COMPENDIUM OF IN VIVO MONITORING IN REAL-TIME MOLECULAR NEUROSCIENCE

Volume 3: Probing Brain Function, Disease and Injury with Enhanced Optical and Electrochemical Sensors

Compendium of In Vivo Monitoring in Real-Time Molecular Neuroscience

Editors: George S. Wilson *(University of Kansas, USA)*
Adrian C. Michael *(University of Pittsburgh, USA)*

Published

Volume 1: Fundamentals and Applications
Volume 2: Microdialysis and Sensing of Neural Tissues
Volume 3: Probing Brain Function, Disease and Injury with Enhanced Optical and Electrochemical Sensors

COMPENDIUM OF IN VIVO MONITORING IN REAL-TIME MOLECULAR NEUROSCIENCE

Volume 3: Probing Brain Function, Disease and Injury with Enhanced Optical and Electrochemical Sensors

Editors

George S Wilson
University of Kansas, USA

Adrian C Michael
University of Pittsburgh, USA

World Scientific

NEW JERSEY • LONDON • SINGAPORE • BEIJING • SHANGHAI • HONG KONG • TAIPEI • CHENNAI • TOKYO

Published by

World Scientific Publishing Co. Pte. Ltd.
5 Toh Tuck Link, Singapore 596224
USA office: 27 Warren Street, Suite 401-402, Hackensack, NJ 07601
UK office: 57 Shelton Street, Covent Garden, London WC2H 9HE

Library of Congress Cataloging-in-Publication Data
Compendium of in vivo monitoring in real-time molecular neuroscience / editors, George S. Wilson, University of Kansas, USA, Adrian C. Michael, University of Pittsburgh, USA.
volumes cm
Includes bibliographical references and index.
Contents: volume 1. Fundamentals and applications
ISBN 978-9814619769 (v. 1 : hardcover : alk. paper)
1. Neurochemistry. 2. Electrochemical analysis. I. Wilson, George S., 1939– II. Michael, Adrian C.
QP356.3.C64 2014
612.8'042--dc23
2014029387

British Library Cataloguing-in-Publication Data
A catalogue record for this book is available from the British Library.

Compendium of In Vivo Monitoring in Real-Time Molecular Neuroscience
Volume 3
ISBN 978-981-120-622-1

For any available supplementary material, please visit
https://www.worldscientific.com/worldscibooks/10.1142/11441#t=suppl

Typeset by Stallion Press
Email: enquiries@stallionpress.com

Printed in Singapore

PREFACE

The importance of global interactions within the brain has inspired the development of methods designed to measure synchronization of neurobiological events. This would include metabolic coupling of neurons and astrocytes (Magistretti, PJ *et al.*, *Nature Reviews Neuroscience*, 2018, 19: 235–249), cellular homeostasis through Na^{+}/K^{+} transport, and the role of various neurotransmitters and neuromodulators. At the same time, it is necessary to develop techniques with sufficiently high spatial resolution such that processes at the level of an individual neuron can be studied.

Accordingly, this volume places heavy emphasis on the development of techniques capable of high temporal and spatial resolution. Optogenetics plays an increasingly important role in this regard as Lin Tian's group demonstrates the selective binding of dopamine to a receptor with a time resolution of 10–100 ms. Neuromodulator peptides can be studied by coupling to a G-protein-coupled receptor that generates a cascade of steps resulting in a fluorescence change as described by Slesinger. Alternatively, monoamine-secreting vesicles can be loaded with fluorescently-labeled neurotransmitter (fluorescent false neurotransmitter) and its release by exocytosis can be studied (Sulzer). Electrical and optogenetic methods were used by Rayport to construct connectome maps of dopamine neurons. Johnson discusses the instrumentation required for optogenetics which can be employed in any part of the brain, not just in brain slices or on the surface. EEG measurements are also described as an approach to assessing overall brain activity.

Enhancements in electrochemical methodology are also noted as smaller electrodes afford higher spatial and temporal resolution. Electrochemical imaging of exocytotic fusion events with 1 ms time resolution has been demonstrated by Lindau. Micro/nanoelectrode arrays can measure time-dependent concentration events in response to stimuli along with neural spikes and local field action potentials as demonstrated by Cai. Electrode surface modification has been used by Yang

Tian to selectively monitor reactive oxygen species. NanoITIES, a less familiar electrochemical technique, has been used by Shen to insert a 15 nm electrode into a 100 nm synaptic cleft to monitor acetylcholine release from a single synapse.

Phillips has demonstrated that Fast Scan Cyclic Voltammetry (FSCV) electrodes can be used over a period of three weeks for the study of addictive behavior, in contrast to the usual short term measurements. The influence of dopamine and norepinephrine on behavior during alcohol consumption was studied via microdialysis by Gonzales.

Several contributions have focused on the link between the levels of key neurotransmitters and neuromodulators in the brain and various diseases. Rebec described the link in Huntington's disease as related to ascorbate release and glutamate uptake. Thompson has developed capability for monitoring K^+ and exploring its link to neuromuscular diseases. Kiyatkin has focused on energy metabolism in the brain by monitoring oxygen and glucose levels in response to various stimuli and Marinesco monitors brain injury via various energy criteria for brain energy metabolism. Mateo has measured dopamine and its role in basal ganglia disorders.

The cover figure is a simulated structure of a circularly permuted (cp)GFP-based sensor for neuromodulators. These sensors are engineered by inserting cpGFP into G-protein coupled receptors or E.coli periplasmic binding proteins. Ligand binding induces conformational changes which modulates the core conformation and leads to fluorescence changes. Further details may be found in Chapter 1.

George S. Wilson
Adrian C. Michael

LIST OF CONTRIBUTORS

Clélia Allioux
INSERM U1028, CNRS UMR5292,
Lyon Neuroscience Research Center — CRNL, Team TIGER
and AniRA-Neurochem Technological Platform, Lyon, France
Chapter 16

Daniel V. Aillon
Pinnacle Technology, Inc., USA
Chapter 5

Baptiste Balança
INSERM U1028, CNRS UMR5292,
Lyon Neuroscience Research Center — CRNL,
Team TIGER and Université Claude Bernard Lyon 1, Lyon, France
and Hospices Civils de Lyon, Lyon, France
Chapter 16

Xinxia Cai
State Key Laboratory of Transducer Technology
Institute of Electronics, Chinese Academy of Sciences
Beijing, 100190, P.R. China
and University of Chinese Academy of Sciences
Beijing, 100049, P.R. China
Chapter 7
xxcai@mail.ie.ac.cn

Ran Chen
Department of Chemistry
University of Illinois at Urbana-Champaign, USA
Chapter 9
mshen233@illinois.edu

Nao Chuhma
Department of Psychiatry
Columbia University
and Department of Molecular Therapeutics
New York State Psychiatric Institute, USA
Chapter 4

Daniel Eskenazi
Department of Psychiatry
Columbia University
and Department of Molecular Therapeutics
New York State Psychiatric Institute, USA
Chapter 4

Qinghua Fang
School of Applied & Engineering Physics
Cornell University, NY, USA
and Laboratory for Nanoscale Cell Biology
Max-Planck-Institute for Biophysical Chemistry
Göttingen, Germany
Chapter 6

Ryan D. Farero
Center for Excellence in Neurobiology of Addiction, Pain and Emotion
Graduate Program in Neuroscience
Department of Psychiatry & Behavioral Sciences
and Department of Pharmacology
University of Washington, USA
Chapter 10

Rueben A. Gonzales
Division of Pharmacology and Toxicology
College of Pharmacy
The University of Texas at Austin, USA
Chapter 13
rgonzales@austin.utexas.edu

Meng Huang
School of Applied & Engineering Physics
Cornell University, NY, USA
Chapter 6

Saul Jaime
Division of Pharmacology and Toxicology
College of Pharmacy, The University of Texas at Austin, USA
Chapter 13
rgonzales@austin.utexas.edu

Eugene A. Kiyatkin
In Vivo Electrophysiology Unit
Behavioral Neuroscience Branch
National Institute on Drug Abuse — Intramural Research Program
National Institutes of Health, Baltimore, MD, USA
Chapter 12
ekiyatki@intra.nida.nih.gov

Ziyue Li
State Key Laboratory of Transducer Technology
Institute of Electronics, Chinese Academy of Sciences
Beijing, 100190, P.R. China
and University of Chinese Academy of Sciences
Beijing, 100049, P.R. China
Chapter 7

David A. Johnson
Pinnacle Technology, Inc., USA
Chapter 5
davej@pinnaclet.org

Daniel M. Kircher
Department of Neuroscience
Friedman Brain Institute
Icahn School of Medicine at Mount Sinai
New York, NY 10029, USA
Chapter 2

David Kleinfeld
Department of Physics
and Section of Neurobiology
University of California, San Diego
La Jolla, California, USA
Chapter 2

Macit Emre Lacin
Department of Neuroscience
Friedman Brain Institute
Icahn School of Medicine at Mount Sinai
New York, NY 10029, USA
Chapter 2

Manfred Lindau
School of Applied & Engineering Physics
Cornell University, NY, USA
and Laboratory for Nanoscale Cell Biology
Max-Planck-Institute for Biophysical Chemistry
Göttingen, Germany
Chapter 6
ML95@cornell.edu

Stéphane Marinesco
INSERM U1028, CNRS UMR5292,
Lyon Neuroscience Research Center — CRNL, Team TIGER
and AniRA-Neurochem Technological Platform, Lyon, France
and Université Claude Bernard Lyon 1, Lyon, France
Chapter 16
Stephane.marinesco@univ-lyon1.fr

Yolanda Mateo
Laboratory for Integrative Neuroscience
Division of Intramural Clinical and Biological Research
National Institute on Alcohol Abuse and Alcoholism, USA
Chapter 11
mateoy@mail.nih.gov

Anne Meiller
INSERM U1028, CNRS UMR5292,
Lyon Neuroscience Research Center — CRNL, Team TIGER
and Université Claude Bernard Lyon 1, Lyon, France
Chapter 16

Susana Mingote
Department of Psychiatry
Columbia University
and Department of Molecular Therapeutics
New York State Psychiatric Institute, USA
Chapter 4

Grace O. Mizuno
Department of Biochemistry and Molecular Medicine
School of Medicine
University of California, Davis, USA
Chapter 1

Erik Naylor
Pinnacle Technology, Inc., USA
Chapter 5

Paul E. M. Phillips
Center for Excellence in Neurobiology of Addiction, Pain and Emotion
Graduate Program in Neuroscience
Department of Psychiatry & Behavioral Sciences
and Department of Pharmacology
University of Washington, USA
Chapter 10
pemp@uw.edu

Shailendra S. Rathore
School of Applied & Engineering Physics
Cornell University, NY, USA
Chapter 6

Stephen Rayport
Department of Psychiatry
Columbia University
and Department of Molecular Therapeutics
New York State Psychiatric Institute, USA
Chapter 4
sgr1@columbia.edu

George V. Rebec
Program in Neuroscience
Department of Psychological and Brain Sciences
Indiana University, Bloomington, USA
Chapter 14
rebec@indiana.edu

Armando G. Salinas
Laboratory for Integrative Neuroscience, Division of Intramural Clinical and Biological Research, National Institute on Alcohol Abuse and Alcoholism, USA and Department of Bioengineering, George Mason University, USA
Chapter 11

Dalibor Sames
Department of Chemistry
Columbia University, USA
Chapter 3

Mei Shen
Department of Chemistry
University of Illinois at Urbana-Champaign, USA
Chapter 9
mshen233@illinois.edu

Paul A. Slesinger
Department of Neuroscience
Friedman Brain Institute
Icahn School of Medicine at Mount Sinai
New York, NY 10029, USA
Chapter 2
paul.slesinger@mssm.edu

Yilin Song
State Key Laboratory of Transducer Technology
Institute of Electronics, Chinese Academy of Sciences
Beijing, 100190, P.R. China
and University of Chinese Academy of Sciences
Beijing, 100049, P.R. China
Chapter 7

David Sulzer
Departments of Psychiatry, Neurology, Pharmacology
Columbia University Medical Center
New York State Psychiatric Institute, USA
Chapter 3
ds43@cumc.columbia.edu

Michael Thompson
Department of Chemistry
and Institute for Biomaterials and Biomedical Engineering
University of Toronto, Canada
Chapter 15
m.thompson@utoronto.ca

Lin Tian
Department of Biochemistry and Molecular Medicine
School of Medicine
University of California, Davis, USA
Chapter 1
lintian@ucdavis.edu

Yang Tian
Shanghai Key Laboratory of Green Chemistry and Chemical Processes
School of Chemistry and Molecular Engineering
East China Normal University
Dongchuan Road 500, Shanghai 200241, P. R. China
Chapter 8
ytian@chem.ecnu.edu.cn

Elizabeth K. Unger
Department of Biochemistry and Molecular Medicine
School of Medicine
University of California, Davis, USA
Chapter 1

Mixia Wang
State Key Laboratory of Transducer Technology
Institute of Electronics, Chinese Academy of Sciences
Beijing, 100190, P.R. China
and University of Chinese Academy of Sciences
Beijing, 100049, P.R. China
Chapter 7

Wenjing Wei
State Key Laboratory of Transducer Technology
Institute of Electronics, Chinese Academy of Sciences
Beijing, 100190, P.R. China
and University of Chinese Academy of Sciences
Beijing, 100049, P.R. China
Chapter 7

Guihua Xiao
State Key Laboratory of Transducer Technology
Institute of Electronics, Chinese Academy of Sciences
Beijing, 100190, P.R. China
and University of Chinese Academy of Sciences
Beijing, 100049, P.R. China
Chapter 7

Shengwei Xu
State Key Laboratory of Transducer Technology
Institute of Electronics, Chinese Academy of Sciences
Beijing, 100190, P.R. China
and University of Chinese Academy of Sciences
Beijing, 100049, P.R. China
Chapter 7

Limin Zhang
Shanghai Key Laboratory of Green Chemistry and Chemical Processes
School of Chemistry and Molecular Engineering
East China Normal University
Dongchuan Road 500, Shanghai 200241, P. R. China
Chapter 8

Ying Zhao
School of Applied & Engineering Physics
Cornell University, NY, USA
Chapter 6

Samira Ztaou
Department of Psychiatry
Columbia University
and Department of Molecular Therapeutics
New York State Psychiatric Institute, USA
Chapter 4

CONTENTS

CHAPTER 1

REAL TIME MONITORING OF NEUROMODULATORS IN BEHAVING ANIMALS USING GENETICALLY ENCODED INDICATORS

Grace O. Mizuno*, Elizabeth K. Unger* and Lin Tian†

*Department of Biochemistry and Molecular Medicine, School of Medicine, University of California, Davis, USA

†lintian@ucdavis.edu

1.1 Introduction

Behavior is driven by precise spatial and temporal patterns of neural activity. Understanding how the brain processes these complex patterns requires large-scale recording and manipulation of genetically defined neuronal populations in behaving animals. Over the past decade, protein engineering has produced high-quality genetically encoded indicators to detect changes in calcium, voltage, fast neurotransmitters (glutamate/GABA), and vesicular release. In combination with modern microscopy techniques, these sensors, especially genetically encoded calcium indicators (GECIs), have enabled large-scale recordings from genetically defined neurons or glial cells with cellular and sub-cellular resolution. In addition, sensors with non-overlapping spectra now enable multiplexed imaging of calcium and voltage or can be combined with optogenetic manipulations. These advances provide new opportunities for *in vivo* dissection of cellular and circuit activity and its relationship to behavior.

*Equal contribution.

Much of the focus of neuroscience has been on the rapid inputs and outputs of neurons, however the immense importance of neuromodulators in profoundly influencing neuronal and astrocytic function is now coming to light. The functional significance of their ability to modulate global brain processes such as arousal, attention, or emotion is now well recognized. Yet how they control the activity of local neurons in the brain — how they alter circuit dynamics or modulate behavioral states — is largely unknown. Analytical chemistry has provided useful insights about the concentrations of neuromodulators in the brain via *in situ* microdialysis, fast scan cyclic voltammetry (FSCV), and amperometry with enzyme-modified electrochemical sensors; however, these methods face limitations that prevent large-scale measurements of neuromodulator release in behaving animals.

To address these technical barriers, we and other labs have developed a toolbox of genetically encoded indicators that enable direct, specific and long-term imaging of neuromodulatory signals in target circuits during natural behavior. dLight1, provides high molecular specificity, rapid temporal response (10 ms on and 100 ms off), cellular-level spatial resolution, and stable, long-term measurements over hours, days, or even months. Applications of dLight1 have started to lead to new discoveries about the neural mechanisms controlling dopamine release.

Here we focus on recent developments and applications of genetically encoded indicators for neurotransmitters and neuromodulators. We take an in-depth look at the design and screening of these sensors based on various protein scaffolds. We also discuss the novel applications these new optical probes have enabled, as well as practical concerns in awake, behaving animals. In combination with calcium/voltage imaging and optogenetics, the fast-growing toolbox of sensors for chemical transmission is well-poised to permit direct functional analysis of how the spatiotemporal coding of chemical input signaling mediates the plasticity and function of brain circuits.

1.2 Design of genetically encoded indicators of neural activity

Recent advances in the creation and optimization of genetically encoded (i.e. protein-based) sensors for a variety of molecules and cellular states have led to several high-quality optical probes, such as the genetically encoded calcium indicator (GCaMP) family of GECIs and voltage sensors based on either opsins or voltage sensing domains. Meanwhile, advances in the speed, depth of penetration, and spatial resolution of fluorescence microscopy techniques have extended our ability to non-invasively measure neural activity with superior spatiotemporal resolution. Combined with advanced microscopy, calcium imaging using GCaMPs is now routinely employed to enable large-scale, mesoscopic recording of neuronal populations or glia cells at 30 Hz or faster and with subcellular resolution in awake, behaving animals.

Protein-based sensors typically consist of an analyte-binding or sensing domain and a reporter element based on either a single fluorescence protein (FP) or two FPs. In the case of two FP-based sensors, the FPs have overlapping excitation and emission spectra. The conformational changes in the analyte-binding or sensing domain move the two FPs into sufficient proximity for Förster Resonance Energy Transfer (FRET) between the two FPs to occur, and the readout is a ratio of the intensities of the two FPs. In the case of single FP sensors, changes in the cellular environment detected by the analyte binding or sensing domain result in changes in the chromophore environment of the FP leading to an increase or decrease in fluorescence intensity. Single FP based indicators offer several appealing advantages, such as superior sensitivity, enhanced photostability, broader dynamic range, and faster kinetics compared to FRET-based indicators (Piston and Kremers, 2007). They are also relatively small, and thus easier to express in an adeno-associated virus (AAV) and to target to subcellular locations, such as spines and axon terminals. In addition, imaging with single-FP indicators preserves spectrum bandwidth, thus allowing for multiplexed imaging or use alongside optogenetic effectors, such as channelrhodopsin.

The main advantage of these sensors over small molecule-based fluorophores stems from their genetic encoding, which enables reporters to be constructed from proteins that are evolutionarily designed to respond to neural activity. This enables them to be optimized by computational modeling and directed evolution. Then they can be selectively targeted to cells with specific anatomical connectivity. Finally, these sensors can be stably expressed over long periods of time (from days to months), allowing neuroscientists to study how patterns of neural activity change with learning, development, or disease progression. *For reviews please see (Palmer et al., 2011; Broussard et al., 2014a; Lin and Schnitzer, 2016).*

In recognition of the potential of combining genetically encoded indicators with modern microscopy, we and others have now extended the concept of single-FP based sensors to the design of genetically encoded sensors for neurotransmitters and neuromodulators. Genetically encoded indicators are generally categorized by their scaffold: G-protein coupled receptors (GPCR) and bacterial periplasmic binding proteins (PBP).

1.3 dLight1: A GPCR-based dopamine indicator

To enable direct and specific measurements of diverse types of neuromodulators with the necessary spatiotemporal resolution, it is advantageous to design a sensor with molecular specificity, affinity, and kinetics similar to endogenous receptors.

As GPCRs are native neuromodulator targets, we have pioneered a new approach for biosensor design based on protein engineering of the cognate GPCRs. This approach, facilitated by recent advances in GPCR structural biology, allows us to leverage the ligand specificity, affinity, and binding kinetics that has evolved in the GPCR family and provides a universal platform to design sensors for any desired neuromodulator or pharmacological drug. Using this design platform, we engineered dLight1, a suite of intensity-based genetically encoded indicators for dopamine (DA) that work by directly coupling the conformational changes of inert human dopamine receptors to changes in the fluorescence intensity of a circularly permuted GFP (cpGFP) (Patriarchi *et al.*, 2018). The dLight1 family consists of six sensors based on three dopamine receptors (DRD1, DRD2, and DRD4) with broadly tunable affinity and dynamic range to probe dopamine transients across the pM-mM range of concentrations. We have also applied the design strategy of dLight1 to develop a class of intensity-based sensors for serotonin, norepinephrine, melatonin and opioid peptides. Below we use dLight1 as a prototype to discuss GPCR-based sensor design, characterization and applications.

1.3.1 Sensor engineering and screening pipeline

To engineer a GPCR-based sensor, the optimal insertion site of cpGFP first needs to be determined. The ideal insertion site in a GPCR would maximize the coupling of fluorescence changes with conformational changes of the receptor upon ligand binding without interfering with membrane trafficking. Crystallography has yielded detailed structural information on both the active and inactive states of β2 adrenergic receptor (β2AR), a catecholamine receptor, which suggested that intracellular loop 3 (IL3) undergoes significant conformational change upon ligand binding. Given the structural similarity of GPCRs, it was hypothesized that insertion of cpGFP into IL3 of DRDs would alter the chromophore environment of cpGFP resulting in a change in cpGFP fluorescence. To determine the insertion site of cpGFP, DRD1 and DRD4 were first aligned with that of β2AR. The exact site of the cpGFP insertion into IL3 was further optimized based on expression on the membrane. The cpGFP module (LSSLE-cpGFP-LPDQL) from GCaMP was inserted between K232 and K269 of DRD1 (Patriarchi *et al.*, 2018). When expressed in human embryonic kidney cells (HEK), this mutant was well-expressed at the plasma membrane. However, when DA was applied, we observed a decreased fluorescence signal at the cell membrane.

Through engineering GCaMP (Tian *et al.*, 2009) and the glutamate sensing fluorescent reporter, iGluSnFR (Marvin *et al.*, 2013), we learned that both linker regions between the ligand-binding proteins and cpGFP are critical for a sensor's dynamics and kinetics. Therefore, to further optimize and obtain a positive-response sensor, a linker library on each side of the cpGFP (LSSXX-cpGFP-XXDQL) was engineered and variants were screened for their responses in mammalian HEK cells. After screening a library of 585 variants in HEK cells and tuning the dynamic range with mutations in IL3 we obtained a suite of dLight1 sensors with excellent membrane localization and varying dynamic ranges (~100% to 1000% ΔF/F) and affinities to cover a broad range of physiologically relevant concentrations (pM to mM) (Patriarchi *et al.*, 2018).

1.3.2 Sensor characterization: Affinity, specificity, and kinetics

Before applying sensors *in vivo*, it is important to perform systematic characterization of the sensor's intrinsic properties, including affinity, specificity, and kinetics. We used *in situ* ligand titration in HEK cells to determine the apparent affinity and molecular specificity of the sensor (Patriarchi *et al.*, 2018). The apparent affinity combined with the dynamic range of a sensor can be used to predict the sensitivity to physiologically relevant release, thus guiding the selection of sensors for *in vivo* applications. The apparent affinities and dynamic ranges of the dLight1 family based on various DRDs are shown in Table 1.1.

For monoamines, it is critical to confirm ligand specificity as they are structurally very similar. While DA and norepinephrine (NE) only differ by one hydroxyl group, their affinities for dLight1.1 (based off DRD1) are vastly different: the K_d of dLight1.1(DRD1 based sensor) for DA is 330 nM while the K_d of dLight1.1 for NE is 19,850 nM, which is a 40 fold lower affinity (Patriarchi *et al.*, 2018). Negligible responses were observed for other neurotransmitters and neuromodulators, including epinephrine, octopamine, serotonin, GABA, histamine, and acetylcholine.

One of the major concerns using an endogenously expressed receptor as a sensor is the possibility of endogenous signaling interference. To investigate this, the effect of the sensor expression on ligand-induced cyclic AMP (cAMP) production was monitored. dLight1.1 did not display significant cAMP response in HEK cells. Furthermore, when dLight1.1 was introduced into U2OS, a cell line that endogenously expresses DRD1, dLight1.1 did not significantly alter the DA dose-response curve (Patriarchi *et al.*, 2018).

Next, dLight1.1 was characterized in primary rat hippocampal neurons. To do so, the sensor was first cloned into an adeno-associated viral vector driven by a human synapsin or CAG promoter. Similar to in HEK cells, in primary rat hippocampal neurons, dLight1.1 showed excellent membrane localization and a similar affinity for DA (K_d 311 nM) (Patriarchi *et al.*, 2018). In cultured hippocampal slices, dLight1.1 reliably detects sub-micromolar DA concentration changes at dendrites and single dendritic spines.

Table 1.1 The apparent affinity and dynamic range of the dLight1 family.

Version	GPCR	Dynamic Range	Affinity
dLight1.1	DRD1	230 ± 9%	330 ± 30 nM
dLight1.2	DRD1	340 ± 20%	770 ± 10 nM
dLight1.3a	DRD1	660 ± 30%	2300 ± 20 nM
dLight1.3b	DRD1	930 ± 30%	1680 ± 10 nM
dLight1.4	DRD4	170 ± 10%	4.1 ± 0.2 nM
dLight1.5	DRD2	180 ± 10%	110 ± 10 nM

Lastly, we used dLight1.2 (based on DRD1 with additional mutations) to measure the time course of dopamine release triggered by electrical stimulation in acute striatal slices with two photon imaging (Patriarchi *et al.*, 2018). Fast line scan revealed a rapid onset of fluorescence increase (rise $\tau_{½} = 9.5 \pm 1.1$ ms) followed by a plateaued peak (averaged $\Delta F/F = 220 \pm 50\%$) for about 150 ms, which decayed to baseline in about 400 ms (decay $\tau_{1/2} = 90 \pm 11$ ms). Blockade of dopamine reuptake with cocaine significantly prolonged the decay of fluorescence from peak to baseline, but with equivocal effect on response amplitude ($P = 0.056$). Application of the competitive antagonist SKF83566 eliminated the responses, confirming that fluorescent signals are indeed due to DA binding.

1.3.3 Application of dLight1

dLight1 offers fast temporal resolution, matching that of electrochemical methods for detecting monoamines, while also providing subcellular resolution and molecular specificity. dLight1 enables the measurement of DA reuptake versus diffusion which would not be possible with other methods. These properties of dLight1 enable robust and chronic detection of physiologically or behaviorally relevant DA transients, which opens new doors to study how neuromodulators govern the rapid changes in activity and brain state.

One of the major questions surrounding DA is the concentration of DA release and its synaptic and extra-synaptic contribution to circuit function. By utilizing the fast kinetics of dLight1.2, we observed peak DA concentrations of 10–30 μM in acute striatal slices following electrical stimulation: such concentrations were closely associated with sites of release (Patriarchi *et al.*, 2018). Under similar stimulus parameters, DA concentrations reported with FSCV are 1–2 orders of magnitude lower (Yorgason *et al.*, 2017).

dLight1 also permits deep-brain recording of bi-directional dopamine dynamics compatible with optogenetic stimulation or calcium imaging of local neuronal activity in freely behaving mice. dLight1.1, when used together with ChrimsonR (a red-shifted optogenetic actuator), reported an increase of DA in the nucleus accumbens (NAc) when optogenetically activating DAergic neurons in the ventral tegmental area (VTA) (Figure 1.1A) and a decrease of DA when inhibiting DAergic neurons through optogenetic stimulation of GABAergic neurons in the VTA (Figure 1.1B). Furthermore, when performing dual-color imaging, the fluorescence of dLight1.1 and red-shifted calcium indicator, jRGECO1a, increased while consuming sucrose. However, dLight1.1 fluorescence decreased while the fluorescence of red-shifted calcium indicator, jRGECO1a,

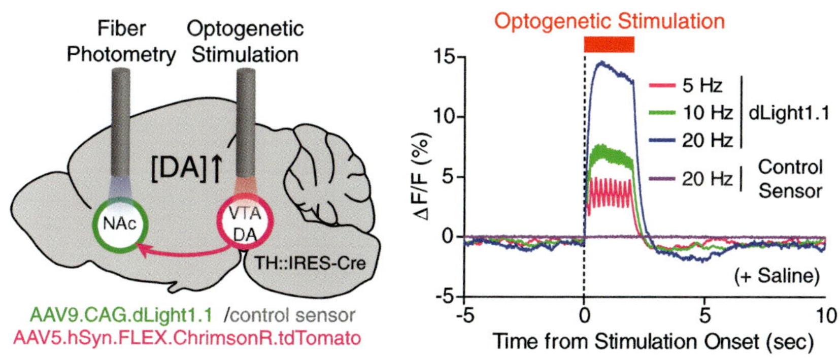

(A)

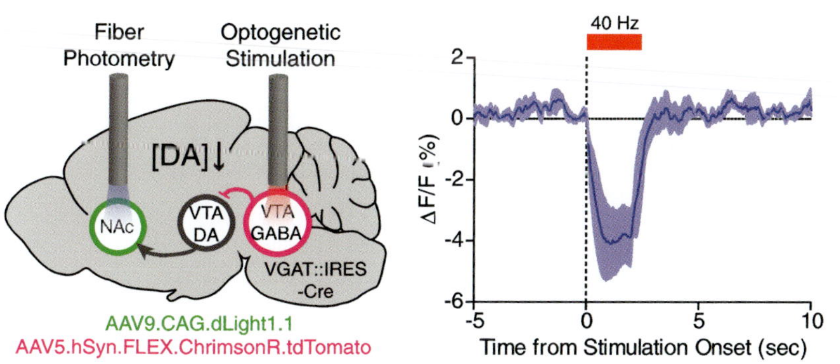

(B)

Figure 1.1 (A) Schematics showing fiber photometry recording of dLight1.1 or control sensor in the NAc while stimulating VTA DA neurons by optogenetics. Averaged fluorescence increase in response to optogenetic stimuli (n = 5 mice). (B) Schematics showing fiber photometry recording of dLight1.1 in the NAc and optogenetic stimulation of VTA GABA neurons that inhibits VTA DA neurons. Averaged fluorescence decrease in response to optogenetic stimulation at 40 Hz (n = 4 mice). Figures reproduced with permission from Patriarchi *et al.*, 2018.

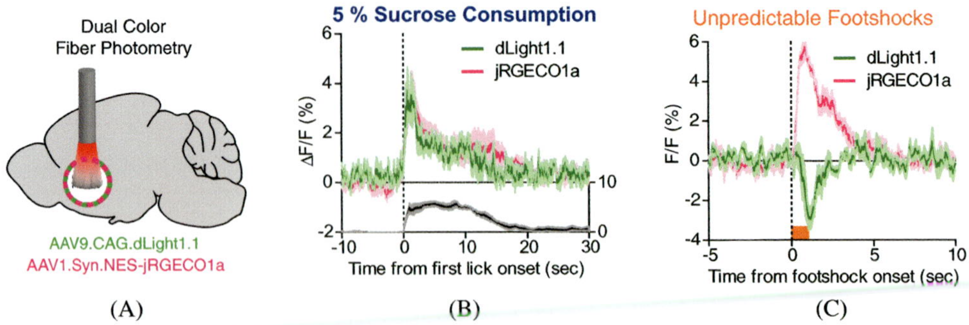

Figure 1.2 **(A)** Dual-color fiber photometry recording of DA release with dLight1.1 and local neuronal activity with jRGECO1a. **(B)** Fluorescence increase or decrease in dLight1.1 (green) and increase in jRGECO1a (red) during unpredictable reward or **(C)** footshock delivery (0.6 mA for 1 s, n = 5 mice). Figures reproduced with permission from Patriarchi *et al.*, 2018.

increased during unexpected footshock (Figure 1.2). These findings demonstrate dLight1 can be used with calcium imaging to correlate postsynaptic local circuit activity with DA input signals. We were also able to use dLight1.1 to chronically measure learning-induced dynamic changes of DA transients within the NAc at sub-second resolution. Mice trained using Pavlovian conditioning displayed an increase of DA response to predictive cues and a decrease to reward consumption (Figure 1.3).

As DA neurons are densely confined in relatively small deep-brain areas and extensively project to distant brain regions to modulate local circuitry, it is extremely important to analyze different spatiotemporal transients across depth. One of the most powerful aspects of genetically encoded sensors such as dLight1 is the ability to allow for imaging cellular activity at spatial scales ranging from subcellular to millimeter. Therefore, we performed two-photon imaging with dLight1.2 in layers 1 and 2 of mouse motor cortex, which revealed a high-resolution (cellular level) dopamine transient map of spatially distributed, functionally heterogeneous dopamine signals during a visuomotor learning task (Figure 1.4). Our results suggest that DA release from mesocortical projections are thus spatially intermingled and depends on motor behavior, reward expectation and consumption.

Dissemination of dLight1 to the neuroscience community has led to new discoveries involved in the neural mechanisms of DA release in modulating motivation, reinforcement, and rewarding and aversive stimuli. Using dLight1, the Lüscher lab demonstrated DA's direct involvement in the reinforcing effect of heroin, a hypothesis severely challenged largely based on pharmacological

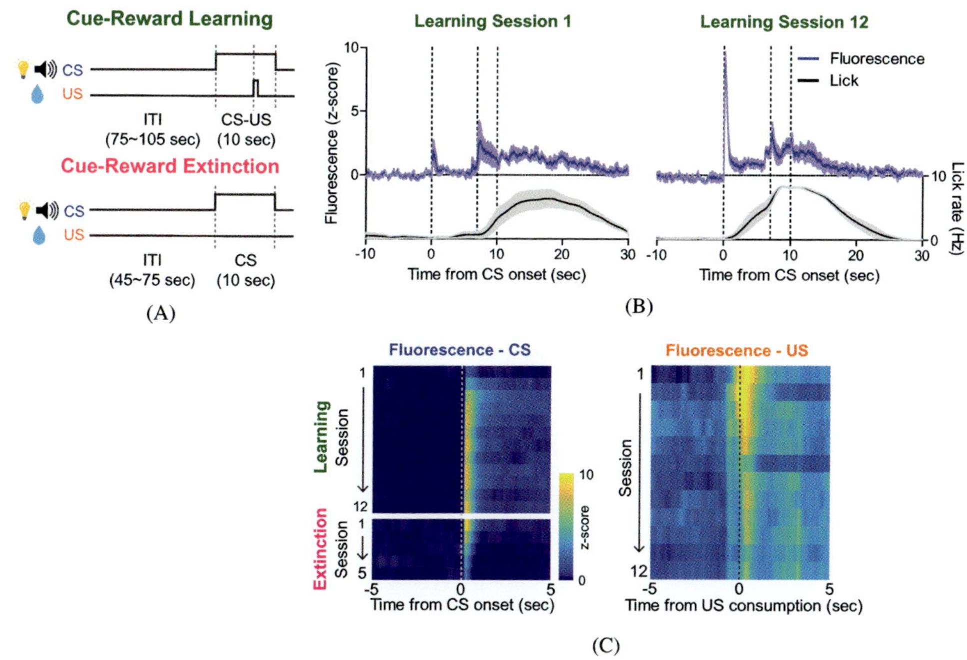

Figure 1.3 **(A)** Pavlovian conditioning procedures involved learning to associate neutral cues (CS; house light and 5 kHz tone) with a sucrose reward (US; 50μL of 5% sucrose), and subsequent extinction. **(B)** dLight1.1 dynamics in response to CS and US in first and last sessions of cue-reward learning, shown in single (gray) and averaged (blue) trials (n=20 trials) from a single animal. **(C)** Evolution of CS- (left) and US-evoked (right panel) average fluorescence. Figures reproduced with permission from Patriarchi *et al.*, 2018.

experiments suggesting DA-independent mechanisms for heroin reinforcement (Corre *et al.*, 2018). Using fiber photometry dLight1.1 in the NAc medial shell resulted in an increase in fluorescence after self-administration of heroin during initial reinforcement within less than a minute through the activation of a subset of DAergic neurons in the medial VTA. The Lammel lab, on the other hand, used dLight1 combined with calcium imaging to reveal an anatomically and functionally distinct DA projection underlying motivated behavior (de Jong *et al.*, 2019). They discovered that DA terminals in the ventral NAc medial shell are excited by unexpected aversive outcomes and to cues that predict them, whereas excitation in response to reward-predictive cues dominated in the NAc lateral shell and was largely absent in the medial shell. Recording of direct and specific DA transients using dLight1 combined with calcium imaging make it possible to monitor functionally and anatomically distinct heterogeneous subpopulations of neurons important for a wide range of behaviors.

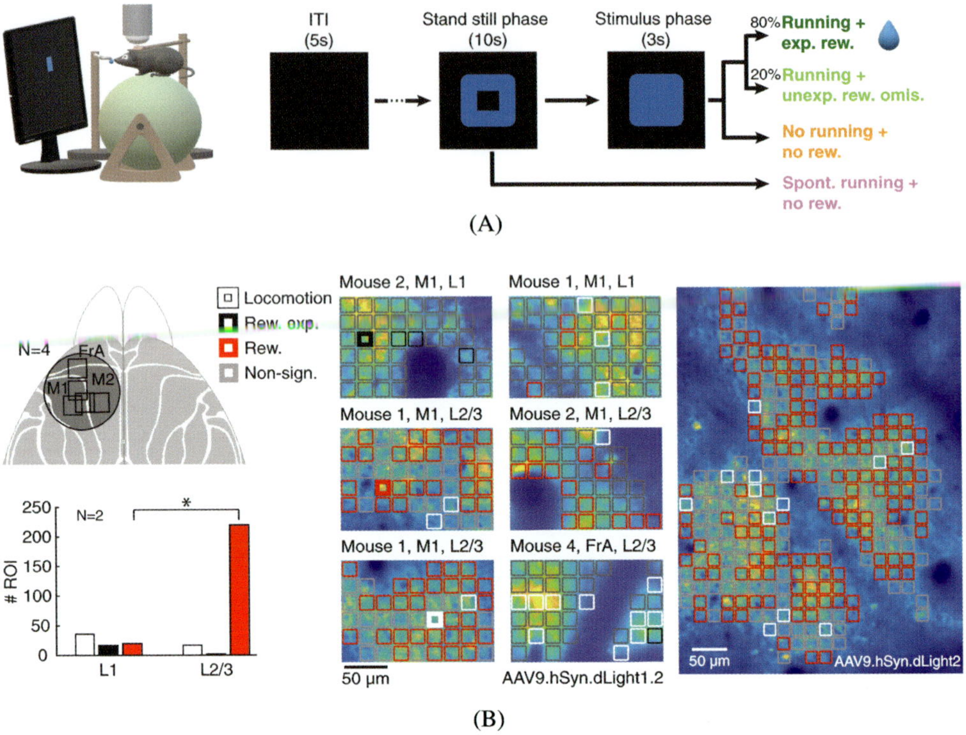

Figure 1.4 Two-photon imaging revealed spatially-distinct, task specific DA transients in motor cortex. **(A)** Schematics of experimental setup. To initiate a trial, mice were required to stand still for 10s following a visual cue (blue square). If mice started to run during the stimulus phase ("Hit trials"), a water reward was given. In 20% of randomly selected "Hit trials" the reward was withheld. If no run was triggered by stimulus presentation, the trials were counted as "Miss trials". Erroneous/spontaneous runs during the stand-still phase ended the trial (no "Go" cue or reward). **(B)** Dorsal view of mouse cortex with the chronic cranial window (circle) and imaging location indicated (square). Heatmap of dLight1.2 expression pattern in layer 2/3 of M1 cortex. The image is overlaid with computationally defined regions of interest (ROIs, ~17 × 17 μm). Colored ROIs indicate the type of fluorescence responses observed during the task. Figures reproduced with permission from Patriarchi *et al.*, 2018.

1.3.4 Scaffold versatility: Other neurotransmitters, neuromodulators and neuropeptides

The abundance of GPCRs for neurotransmitters and neuropeptides is advantageous from a sensor design point of view. We demonstrated the universality of applying the design strategy of dLight1 to develop a class of intensity-based sensors for various neurotransmitters, neuromodulators and neuropeptides, such as serotonin, norepinephrine, melatonin and opioid peptides (Patriarchi *et al.*, 2018). However, while all receptors share similar structural motifs, each sensor must be individually mutagenized and optimized.

1.4 From glutamate to neuromodulators: Bacterial PBP-based single-FP sensors: iMaltSnFR and iGluSnFR

The Looger lab has shown that bacterial PBPs can be converted into sensors *via* the insertion of cpGFP in loops tolerant of insertions, followed by significant direct evolution to discover mutants with desired intrinsic properties. PBPs consist of two lobes connected by a hinge region, which close around the ligand in a Venus flytrap-like motion (Dwyer and Hellinga, 2004). The first PBP based sensor was first developed from the *E. coli* maltodextrin-binding protein (MalE) (Marvin *et al.*, 2011). Using the crystal structures of the ligand-free (open) and the ligand-bound (closed) confirmations, residues sensitive to ligand-dependent structural changes were identified. These residues were hypothesized to be well-suited for modulating the fluorescence of an inserted cpGFP. After sequential rounds of screening, iMaltSnFR was created that had a ΔF/F of greater than 600%. iMaltSnFR was the foundation for the neurotransmitter sensor, iGluSnFR, which reports the presence of the excitatory molecule glutamate, based on the bacterial protein, Gltl (Marvin *et al.*, 2013). iGluSnFR was developed by inserting cpGFP into GltI, which has significant homology to the glutamate-binding domain from ionotropic glutamate receptors. Although Gltl had been previously used to develop FRET-based glutamate sensors, the single-FP based iGluSnFR, including versions recently optimized for their kinetics and affinities, are the first optical sensors with sufficient sensitivity and photostability to permit reliable in vivo optical measurements of glutamate transients in single dendritic spines (Helassa *et al.*, 2018; Marvin *et al.*, 2018).

iGluSnFR has been publicly available for many years now and several groups have used it to study glutamate release from neurons and glia in applications including the hippocampus, the cerebellum, the visual system, Huntington's disease, Rett syndrome, and many more. A wide range of model organisms were used for these studies, including iPSCs, flies, fish and mice. The Allen Institute has even created a mouse for conditional expression of iGluSnFR, which one lab used to do mesoscopic whole brain imaging of glutamate dynamics during stressed and treatment conditions (Xie *et al.*, 2016).

1.4.1 Beyond iGluSnFR: PBP-based single-FP neuromodulator sensors

PBPs make great scaffolds for small molecule sensors because their large conformational change is relatively easy to translate into a change in fluorescence, offering unmatched dynamic ranges. Additionally, because PBPs are bacterial proteins, it is possible to screen out a large number of variants in bacteria. PBPs are versatile and can further be mutated to bind neurotransmitters for which there are no available PBPs. In addition to iGluSnFR, we and other labs are currently creating other neurotransmitter and neuromodulator sensors based on PBP scaffolds (Kazemipour *et al.*, 2018; Looger *et al.*, 2018; Shivange *et al.*, 2019).

1.5 Practical considerations for choosing the most appropriate sensor

For end users, choosing the most appropriate sensor at the start of a project is paramount. The sensor selection should take into consideration technical experimental conditions such as light source, image acquisition method (one-photon, fiber photometry, two-photon, etc.), camera, image analysis algorithms, etc. as well as biological conditions such as expected ligand concentrations and the model organism or cell type. To maximize the signal to noise ratio (SNR), the intrinsic properties of a sensor such as expression, affinity, and dynamic range need to match with the time-course, frequency and concentration of release events in the brain region. Here we discuss, and summarize in Table 1.2, several practical criteria that users may wish to consider, including the properties of the sensor relative to the experimental parameters, and the potential for off-target effects.

1.5.1 Affinity

In order to create sensors with affinities well within the physiological range, GPCR-based scaffolds are an obvious choice, in that they leverage the evolutionary specificity for the ligand of choice. In addition, there exists an array of naturally occurring GPCRs for each ligand, providing the possibility of creating multiple sensors with different affinities, such as dLight1.1-5, all produced from different receptors and with different affinities for DA.

While some PBP-based sensors, such as iGluSnFR, are able to take advantage of evolutionarily designed binding proteins, for the most part neurotransmitters do not exist in bacteria and therefore binding must be engineered into the sensor. Changing the specificity of the binding pocket is an extremely difficult process and one for which techniques need to be refined.

GPCRs have affinities that range from pM to μM. However, a high affinity may be a double-edged sword. While one can be sure the affinity is within the physiological range, there is the distinct possibility of the sensor buffering the ligand concentration and thus altering the properties of the cell. On the other hand, PBP-based sensors have affinities for their ligands that are in the μM range, which is likely to be higher than the local concentration. Thus, it is informative to determine a "sweet spot" at which the expression level maximizes the SNR while minimizing the buffering effect.

Table 1.2

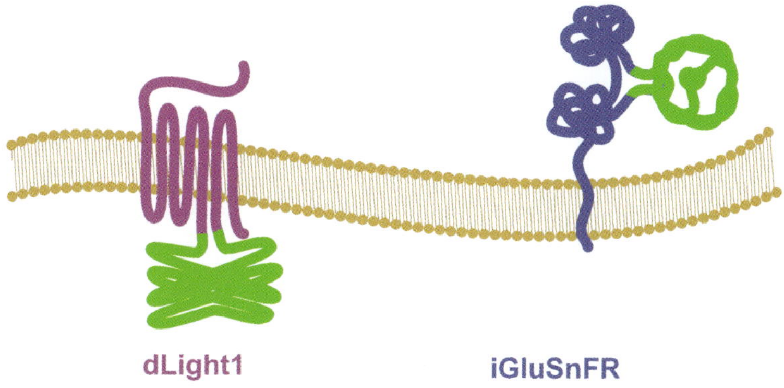

Properties of the sensor		(including ultrafast and affinity variants)
Apparent affinity	100nM–1μM	0.6–53μM
Temporal dynamics	10 ms on, 100 ms off	1 ms on, 2–30 ms off
***In vivo* expression and activity**		
Genetically encoded	Yes	Yes
Model organisms	mouse (... thus far)	mouse (knock-in), fly, fish
Subcellular localization	Obligate transmembrane protein Can be targeted to dendrite or axon	Can be tagged to any compartment
Agonist/Antagonist	Mimics endogenous receptor	Specificity is distinct from endogenous receptors
Downstream signaling	Tested negative for cAMP	No known activity in eukaryotic cells
Buffering of ligand	Potential concern	Kd outside of physiologic range reduces the potential for ligand buffering

1.5.2 Dynamic range

The dynamic range is calculated as the change in fluorescence after ligand addition divided by the fluorescence before ligand addition. Given this, it is possible to improve the dynamic range by either increasing the change in fluorescence, or by decreasing the basal fluorescence. For some applications, it is advantageous to have a very low basal fluorescence, such as photometry, where small signals are easier to detect when there is little or no background. In others, it is advantageous to have moderate basal fluorescence, such as when searching for infected or transfected cells amongst untransfected ones, prior to the addition of ligand. Sensor optimization is thus tailored to its intended application. However, it is always advantageous to have a larger change in fluorescence. Signal detection above noise is easier when the signal is significantly larger than the noise.

Because the affinity of GPCR-based sensors is already optimal, most of the effort of creating the sensor is in optimizing the coupling. This is done by optimizing the cpGFP insertion site and then screening amino acid composition in the linker regions between cpGFP and scaffold proteins (Marvin *et al.*, 2011; Patriarchi *et al.*, 2018). The cpGFP is typically inserted into a loop region between two more structured areas of the protein whereby ligand binding creates a displacement between the two structures. A linker is then added such that steric hindrance from the cpGFP will not interfere with protein function and vice versa. A rigid linker will translate the motion of the binding protein well, but if it is too rigid, it may break the sensor. Similarly, a long linker will provide good separation between the two fused proteins, but if it is too long, the motion of the binding protein may not be translated to the GFP.

The other factor to consider is how much the binding protein moves. A large conformational change is easier to translate into a large dynamic range because there is then a wide range of angles to choose from for the dim and bright states of the sensor. While PBP-based sensors may have lower affinity for their ligands, this shortcoming is readily compensated for in their broad dynamic range due to their venus-fly-trap-like closure upon ligand binding.

1.5.3 Subcellular localization

One of the advantages of genetically encoded indicators over small molecules is the ability to target them to specific sub-cellular locations. To maximize SNR, subcellular targeting has proven to be effective in increasing the sensor's performance in subcellular compartments such as soma, axons and dendrites. For example, axon-targeted GCaMP and soma-targeted voltage indicators showed significantly increased sensitivity enabling robust SNR recordings compared to untargeted ones (Abdelfattah *et al.*, 2018; Broussard *et al.*, 2018).

Compared to GPCR-based sensors, PBP-based sensors offer more flexibility for fusion with a targeting motif (plasma membrane targeted, ER targeted, cytoplasmically expressed, etc). Because PBPs are soluble proteins, there are many fewer restrictions on where the sensor may exist within the cell. GPCRs are obligatory transmembrane proteins. However, adding targeting motifs is no guarantee that proper targeting will occur.

1.5.4 Off-target effects

Sensors are capable of disrupting normal biological functions in three major ways: by buffering the ligand concentration as discussed above, by initiating signal cascades, and by interfering with normal processes.

GPCRs, as endogenous receptors, are capable of initiating signaling cascades. Therefore, it is important to engineer them as inert sensors. PBPs, on the other hand, do not initiate signaling cascades in eukaryotic cells and thus no modification is necessary. However, the potential for off-target effects from foreign proteins, such as changes in electrophysiological, morphologic and circuit properties, must always be considered. Each new study must first determine if expression of the sensor is affecting the process being studied.

Any exogenously expressed protein has the potential to overwhelm protein production machinery or interfere with endogenous signaling pathways, particularly if it is overexpressed (Palmer *et al.*, 2011). Overexpression of these sensors (both PBP and GPCR) also may buffer the ligand (regardless of the affinity) or may begin binding lower affinity targets, thus disrupting circuitry properties. For example, overexpression of GCaMP (which is made from eukaryotic proteins but whose structure is more similar to that of the PBP-based sensors) has long been known to be an issue (Rose *et al.*, 2014). High expression levels of iGluSnFR or dLight1 have not been shown to create any gross morphological abnormalities, and bright cells respond similarly to dim cells (Marvin *et al.*, 2013; Patriarchi *et al.*, 2018). Also, in cells, dLight1 did not signal via G-protein signaling. We have observed, however, that punctate expression patterns occasionally emerge, and in these cases (either GPCR or PBP) the cell appears not to respond to normal stimuli. To balance all these effects, expression level should always be as low as possible while still obtaining a detectable signal. Very low expression will preclude sensor visualization and response detection due to a limited photon budget. To accomplish this, users may wish to explore multiple promoters such as synapsin, CAG or CaMKII, regulatory sequences, transduction methods (e.g. different viral serotypes, electroporation techniques, or transgenic lines) and examine expression level over time. It is also important to compare physiological properties of cells with and without sensor expression such as resting potential, membrane resistance and capacitance and signaling cascades, such as cAMP production.

1.6 Outlook

More often than not, the barriers to research are technical rather than intellectual. The development and application of tools such as genetically encoded indicators expand the horizons of what is even possible to explore in neuroscience.

Continued sensor development is likely to progress in the following ways: (1) refinement of the properties of existing sensors, including kinetics, affinity, specificity, and subcellular localization as defined by the individual needs of the user, (2) expansion of the colors available for each tool, (3) expansion of the molecular targets of sensors, and (4) adaptation of the sensors for different model organisms and experimental platforms. We and other labs are actively engaged in all four of these areas: we are refining the subcellular localization of iGluSnFR, while others are adjusting the kinetics; we and other labs are developing red and far red versions of our existing sensors; we and other labs are expanding our GPCR- and PBP-based sensors to recognize other neuromodulators and small molecule peptides (Feng *et al.*, 2018; Jing *et al.*, 2018; Sun *et al.*, 2018); and we are creating stable cell lines while other labs and organizations are making worms, fish, flies and mice that conditionally express sensors.

The application of these new sensors to existing questions in neuroscience can be potentially transformative. We and other labs devoted to tool development are working to enable this expansion and ultimately gain a deeper understanding of the brain.

Acknowledgments

This work was supported by funding to L.T. (BRAIN Initiative U01NS090604 and U01NS013522, DP2MH107056 and R21NS095325 from the National Institutes of Health), to G.M. (ARCS Scholarship) and E.U. (Mistletoe Research Fellowship).

References

Abdelfattah AS *et al.* (2018) Bright and photostable chemigenetic indicators for extended *in vivo* voltage imaging. *bioRxiv*, 436840.

Broussard GJ, Liang R, Tian L (2014a) Monitoring activity in neural circuits with genetically encoded indicators. *Front Mol Neurosci,* 7: 97.

Broussard GJ, Liang Y, Fridman M, Unger EK, Meng G, Xiao X, Ji N, Petreanu L, Tian L (2018) *In vivo* measurement of afferent activity with axon-specific calcium imaging. *Nat Neurosci*, 21: 1272–1280.

Corre J, van Zessen R, Loureiro M, Patriarchi T, Tian L, Pascoli V, Lüscher C (2018) Dopamine neurons projecting to medial shell of the nucleus accumbens drive heroin reinforcement Monteggia LM, Dulac C, Wolf ME, eds. *eLife*, 7: e39945.

de Jong JW, Afjei SA, Pollak Dorocic I, Peck JR, Liu C, Kim CK, Tian L, Deisseroth K, Lammel S (2019) A neural circuit mechanism for encoding aversive stimuli in the mesolimbic dopamine system. *Neuron*, 101:133–151.e7.

Dwyer MA, Hellinga HW (2004) Periplasmic binding proteins: a versatile superfamily for protein engineering. *Curr Opin Struct Biol*, 14: 495–504.

Feng J, Zhang C, Lischinsky J, Jing M, Zhou J, Wang H, Zhang Y, Dong A, Wu Z, Wu H, Chen W, Zhang P, Zou J, Hires A, Zhu J, Cui G, Lin D, Du J, Li Y (2018) A genetically encoded fluorescent sensor for rapid and specific *in vivo* detection of norepinephrine. *bioRxiv*, 449546.

Helassa N, Dürst CD, Coates C, Kerruth S, Arif U, Schulze C, Wiegert JS, Geeves M, Oertner TG, Török K (2018) Ultrafast glutamate sensors resolve high-frequency release at Schaffer collateral synapses. *PNAS*, 115: 5594–5599.

Jing M *et al.* (2018) A genetically encoded fluorescent acetylcholine indicator for *in vitro* and *in vivo* studies. *Nat Biotechnol*, 36: 726–737.

Kazemipour A, Novak O, Flickinger D, Marvin JS, King J, Borden P, Druckmann S, Svoboda K, Looger LL, Podgorski K (2018) Kilohertz frame-rate two-photon tomography. *bioRxiv*, 357269.

Lin MZ, Schnitzer MJ (2016) Genetically encoded indicators of neuronal activity. *Nat Neurosci*, 19: 1142–1153.

Looger LL, Marvin JS, Shimoda Y, Magloire V, Leite M, Kawashima T, Jensen TP, Knott EL, Novak O, Podgorski K, Leidenheimer NJ, Rusakov DA, Ahrens MB, Kullmann DM (2018) A genetically encoded fluorescent sensor for *in vivo* imaging of GABA. *bioRxiv*, 322578.

Marvin JS *et al.* (2018) Stability, affinity, and chromatic variants of the glutamate sensor iGluSnFR. *Nat Methods*, 15: 936.

Marvin JS, Borghuis BG, Tian L, Cichon J, Harnett MT, Akerboom J, Gordus A, Renninger SL, Chen T-W, Bargmann CI, Orger MB, Schreiter ER, Demb JB, Gan W-B, Hires SA, Looger LL (2013) An optimized fluorescent probe for visualizing glutamate neurotransmission. *Nat Methods*, 10: 162–170.

Marvin JS, Schreiter ER, Echevarría IM, Looger LL (2011) A genetically encoded, high-signal-to-noise maltose sensor. *Proteins*, 79: 3025–3036.

Palmer AE, Qin Y, Park JG, McCombs JE (2011) Design and application of genetically encoded biosensors. *Trends Biotechnol*, 29: 144–152.

Patriarchi T, Cho JR, Merten K, Howe MW, Marley A, Xiong W-H, Folk RW, Broussard GJ, Liang R, Jang MJ, Zhong H, Dombeck D, Zastrow M von, Nimmerjahn A, Gradinaru V, Williams JT, Tian L (2018) Ultrafast neuronal imaging of dopamine dynamics with designed genetically encoded sensors. *Science*, eaat4422.

Piston DW, Kremers G-J (2007) Fluorescent protein FRET: the good, the bad and the ugly. *Trends Biochem Sci*, 32: 407–414.

Rose T, Goltstein PM, Portugues R, Griesbeck O (2014) Putting a finishing touch on GECIs. *Front Mol Neurosci*, 7: 88.

Shivange AV, Borden PM, Muthusamy AK, Nichols AL, Bera K, Bao H, Bishara I, Jeon J, Mulcahy MJ, Cohen B, O'Riordan SL, Kim C, Dougherty DA, Chapman ER, Marvin JS, Looger LL, Lester HA (2019) Determining the pharmacokinetics of nicotinic drugs in the endoplasmic reticulum using biosensors. The Journal of General Physiology:jgp.201812201.

Sun F *et al.* (2018) A genetically encoded fluorescent sensor enables rapid and specific detection of dopamine in flies, fish, and mice. *Cell*, 174: 481–496.e19.

Tian L, Hires SA, Mao T, Huber D, Chiappe ME, Chalasani SH, Petreanu L, Akerboom J, McKinney SA, Schreiter ER, Bargmann CI, Jayaraman V, Svoboda K, Looger LL (2009) Imaging neural activity in worms, flies and mice with improved GCaMP calcium indicators. *Nat Methods*, 6:875–881.

Xie Y, Chan AW, McGirr A, Xue S, Xiao D, Zeng H, Murphy TH (2016) Resolution of high-frequency mesoscale intracortical maps using the genetically encoded glutamate sensor iGluSnFR. *J Neurosci*, 36: 1261–1272.

Yorgason JT, Zeppenfeld DM, Williams JT (2017) Cholinergic interneurons underlie spontaneous dopamine release in nucleus accumbens. *J Neurosci*, 37: 2086–2096.

CHAPTER 2

CNIFERS: CELL-BASED BIOSENSORS WITH NANOMOLAR SENSITIVITY TO *IN VIVO* CHANGES IN NEUROMODULATION

Daniel M. Kircher[*,§], Macit Emre Lacin[*,§], David Kleinfeld[†,‡] and Paul A. Slesinger[*,¶]

*Department of Neuroscience, Friedman Brain Institute, Icahn School of Medicine at Mount Sinai, New York, NY 10029, USA

†Department of Physics, University of California, San Diego, La Jolla, California, USA

‡Section of Neurobiology, University of California, San Diego, La Jolla, California, USA

¶paul.slesinger@mssm.edu

2.1 Introduction

Neuromodulators, such as dopamine, norepinephrine, serotonin, endocannabinoids and neuropeptides, all produce profound effects on circuits that control animal behavior. Perturbation of neuromodulation is implicated in several different brain disorders that include substance abuse and dependence, depression, and severe anxiety. A major obstacle to studying the effect of neuromodulators in the brain has been the inability to detect their release *in vivo* with sufficient sensitivity, chemical specificity, and spatial and temporal resolution. Here, we describe the development of new tools for detecting neuropeptides and other neuromodulators.

§Equal contribution.

2.2 Neuropeptides

Neuropeptides are small proteins that signal like neurotransmitters in the nervous system. Except for a few exceptions, such as FMRFamide-dependent opening of Na^+-channels, neuropeptides typically signal through G protein-coupled receptors (GPCRs) (Linguegla *et al.*, 1995), have a slower time course than ionotropic receptors, and produce persistent signaling. Peptide signaling through the GPCRs is amplified by G proteins, which stimulate/inhibit enzymes that generate second messengers like cAMP, IP_3, and Ca^{2+} that in turn modulate downstream kinases, e.g., protein kinase A (PKA), protein kinase C (PKC) (Sakurai *et al.*, 1998; Thibonnier *et al.*, 1998). These signaling pathways can also modulate the response of neurons through activation or inhibition of multiple ion channels, including voltage-gated calcium channels and potassium channels (Borgland *et al.*, 2008; Li and van den Pol, 2008).

Neuropeptides are synthesized as an inactive precursor on the rough endoplasmic reticulum in the soma, transferred to the Golgi apparatus for packaging, and then transported down the axon (Sossin *et al.*, 1989) (see Figure 2.1). The active peptide is generated by proteolytic cleavage of the precursor in large dense core

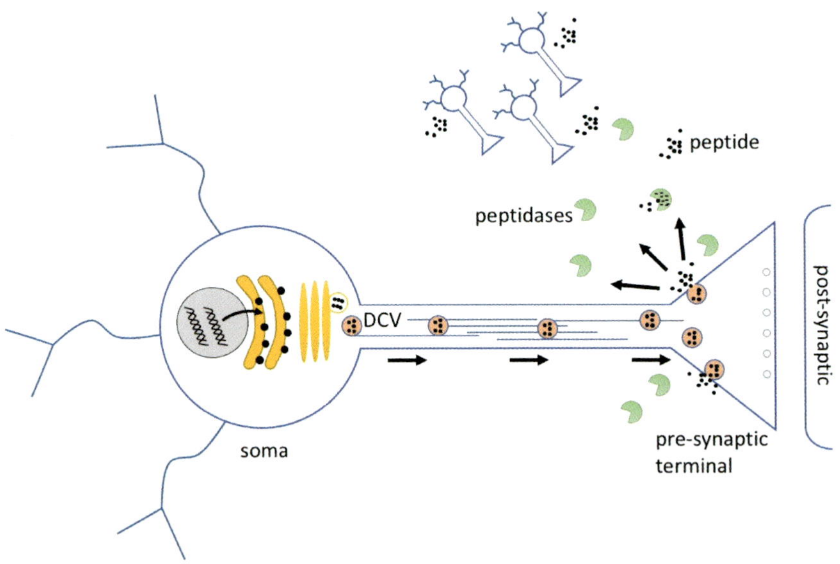

Figure 2.1 Cartoon of peptide synthesis and release in neurons. Neuropeptides are synthesized from mRNA in the soma on the rough ER as an inactive precursor, transferred to the Golgi apparatus for sorting, and transported down the axon in large dense core vesicles (DCVs). Active neuropeptide is generated by the proteolytic cleavage of the precursor peptide in DCVs. Neuropeptides act on receptors located on neurons several hundred micrometers away from release sites. Peptide half-life and diffusion are controlled, in part, by extrasynaptic peptidases, which catabolize the released peptides. In contrast to primary neurotransmitters, peptides are not recycled for re-use.

vesicles (DCVs). In general, DCVs contain multiple peptides and coexist with small vesicles housing small molecule transmitters (Chou *et al.*, 2001; Goodman *et al.*, 2007). In contrast to primary neurotransmitters, peptide-containing DCVs are mostly excluded from the active zone and found at greater distances from the synapse (Kreiner *et al.*, 1986; Fisher *et al.*, 1988). As most voltage-gated Ca^{2+} channels are localized to the active zone, peptide release is thought to require longer and higher frequency firing compared to small vesicle release (Sudhof, 2012).

The extrasynaptic localization of DCVs raises questions on the nature of neuropeptide signaling, particularly how far neuropeptides travel through the extracellular space. Minute-long half-lives of the neuropeptides and high-affinity of the neuropeptide receptors to the ligands (K_d ~ nM) have led researchers to postulate that peptides signal through volume transmission and act on extra-synaptic targets located several hundred micrometers away (Mens *et al.*, 1983; Solano *et al.*, 1996; Schloos *et al.*, 1997; Ludwig and Leng, 2006).

Perturbations in neuropeptide signaling have been implicated in several brain disorders. For example, alterations in neuropeptides appear to play a role in schizophrenia (Caceda *et al.*, 2007; LaCrosse and Olive, 2013). The hypothalamic neuropeptides oxytocin and vasopressin appear to be critically important for the regulation of social and cognitive processes that are abnormal in schizophrenia. The genes that encode oxytocin and vasopressin are associated with schizophrenia, with reduced pituitary and brain levels for both peptides found in individuals with the disease (Frank *et al.*, 2005). In contrast, oxytocin and vasopressin in the hypothalamus appear to be upregulated in individuals who suffer from major depressive disorder (Purba *et al.*, 1996).

Both of these neuropeptides can potentiate the effects of the corticotropin releasing factor (CRF) peptide and thus the activation of the hypothalamic pituitary adrenal (HPA) axis (Raadsheer *et al.*, 1994). Disruption in the stress related peptide CRF has also been implicated in drug addiction (Koob, 1999; Logrip *et al.*, 2011). Many drugs with abuse potential produce potentiation of the CRF system and likely drive atypical activity in response to stressors and contribute to relapse of drug abuse (Kreek and Koob, 1998; Roberto *et al.*, 2017).

2.3 Challenges of studying neuromodulators

Contemporary techniques for studying neuromodulators, such as microdialysis, have provided information on the extracellular concentration of neuromodulators *in vivo* (Day *et al.*, 2001). Microdialysis has been used to measure dopamine release in the medial prefrontal cortex (mPFC) in response to alcohol administration (Schier *et al.*, 2013), where alcohol administration produced significant peak increases in dopamine in the dialysate from mPFC. Microdialysis has also been used to study other neuromodulators such as the neuropeptide oxytocin. Oxytocin is released in response to social stimuli in the dorso-lateral septum, and is blocked by social fear conditioning (Zoicas *et al.*, 2014). Although microdialysis is suitable for the accurate identification of different neuromodulators (Ji *et al.*, 2008; Greco *et al.*, 2013), there are some disadvantages. First, the collection of large CSF samples and the insertion of the microdialysis probe can disrupt neuromodulator activity near the probe track (Wang and Michael, 2012). Second, the temporal resolution is poor, with the sample collection typically occurring at 10 min intervals or longer, and not in real-time (Mingote *et al.*, 2004).

An alternative method for measuring the extracellular concentration of neuromodulators is fast-scan cyclic voltammetry (FSCV) (Robinson and Wightman, 2007). Compared to microdialysis, FSCV improves the spatial and temporal resolution as well as having nanomolar sensitivity. However, the number of neuromodulators that have been successfully examined with FSCV is limited (e.g., DA, serotonin, and NE). This technique is further limited by the inability to distinguish chemically similar compounds. For example, DA and NE differ by only one hydroxyl, and produce similar cyclic voltammetry signatures (Robinson *et al.*, 2003), making examinations of neuromodulator concentrations in brain regions where both are expressed a challenge.

Measuring changes in neuronal activity with single-cell electrophysiology is another technique for studying the effect of neuropeptides in the brain. Exposure to neuropeptides induces changes in excitatory or inhibitory post synaptic currents (amplitude or frequency), action potential firing, cell excitability, or current/voltage relationships (Radhakrishnan and Henry, 1995; Raggenbass, 2001; Jobst *et al.*, 2004; Gilpin and Roberto, 2012). However, electrophysiological measurements of neuropeptide release are hampered by the small amount of neuropeptide released by a single axonal bouton in the CNS (Dreifuss, 1975; Nordmann and Morris, 1984), which is likely due to a requirement for elevations in cytoplasmic Ca^{2+} that are higher than needed for fast, synaptic release (Tallent, 2007). In addition, the response of receptive cells is relatively slow (Rosenbaum *et al.*, 2009; Hazell *et al.*, 2012), taking many seconds to minutes, making it a challenge to correlate peptide release with a post synaptic response, or to use quantal analysis of single or multiple exocytotic events (Hazell *et al.*, 2012). Unlike fast neurotransmitters, such as glutamate or GABA, the rise time for neuromodulators is much slower making it harder to segregate individual events (van den Pol, 2012).

2.4 Cell-based neurotransmitter fluorescent engineered reporters

To overcome limitations with detecting release of neuromodulators *in vivo*, we have developed a new optical approach for detecting the release of neuromodulators. This new technology is termed <u>c</u>ell-based <u>n</u>eurotransm<u>i</u>tter <u>f</u>luorescent <u>e</u>ngineered <u>r</u>eporters (CNiFERs).

CNiFERs leverage fluorescence resonance energy transfer (FRET) to translate transmitter binding into an optical signal. CNiFER cells are engineered from HEK293 cells by stably expressing a specific GPCR for the neuromodulator of interest and a FRET-based calcium sensor, e.g., TN-XXL (Nguyen *et al.*, 2010; Muller *et al.*, 2014; Lacin *et al.*, 2016). For Gq-coupled GPCRs, binding of the neuromodulator of interest to the GPCR expressed on the surface of HEK293 cells leads to an increase in intracellular Ca^{2+} which is detected by the FRET-based Ca^{2+} sensor, i.e. decrease in CFP (FRET donor) emission and an increase of YFP (FRET acceptor) emission (Figure 2.2). CNiFERs can be implanted directly into target brain regions where they can detect release of physiological concentrations of modulators with a temporal resolution of seconds and spatial resolution of less than 100 μm (Nguyen *et al.*, 2010; Muller *et al.*, 2014) (Figures 2.3 B–D). The dual wavelength measurement of the

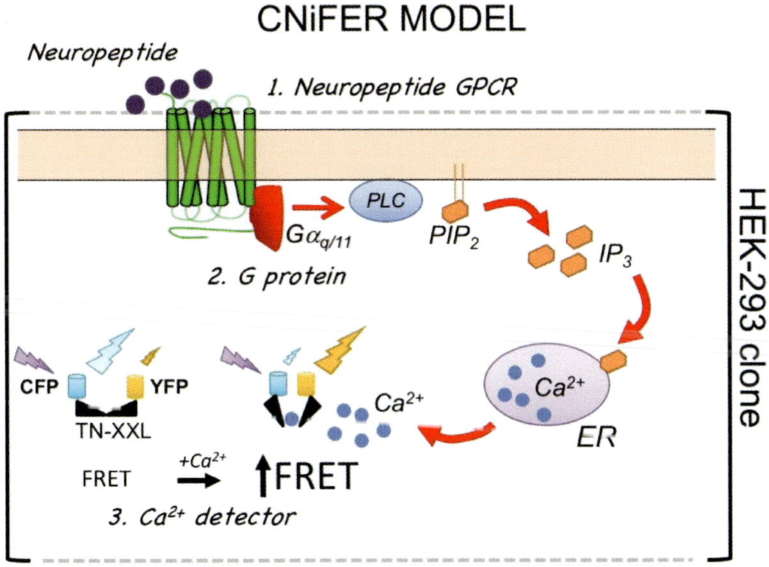

Figure 2.2 Design of peptide CNiFERs. (1) The GPCR for the neuropeptide of interest is expressed on the plasma membrane of HEK-293 cells. (2) Neuropeptide binding produces a conformational change and activation of the G_q G protein, which in turn activates phospholipase C (PLC). A chimeric protein (G_{qi5} or G_{qs5}) can also be expressed to redirect $G_{i/o}$ or G_s-coupled receptors, respectively, to the PLC pathway. Cleavage of PIP_2 into diacyl glycerol and IP_3 by PLC leads to the release of intracellular Ca^{2+} from the endoplasmic reticulum. (3) Ca^{2+} binds to the genetically encoded Ca^{2+} sensor which undergoes FRET. Two-photon excitation of the CNiFER in the absence of the neuropeptide agonist exhibits high CFP and low YFP fluorescence. Neuropeptide binding increases intracellular Ca^{2+} and produces an increase in FRET, detected as a decrease in CFP emission and increase in the YFP emission with CFP excitation.

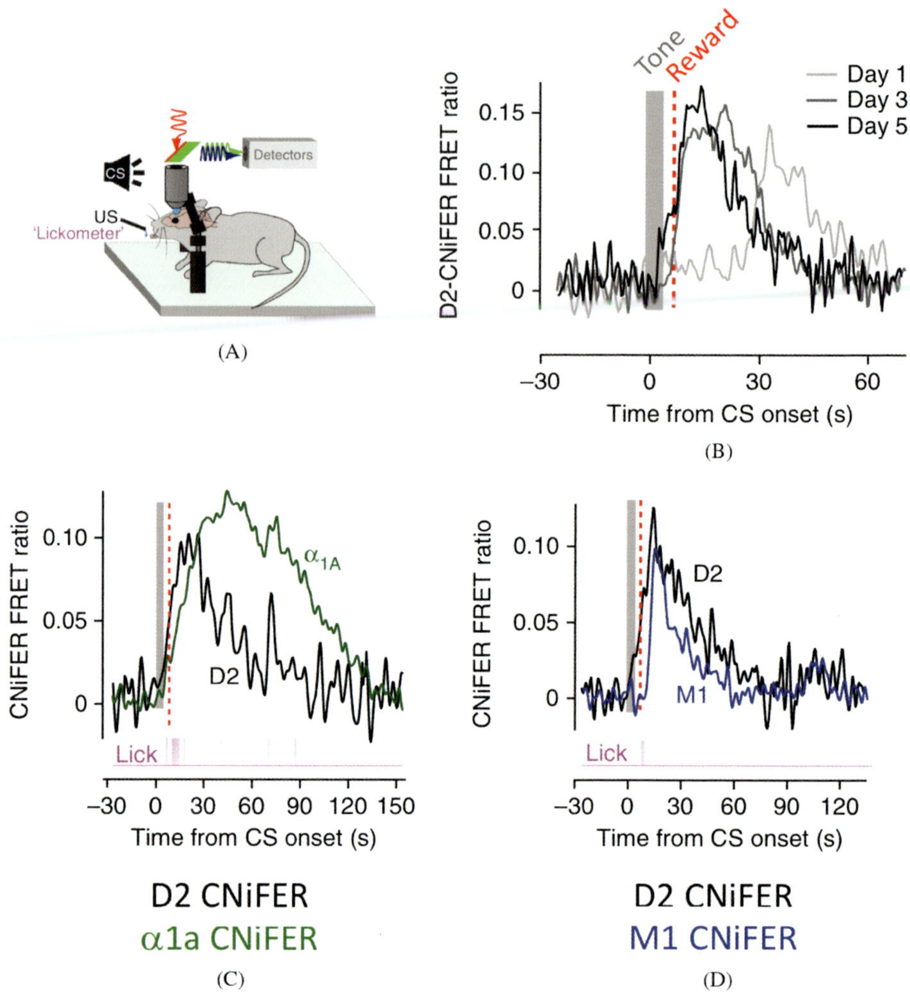

Figure 2.3 Example of CNiFER detection of DA, NE and ACh *in vivo.* (A) Procedure to measure licking behavior and CNIFER fluorescence in head-restrained mice during classical conditioning. (B) Single trial examples of the D2-CNiFER response in the same animal at days 1, 3 and 5 of training. (C) Simultaneous measurement of D2- and α1A-CNiFER FRET, or D2- and M1-CNiFER FRET, in conjunction with bouts of licking, during a single conditioning trial (modified from Muller *et al.*, 2014).

The FRET-based Ca^{2+} sensor provides an intrinsic signal for correcting possible motion-induced artifacts.

2.4.1 Creation of cell-based neurotransmitter fluorescent engineered reporters (CNiFERs)

For each neuromodulator, an appropriate GPCR is stably expressed in a clonal HEK293 cell with a Ca^{2+} sensor, i.e., TN-XXL or equivalent Ca^{2+} sensor

(Figure 2.2). For $G_{i/o}$- or G_s-coupled receptors, a chimeric G protein G_{qi5} or G_{qs5}, respectively, is also expressed in the HEK293 cell to redirect GPCR signaling to the G_q-Ca^{2+} pathway. The calculated fractional change in fluorescence, $\Delta F/F$, for each signal is used to obtain the FRET ratio, denoted $\Delta R/R$. Each clonal CNiFER is assessed for potency (e.g. EC_{50}), efficacy (i.e., largest increase in FRET ratio), and lowest background responses to other neurotransmitter/neuromodulators (Nguyen *et al.*, 2010; Muller *et al.*, 2014; Lacin *et al.*, 2016). The control CNiFER contains the same Ca^{2+} detection components but lacks the specific GPCR. The CNiFER approach is easily adapted for any neurotransmitter, neuromodulator or neuropeptide that signals through a specific GPCR. Thus, a library of CNiFERs can be created for assessing the release of a large number of different neuromodulators. In addition, implanting two different CNiFER types in the brain affords the opportunity for multiplexing, and simultaneous detection of the release of two different neuromodulators (Figures 2.3C and 2.3D) (Nguyen *et al.*, 2010; Muller *et al.*, 2014; Lacin *et al.*, 2016).

The creation of dopamine and norepinephrine CNiFERs provides a good example of simultaneous detection of chemically similar neurotransmitters (Muller *et al.*, 2014). The dopamine CNiFER was created with the D2 dopamine receptor (D2) and norepinephrine CNiFER with the α_{1A} receptor. The D2 and α_{1A} CNiFERs have an EC_{50} of 2.5 nM and 19 nM, respectively, suitable for detection of endogenously released transmitter (Muller *et al.*, 2014). We injected the D2 CNiFER and α_{1A} CNiFER into the cortex, approximately 200 μm apart, and measured the release of DA and NE in awake mice undergoing Pavlovian conditioning, i.e. repeated pairing of a tone (CS) with a sugar reward (Figures 2.3A and 2.3B). We also compared the release of DA with ACh, using a muscaranic M1 CNiFER (Nguyen *et al.*, 2010) and D2 CNiFER for detecting DA (Figure 2.3C). This multiplexing of two different CNiFERs in the brain allows for monitoring the release of two different neuromodulators simultaneously. As predicted, the timing of DA release shifts from the reward to the CS with training. Interestingly, the timing of NE release is highly variable whereas ACh release is tightly correlated with licking (Muller *et al.*, 2014).

Multiplexing of CNiFERs could also be used to compare a single transmitter interacting with two different receptors. For example, dopamine signals through both D1 and D2 receptors, which have different affinities for DA (Lobo and Nestler, 2011). Endogenous dopamine release in a particular region may preferentially activate one CNiFER (D2 *versus* D1) over the other. Another example of multiplexing CNiFERs is to co-implant one CNiFER for a traditional neuromodulator, such as DA, with another CNiFER that expresses a GPCR for a neuropeptide, such as for the kappa opioid receptor to measure the neuropeptide dynorphin (Dyn). DA and Dyn CNiFERs could be implanted into the striatum of a mouse to measure release dynamics in response to exposure to a drug of abuse such as alcohol (Resendez *et al.*, 2016).

The CNiFER design allows for measuring neuromodulator release on the time scale of seconds, mimicking the activation of natively expressed GPCRs. This temporal response of CNiFERs is a significant improvement over the response time of minutes to hours for microdialysis. The fast response time, high affinity, and the ability to generate a library of sensors capable of measuring any modulator that signals through a GPCR, makes CNiFERs well suited for examining the temporal dynamics of local neuropeptide release. A limitation with CNiFERs, on the other hand, is the inability to directly capture the activity of neuromodulators released within the synapse.

2.5 Future for CNiFER-based research

CNiFERs were first created with HEK293 cells, because HEK293 cells typically have low expression of endogenous GPCRs and are easily transduced to create clonal lines. Although HEK293 cells are human cells, implantation into mice does not appear to produce a significant inflammatory response (Muller *et al.*, 2014) when mice are injected with an immunosuppressant (e.g., cyclosporine). Under these conditions, CNiFERs can remain functional for up to a week. However, monitoring neuromodulator release over longer periods might require repeated CNiFER injections. An alternative approach would be to create CNiFERs using rodent cell lines, such as mouse embryonic fibroblasts or chinese hamster ovary (CHO) cells. We have tested CHO cells and measured FRET responses comparable to HEK293 cells (unpublished data).

Currently, CNiFERs are implanted into cortex and monitored *in vivo* using two-photon laser scanning microscopy (TPLSM). However, neuropeptide-producing cells and terminals are widely distributed in the brain, including subcortical regions (Lein *et al.*, 2007; Sakurai, 2014; Acevedo-Rodriguez *et al.*, 2015). Standard TPLSM allows for imaging up to 1 mm in depth and cannot easily image areas below the cortex. In addition, animals are head-fixed for most TPLSM setups, limiting the range of behaviors that can be studied during imaging. Recent advances in imaging technologies have made it possible to monitor subcortical activity in freely moving animals (Gosh *et al.*, 2011; Cai *et al.*, 2016). For example, miniature microscopes (e.g. endoscopes or Miniscopes) contain all the electronic and optical components necessary for imaging and are directly mounted to a mouse skull and tethered to a cable. The head-mounted scopes are usually coupled with a gradient-refractive-index (GRIN) lens that is implanted near the target and relays the fluorescent signal to the scope (Girven and Sparta, 2017). Although currently available Miniscopes are designed for single-wavelength measurements, it should be possible to create a two-channel Miniscope without adding too much weight. Another approach for subcortical measurements of CNiFERs is to use fiber photometry. Fiber photometry is commonly used to measure Ca^{2+} transients using genetically-encoded Ca^{2+} sensors (e.g., GCaMPs) and involves insertion of an optical fiber and coupling to photomultiplier tubes (PMTs) (Girven and Sparta, 2017). Fibers have the advantage of producing less tissue damage because of the small size, ~0.2 mm, compared to GRIN lenses which typically are 0.5–1 mm in diameter (Girven and Sparta, 2017). Fiber photometry measurements lack spatial information, due to summation of the light signal from all cells. However, the average FRET activity is measured in multiple CNiFER cells within an implant, making fiber photometry well suited for CNiFER imaging.

An alternate approach to cell-based CNiFERs is the development of genetically encoded neuromodulator sensors that are capable of capturing the activity of neuromodulators within synapses. The first generation of genetically encoded neurotransmitter detectors were based on mutations of periplasmic binding proteins and detected glutamate release (iGluSnFR (Marvin *et al.*, 2013), EOS (Namiki *et al.*, 2007), Snifit (Brun *et al.*, 2011), FLIPE (Okumoto *et al.*, 2005), albeit with microMolar sensitivity. Recently, two groups have developed sensors for detecting DA based on coupling steric changes in a GPCR for DA, which occur upon binding of DA, into producing changes in the optical properties of a closely coupled fluorescent protein (FP). Patriarchi *et al.* (2018) created a fusion protein that directly couples dopamine binding induced conformational changes in an inert human dopamine receptor to changes in the fluorescence intensity of a circularly permuted green fluorescent protein (cpGFP) (termed dLight1). Sun *et al.* (2018) used a similar strategy but inserted cpEGFP in the intracellular loop of the human dopamine receptor (termed $GRAB_{DA}$). One advantage of these sensors is the ability to express the sensor in a particular neuronal population of interest and capturing the temporal dynamics of neuromodulator activity within the synapse (Patriarchi *et al.*, 2018; Sun *et al.*, 2018). Additionally, the wide expression enables detection of neuromodulator concentrations across the cortical mantle. Yet, with one exception (Sun *et al.*, 2018) the *in vivo* sensitivity of molecular sensors have yet to achieve the nanoMolar sensitivity needed to study neuromodulation *in vivo*.

An advantage of using CNiFERs is the ability to multiplex and study the simultaneous release of multiple neuromodulators. Currently, the simultaneous imaging of two different transmitters is accomplished by injecting two types of CNiFERs within a 100 μm of each other. Because the two sensors are spatially distinct, it is possible to measure FRET in two implants simultaneously. Another approach is to mix CNiFERs of different colors in the same implant. For example, choosing a calcium sensor that emits at a longer wavelength than the CFP and

YFP of TN-XXL, such as with a red Ca^{2+} sensor (Akerboom *et al.*, 2013). A CNiFER with a longer wavelength calcium sensor also has the advantage of imaging deeper in the brain (Akerboom *et al.*, 2013; Tischbirek *et al.*, 2015) and the longer wavelength excitation two-photon light necessary to excite a red or near-infrared fluorophore scatters less, produces less phototoxic damage, and penetrates deeper (Yaroslavsky *et al.*, 2002; Tischbirek *et al.*, 2015). Moreover, when CNiFER imaging is coupled to optogenetics, it enables continuous imaging of CNiFERs during blue light stimulation of channel-rhodopsin.

In summary, the methods for optically detecting the release of neurotransmitters or neuromodulators *in vivo* in awake animals are rapidly advancing, providing neuroscientists with state-of-the-art tools for probing the function of neuromodulators on brain circuits.

Acknowledgments

Work presented in this chapter was supported by research grants through the US National Institute of Mental Health BRAIN Initiative R01 (MH111499), National Institute on Drug Abuse (NIDA) (DA029706; DA037170), National Institute of Biomedical Imaging and Bioengineering (NIBIB) (EB003832), Hoffman-La Roche (88610A), and the "Neuroscience Related to Drugs of Abuse" training grant through NIDA (DA007315).

References

Acevedo-Rodriguez A, Mani SK, Handa RJ (2015) Oxytocin and estrogen receptor beta in the brain: An overview. *Front Endocrinol (Lausanne)*, 6: 160.

Akerboom J *et al.* (2013) Genetically encoded calcium indicators for multi-color neural activity imaging and combination with optogenetics. *Front Mol Neurosci*, 6: 2.

Borgland SL, Storm E, Bonci A (2008) Orexin B/hypocretin 2 increases glutamatergic transmission to ventral tegmental area neurons. *Eur J Neurosci*, 28: 1545–1556.

Brun MA, Griss R, Reymond L, Tan KT, Piguet J, Peters RJ, Vogel H, Johnsson K (2011) Semisynthesis of fluorescent metabolite sensors on cell surfaces. *J Am Chem Soc*, 133: 16235–16242.

Caceda R, Kinkead B, Nemeroff CB (2007) Involvement of neuropeptide systems in schizophrenia: Human studies. *Int Rev Neurobiol*, 78: 327–376.

Cai DJ *et al.* (2016) A shared neural ensemble links distinct contextual memories encoded close in time. *Nature*, 534: 115–118.

Chou TC, Lee CE, Lu J, Elmquist JK, Hara J, Willie JT, Beuckmann CT, Chemelli RM, Sakurai T, Yanagisawa M, Saper CB, Scammell TE (2001) Orexin (hypocretin) neurons contain dynorphin. *J Neurosci*, 21: RC168.

Day JC, Kornecook TJ, Quirion R (2001) Application of *in vivo* microdialysis to the study of cholinergic systems. *Methods*, 23: 21–39.

Dreifuss JJ (1975) A review on neurosecretory granules: Their content and mechanisms of release. *AnnNY Acad Sci*, 248: 184–201.

Fisher JM, Sossin W, Newcomb R, Scheller RH (1988) Multiple neuropeptides derived from a common precursor are differentially packaged and transported. *Cell*, 54: 813–822.

Frank M, Thumer L, Lohse MJ, Bunemann M (2005) G Protein activation without subunit dissociation depends on a Gαi-specific region. *J Biol Chem*, 280: 24584–24590.

Ghosh KK, Burns LD, Cocker ED, Nimmerjahn A, Ziv Y, Gamal AE, Schnitzer MJ (2011) Miniaturized integration of a fluorescence microscope. *Nat Methods*, 8(10): 871–878.

Gilpin NW, Roberto M (2012) Neuropeptide modulation of central amygdala neuroplasticity is a key mediator of alcohol dependence. *Neurosci Biobehav Rev*, 36: 873–888.

Girven KS, Sparta DR (2017) Probing deep brain circuitry: New advances in *in vivo* calcium measurement strategies. *ACS Chem Neurosci*, 8: 243–251.

Goodman RL, Lehman MN, Smith JT, Coolen LM, de Oliveira CV, Jafarzadehshirazi MR, Pereira A, Iqbal J, Caraty A, Ciofi P, Clarke IJ (2007) Kisspeptin neurons in the arcuate nucleus of the ewe express both dynorphin A and neurokinin B. *Endocrinology*, 148: 5752–5760.

Greco S, Danysz W, Zivkovic A, Gross R, Stark H (2013) Microdialysate analysis of monoamine neurotransmitters — A versatile and sensitive LC–MS/MS method. *Analy Chim Acta*, 771: 65–72.

Hazell GGJ, Hindmarch CC, Pope GR, Roper JA, Lightman SL, Murphy D, O'Carroll A-M, Lolait SJ (2012) G protein-coupled receptors in the hypothalamic paraventricular and supraoptic nuclei — serpentine gateways to neuroendocrine homeostasis. *Front Neuroendocrin*, 33: 45–66.

Ji C, Li W, Ren X-d, El-Kattan AF, Kozak R, Fountain S, Lepsy C (2008) Diethylation labeling combined with UPLC/MS/MS for simultaneous determination of a panel of monoamine neurotransmitters in rat prefrontal cortex microdialysates. *Anal Chem*, 80: 9195–9203.

Jobst EE, Enriori PJ, Cowley MA (2004) The electrophysiology of feeding circuits. *Trends Endocrinol Metab*, 15: 488–499.

Koob GF (1999) Stress, corticotropin-releasing factor, and drug addiction. *Ann NY Acad Sci*, 897: 27–45.

Kreek MJ, Koob GF (1998) Drug dependence: Stress and dysregulation of brain reward pathways. *Drug Alcohol Depend*, 51: 23–47.

Kreiner T, Sossin W, Scheller RH (1986) Localization of Aplysia neurosecretory peptides to multiple populations of dense core vesicles. *J Cell Biol*, 102: 769–782.

Lacin E, Muller A, Fernando M, Kleinfeld D, Slesinger PA (2016) Construction of cell-based neurotransmitter fluorescent engineered reporters (CNiFERs) for optical detection of neurotransmitters *in vivo*. *J Vis Exp*, May 12; (111).

LaCrosse AL, Olive MF (2013) Neuropeptide systems and schizophrenia. *CNS Neurol Disord Drug Targets*, 12: 619–632.

Lein ES *et al.* (2007) Genome-wide atlas of gene expression in the adult mouse brain. *Nature*, 445: 168–176.

Li Y, van den Pol AN (2008) Mu-opioid receptor-mediated depression of the hypothalamic hypocretin/orexin arousal system. *J Neurosci*, 28: 2814–2819.

Lingueglia E, Champigny G, Lazdunski M, Barbry P (1995) Cloning of the amiloride-sensitive FMRFamide peptide-gated sodium channel. *Nature*, 378: 730–733.

Lobo MK, Nestler EJ (2011) The striatal balancing act in drug addiction: Distinct roles of direct and indirect pathway medium spiny neurons. *Front Neuroanat*, 5: 41.

Logrip ML, Koob GF, Zorrilla EP (2011) Role of corticotropin-releasing factor in drug addiction: Potential for pharmacological intervention. *CNS Drugs*, 25: 271–287.

Ludwig M, Leng G (2006) Dendritic peptide release and peptide-dependent behaviours. *Nat Rev Neurosci*, 7: 126–136.

Marvin JS, Borghuis BG, Tian L, Cichon J, Harnett MT, Akerboom J, Gordus A, Renninger SL, Chen TW, Bargmann CI, Orger MB, Schreiter ER, Demb JB, Gan WB, Hires SA, Looger LL (2013) An optimized fluorescent probe for visualizing glutamate neurotransmission. *Nat Methods*, 10: 162–170.

Mens WB, Witter A, van Wimersma Greidanus TB (1983) Penetration of neurohypophyseal hormones from plasma into cerebrospinal fluid (CSF): Half-times of disappearance of these neuropeptides from CSF. *Brain Res*, 262: 143–149.

Mingote S, de Bruin JPC, Feenstra MG (2004) Noradrenaline and dopamine efflux in the prefrontal cortex in relation to appetitive classical conditioning. *J Neurosci*, 24: 2475–2480.

Muller A, Joseph V, Slesinger PA, Kleinfeld D (2014) Cell-based reporters reveal *in vivo* dynamics of dopamine and norepinephrine release in murine cortex. *Nat Methods*, 11: 1245–1252.

Namiki S, Sakamoto H, Iinuma S, Iino M, Hirose K (2007) Optical glutamate sensor for spatiotemporal analysis of synaptic transmission. *Eur J Neurosci*, 25: 2249–2259.

Nguyen QT, Schroeder LF, Mank M, Muller A, Taylor P, Griesbeck O, Kleinfeld D (2010) An *in vivo* biosensor for neurotransmitter release and in situ receptor activity. *Nat Neurosci*, 13: 127–132.

Nordmann JJ, Morris JF (1984) Method for quantitating the molecular content of a subcellular organelle: Hormone and neurophysin content of newly formed and aged neurosecretory granules. *Proc Natl Acad Sci USA*, 81: 180–184.

Okumoto S, Looger LL, Micheva KD, Reimer RJ, Smith SJ, Frommer WB (2005) Detection of glutamate release from neurons by genetically encoded surface-displayed FRET nanosensors. *Proc Natl Acad Sci USA*, 102: 8740–8745.

Patriarchi T, Cho JR, Merten K, Howe MW, Marley A, Xiong WH, Folk RW, Broussard GJ, Liang R, Jang MJ, Zhong H, Dombeck D, von Zastrow M, Nimmerjahn A, Gradinaru V, Williams JT, Tian L (2018) Ultrafast neuronal imaging of dopamine dynamics with designed genetically encoded sensors. *Science*, 360: pii: eaat4422.

Purba JS, Hoogendijk WJ, Hofman MA, Swaab DF (1996) Increased number of vasopressin- and oxytocin-expressing neurons in the paraventricular nucleus of the hypothalamus in depression. *Arch Gen Psychiatry*, 53: 137–143.

Raadsheer FC, Hoogendijk WJ, Stam FC, Tilders FJ, Swaab DF (1994) Increased numbers of corticotropin-releasing hormone expressing neurons in the hypothalamic paraventricular nucleus of depressed patients. *Neuroendocrinology*, 60: 436–444.

Radhakrishnan V, Henry JL (1995) Electrophysiology of neuropeptides in the sensory spinal cord. *Prog Brain Res*, 104: 175–195.

Raggenbass M (2001) Vasopressin- and oxytocin-induced activity in the central nervous system: Electrophysiological studies using *in-vitro* systems. *Prog Neurobiol*, 64: 307–326.

Resendez SL, Jennings JH, Ung RL, Namboodiri VM, Zhou ZC, Otis JM, Nomura H, McHenry JA, Kosyk O, Stuber GD (2016) Visualization of cortical, subcortical and deep brain neural circuit dynamics during naturalistic mammalian behavior with head-mounted microscopes and chronically implanted lenses. *Nat Protoc*, 11: 566–597.

Roberto M, Spierling SR, Kirson D, Zorrilla EP (2017) Corticotropin-releasing factor (CRF) and addictive behaviors. *Int Rev Neurobiol*, 136: 5–51.

Robinson DL, Venton BJ, Heien ML, Wightman RM (2003) Detecting subsecond dopamine release with fast-scan cyclic voltammetry *in vivo*. *Clin Chem*, 49: 1763–1773.

Robinson DL, Wightman RM (2007) Rapid Dopamine Release in Freely Moving Rats. In: Electrochemical Methods for Neuroscience (Michael AC, Borland LM, eds). Boca Raton (FL).

Rosenbaum DM, Rasmussen SGF, Kobilka BK (2009) The structure and function of G-protein-coupled receptors. *Nature*, 459: 356–363.

Sakurai T (2014) The role of orexin in motivated behaviours. *Nat Rev Neurosci*, 15: 719–731.

Sakurai T *et al.* (1998) Orexins and orexin receptors: a family of hypothalamic neuropeptides and G protein-coupled receptors that regulate feeding behavior. *Cell*, 92: 573–585.

Schier CJ, Dilly GA, Gonzales RA (2013) Intravenous ethanol increases extracellular dopamine in the medial prefrontal cortex of the long-evans rat. *Alcohol Clin Exp Res*, 37: 740–747.

Schloos J, Raulf F, Hoyer D, Bruns C (1997) Identification and pharmacological characterization of somatostatin receptors in rat lung. Br *J Pharmacol*, 121: 963–971.

Solano RM, Carmena MJ, Carrero I, Cavallaro S, Roman F, Hueso C, Travali S, Lopez-Fraile N, Guijarro LG, Prieto JC (1996) Characterization of vasoactive intestinal peptide/pituitary adenylate cyclase-activating peptide receptors in human benign hyperplastic prostate. *Endocrinology*, 137: 2815–2822.

Sossin WS, Fisher JM, Scheller RH (1989) Cellular and molecular biology of neuropeptide processing and packaging. *Neuron*, 2: 1407–1417.

Sudhof TC (2012) The presynaptic active zone. *Neuron*, 75: 11–25.

Sun F *et al.* (2018) A genetically encoded fluorescent sensor enables rapid and specific detection of dopamine in flies, fish, and mice. *Cell*, 174: 481–496.

Tallent MK (2007) Presynaptic inhibition of glutamate release by neuropeptides: Use-dependent synaptic modification. In: *Inhibitory Regulation of Excitatory Neurotransmission*, pp. 177–200. Berlin: Springer Berlin Heidelberg.

Thibonnier M, Berti-Mattera LN, Dulin N, Conarty DM, Mattera R (1998) Signal transduction pathways of the human V1-vascular, V2-renal, V3-pituitary vasopressin and oxytocin receptors. *Prog Brain Res*, 119: 147–161.

Tischbirek C, Birkner A, Jia H, Sakmann B, Konnerth A (2015) Deep two-photon brain imaging with a red-shifted fluorometric Ca^{2+} indicator. *Proc Natl Acad Sci USA*, 112: 11377–11382.

van den Pol AN (2012) Neuropeptide transmission in brain circuits. *Neuron*, 76: 98–115.

Wang Y, Michael AC (2012) Microdialysis probes alter presynaptic regulation of dopamine terminals in rat striatum. *J Neurosci Methods*, 208: 34–39.

Yaroslavsky AN, Schulze PC, Yaroslavsky IV, Schober R, Ulrich F, Schwarzmaier HJ (2002) Optical properties of selected native and coagulated human brain tissues *in vitro* in the visible and near infrared spectral range. *Phys Med Biol*, 47: 2059–2073.

Zoicas I, Slattery DA, Neumann ID (2014) Brain oxytocin in social fear conditioning and its extinction: Involvement of the lateral septum. *Neuropsychopharm*, 39: 3027–3035.

CHAPTER 3

FLUORESCENT FALSE NEUROTRANSMITTERS

David Sulzer*,‡ and Dalibor Sames†

*Departments of Psychiatry, Neurology, Pharmacology, Columbia University Medical Center, New York State Psychiatric Institute, USA

†Department of Chemistry, Columbia University, USA

‡ds43@cumc.columbia.edu

3.1 Introduction

The monoaminergic neurotransmitters, particularly dopamine (DA), norepinephrine (NE), and serotonin (5-hydroxytryptamine: 5HT), are widely dispersed throughout the central nervous system and are involved in a multitude of important physiological roles. While many of the axonal release sites of monoamine neurons can be found to have small morphological synaptic specializations, they engage in "social" neurotransmission that entails the diffusion of neurotransmitter to distal extrasynaptic receptors on many cells (Agnati *et al.*, 1986; Sulzer and Pothos, 2000; Sulzer *et al.*, 2016). Thus mechanisms to observe the presynaptic activity of these axons are particularly relevant.

As the monoamines overflow beyond their "synapse", their release and reuptake are effectively measured by extracellular and obviously extrasynaptic carbon fiber microelectrodes in brain tissue. Indeed, most of the important biological insights of DA neurotransmission in particular that have been made since the 1960s have used amperometry and fast scan cyclic voltammetry (FSCV). These approaches, as well as electrochemical detection by microdialysis, are discussed in other chapters in this series.

More recently, a variety of optical approaches based on monoamine receptors that can act as fluorescent reporters have been introduced, including CNifers (Muller *et al.*, 2014), iTango (Lee *et al.*, 2017) and dLight1 (Patriarchi *et al.*, 2018). These are rapidly becoming highly attractive for many labs, particularly as there has been a wide adoption of optical recording methods over the past decade.

Another approach for analyzing release of DA is provided by native or ectopically expressed GIRK channels, which produce an inhibitory ionotropic current when DA activates D2 DA receptors that can be measured by patch clamp (Ford *et al.*, 2009; Marcott *et al.*, 2014).

Each of these methods provide complementary and useful insights, but due to the "social" nature of monoamine transmission, they measure release from many presynaptic sites. There is, however, abundant evidence from electrochemical recordings *in vivo* of striking differences in the presynaptic activity of monoamine axons, including "hot" and "cold" spots of release coupled to specific behaviors in the striatum despite the local presence of DA axons (Owesson-White *et al.*, 2016). To analyze differences between monoamine synapses, an optical approach with very high spatial resolution is required.

Fluorescence microscopy provides sufficient spatial and temporal resolution to measure release from individual synapses. While endocytic probes, such as FM1-43, measure the fusion of synaptic vesicles, they do not measure transmitter release. The genetic fluorescent synaptopHluorin probes likewise measure membrane fusion by detecting changes in pH during fusion and recycling, but do not indicate the vesicle content release or distribution.

To analyze the local activity and modulation of monoamine transmission at specific sites, we developed a set of "fluorescent false neurotransmitters" (FFNs), with properties similar to native neurotransmitters. While in principle applicable to any transmitter, to date these are most effective for the monoamines. In contrast to the other optical approaches, FFNs can report release from individual monoamine synapses. The use of FFNs indicates vast differences between monoamine release from individual sites, even from neighboring sites on the same axon (Pereira *et al.*, 2016), and differences in the properties of secretory vesicles, even within the same cell (Lee *et al.*, 2010).

FFNs can further provide optical measures to estimate monoamine uptake by plasma membrane and vesicular transporters. As we discuss, another advantage is the ability to measure transmission in the many brain regions that have very low levels of monoamines, due to the development of pH-dependent FFNs.

3.2 The classical use of fluorescent monoamine derivatives in fixed tissue

The initial use of fluorescent monoamine derivatives provided an important chapter for neuroscience including the first means to visualize neurotransmitters within cells. This use of fluorescent monoamine derivatives long preceded the use of immunocytochemistry to identify monoamine neurons and their projections.

In the early 1960s, Bengt Falck and Nils-Ake Hillarp at the University of Lund, inspired by Arvid Carlsson's laboratory's recent discovery of high levels of monoamines in the brain by production of fluorescent trihydroxyindole derivatives (Bertler *et al.*, 1958; Carlsson and Waldeck, 1958), developed a fluorescent method to label monoamines in brain tissue. They succeeded in producing fluorescent derivatives of endogenous monoamines by exposing freeze-dried tissue to formaldehyde vapor (Falck and Hillarp, 1959; Carlsson *et al.*, 1962; Falck, 1962). The team found that this occurred by conversion of DA and NE to the intensely fluorescent 1,2,3,4-tetrahydroisoquinoline derivatives (the original 1961 paper was reprinted in 1982) (Falck *et al.*, 1982).

The Falck-Hillarp technique produced stunning fluorescent DA, NE, and 5HT axonal images in brain sections, and was adapted by Hillarp, Annica Dahlstrom, and Kjell Fuxe to identify the monoaminergic nuclei (Hillarp *et al.*, 1966). The nomenclature they developed (e.g., A10 for ventral tegmental DA neurons, A9 for substantia nigra DA neurons) is still used.

The Falck-Hillarp approach was subsequently advanced by the glyoxylic acid method of Olle Lindvall and Anders Bjorklund, which converted DA and NE to 2-carboxymethyl-dihydroisoquinoline derivatives, providing improved sensitivity for imaging DA axons (Lindvall and Bjorklund, 1974).

3.3 The vesicular monoamine transporter accumulates a wide variety of substrates

The basis for the many substrates that exist for monoamine neurosecretory vesicles is due to the "promiscuity" of the vesicular monoamine transporter (VMAT). Indeed, VMAT1 was originally cloned by Robert Edward's laboratory due to its ability to accumulate the neurotoxin substrate MPP^+, which protected the cells from its toxicity (Liu *et al.*, 1992a, 1992b).

The initial studies from the Edwards lab demonstrated that VMAT1 and its brain counterpart, VMAT2, accumulate the endogenous monoamines DA, NE, 5HT, epinephrine, and histamine (Liu *et al.*, 1992a, 1992b; Peter *et al.*, 1994). Shimon Schuldiner's lab followed this to demonstrate that a wide variety of compounds bound to the substrate binding site of VMAT1 (Gros and Schuldiner,

2010) and that the fluorescent compound, rhodamine, is a fluorescent substrate (Yelin and Schuldiner, 1995), although it has not been used as an FFN.

3.4 Native monoamine fluorescence in living tissue

The lab of Watt Webb, who is credited with the development of multiphoton microscopy, realized that 5HT itself could act as a fluorescent compound with three photon excitation of its native UV fluorescence (Maiti *et al.*, 1997; Williams *et al.*, 1999). The use of three photon excitation provided important information on the modes of vesicle exocytosis by large non-synaptic cell body secretory vesicles in mast cells, including, to our knowledge, the first optical demonstration of kiss and run vesicle fusion as well as full fusion of secretory vesicles by the same cell (Williams and Webb, 2000).

This approach has proved difficult to use in neurons, although it has been used to analyze uptake and release from 5HT neuronal cell bodies of the dorsal raphe (Colgan *et al.*, 2004). Due, however, to low signal and neurotoxicity from the high excitation energy, to our knowledge, this approach has not been used successfully in axons.

3.5 The concept of false transmitters

The development of FFNs was inspired from classic work by Irwin Kopin and colleagues who introduced the concept of "false transmitters," originally from studies of the peripheral secretory glands including the adrenal (Kopin *et al.*, 1964; Kopin, 1968). In this scenario, a molecule other than the native monoamine is accumulated within a secretory vesicle and is then available for release by exocytosis.

The paradigmatic false transmitter is tyramine, which is normally broken down by monoamine oxidase. It is thought that dietary tyramine of microbial origin found in red wine, beer, cheese, and sausage promotes hypertension in patients taking monoamine oxidase inhibitors by a false transmitter action that displaces peripheral epinephrine and norepinephrine from secretory vesicles, some of which is released and activates adrenergic receptors.

Due to the ability of VMAT to transport a broad range of substrates, there are many other false transmitters for monoamine systems, including amphetamine and its derivatives including fenfluramine, "ecstasy" (3,4-methylenedioxymethamphetamine), and methamphetamine (Sulzer *et al.*, 2005), and phenylephrine which provides a PET tracer for release from peripheral neurons (Cheng, 2004), the toxins N-methyl-4-phenylpyridinium (MPP^+), the osmophilic monoamine synaptic vesicle marker 5-hydroxydopamine, isoprenaline, metaraminol, and a variety of other compounds.

Fewer false transmitters for glutamate synapses have been reported, including 3-aminoglutarate, which was identified as a substrate for the vesicular glutamate 1 transporter. (Foster *et al.*, 2015; Wu *et al.*, 2015), and threo-beta-hydroxyaspartate (Fleck *et al.*, 2001).

For cholinergic vesicles, monoethylcholine (Van der Kloot and Naves, 1996) carbachol (Baux and Tauc, 1983) and N-aminodeanol (Newton *et al.*, 1985) and additional choline derivatives are reported to act as false transmitters.

3.6 Loading of native monoamines as false transmitters to measure secretion

Native monoamines have been loaded into cells to provide a means for detection, although this has to date been accomplished for electrochemistry rather than optical measures.

DA or 5HT has been loaded into a broad set of non-monoamine cells, including cerebellar granule cells and even yeast, to provide detection by amperometry. It has been suggested that this is due to trapping by protonation in the low pH vesicle compartment (Kim *et al.*, 2000; Koh *et al.*, 2000).

NE is effective as a false transmitter for loading glutamatergic retinal bipolar neurons, enabling amperometric detection of quantal release (Grabner and Zenisek, 2013).

The ability to load monoamines into non-monoamine cell vesicles is greatly enhanced by inducing VMAT2 expression. For these cells, providing a source of DA via tyrosine hydroxylase expression or addition of L-DOPA. was sufficient to convert nominally non-monoamine AtT-20 cells and even hippocampal neurons to DA releasing cells from which exocytosis could be measured by amperometry (Li *et al.*, 2005; Pothos *et al.*, 1999, 2000).

3.7 Early attempts at developing FFNs

A preliminary FFN was the 5HT derivative, 5,7-dihydroxytryptamine (5,7-DHT), which was explored as a neurotoxin for 5HT neurons (Baumgarten and Lachenmayer, 1972; Bjorklund *et al.*, 1974). 5,7-DHT is an analog of another widely used serotonergic neurotoxin, 5,6-dihydroxytryptamine, but 5,7-DHT was noted to condense with formaldehyde to form a fluorescent compound, and so provided a means to map serotonin neurons analogous to the Falck-Hillarp approach.

In living cells, 5,7-DHT is accumulated by both 5HT and DA neurons, although there was some debate on specificity (Dacey, 1988; Silva *et al.*, 1988; Cardozo, 1993; Steensen *et al.*, 1995). While there is an interesting study of secretory vesicle exocytosis and redistribution in cell bodies of serotonin neurons (Colgan *et al.*, 2004), 5,7-DHT has not been used successfully to study neurotransmission from axons.

There have been attempts to modify MPP^+ for use as FFNs. Two derivatives were effective as plasma membrane DA (DAT), NE (NET) and 5HT (SERT) substrates and effectively labeled monoamine neurons in culture; ASP^+ (4-(4-dimethylaminostyryl)-1-methylpyridinium) (Schwartz *et al.*, 2003; Oz *et al.*, 2010) and APP^+ (4-(4-dimethylamino)phenyl-1-methylpyridinium), (Bernstein *et al.*, 2012), the latter of which was shown to be a VMAT2 substrate (Tucker *et al.*, 2015). While these compounds are effective for detecting monoamine neurons and axons, APP^+ is not an effective FFN, probably because it is mostly accumulated in mitochondria, and so there is little or no destaining when it is released by exocytosis (Karpowicz *et al.*, 2013). Similar findings are reported for ASP^+ in serotonin neurons derived from stem cells (Lau *et al.*, 2015).

Other attempts to load synaptic and secretory vesicles for fluorescent study of exocytosis have used weak base vital dyes including acridine orange in large secretory vesicles (Steyer *et al.*, 1997; Williams and Webb, 2000), although this and other such compounds have not been useful for small synaptic vesicles, probably due to the low signal. However, a fluorescent antipsychotic drug, cyamemazine, has been used successfully to observe uptake and release from DA synaptic vesicles (Tucker *et al.*, 2015), apparently due to acidic trapping and not as a VMAT substrate.

Ricardo Borges's laboratory has explored the use of beta blockers, particularly propranolol, and produced a variety of fluorescent derivatives that are effectively accumulated into chromaffin vesicles and may provide effective FFNs (Beltran *et al.*, 2011).

3.8 Current FFN development

Prompted by the necessity for differentiating presynaptic activity at the level of individual release sites, and guided by the insights above, we developed a research program to develop effective FFNs. This is an ongoing effort and here we focus on those that have been published and characterized.

3.8.1 FFN511, a relatively non-selective probe in the striatum

To our knowledge, the first successful FFN for demonstrating synaptic vesicle uptake and release was FFN511 (Figure 3.1), which was based on the exploration of modified coumarins as fluorescent substrates for VMAT2 (IC50 ~ 1 μM) (Gubernator *et al.*, 2009). This provided the first means for optical imaging of

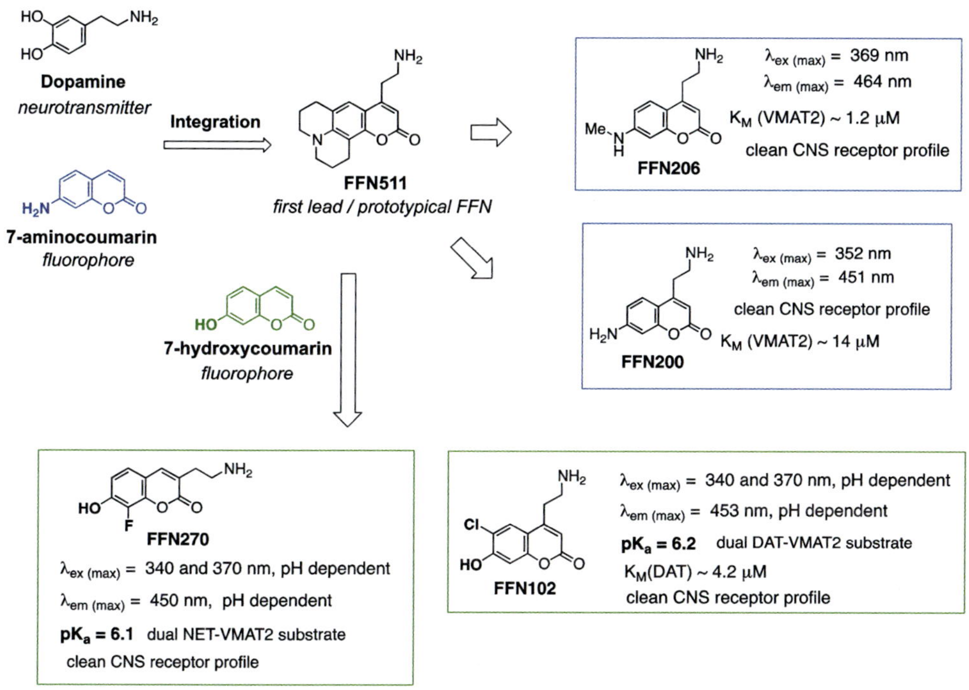

Figure 3.1 FFN probes were designed as fluorescent substrates for relevant monoamine transporters by integrating the key structural features of monoamine neurotransmitters and organic fluorophores. Dopaminergic and noradrenergic FFNs were developed and used in a number of experimental systems ranging from cell culture to living rodents. FFN102, FFN200, and FFN270 led to the discovery of silent monoaminergic synapses.

synaptic vesicular content uptake and release at individual presynaptic boutons in brain tissue. The uptake of FFN511 was inhibited by the VMAT inhibitors reserpine and tetrabenazine. FFN511 destained on exposure to amphetamine. Most importantly, the varicosities were destained in a calcium-dependent manner by evoked depolarization. However, a significant drawback of FFN511 was that much of the signal did not colocalize with the axonal GFP signal from mice that expresses GFP with a tyrosine hydroxlase promoter (TH-GFP). The much better colocalization with subsequently designed FFNs, such as FFN102, indicates that FFN511 also labels additional synapses that do not have VMAT and that it may be trapped in some other types of synaptic vesicles as a lipophilic weak base. Thus, FFN511, while the first compound introduced of this family, is relatively nonspecific for monoamine synaptic vesicles.

FFN511 has been used successfully to image secretion from prolactin-sensitive DA neurons of the hypothalamus (Dulcis *et al.*, 2013; Romano *et al.*, 2013), enterochromaffin cells (Raghupathi *et al.*, 2016), and from cell bodies and neurites of stem cell-derived serotonin neurons (Lau *et al.*, 2015).

3.8.2 FFN200, an effective reporter in cultured neurons

We developed FFN200 (Figure 3.1) as a VMAT2 substrate: it has not been found to act as a substrate for any of the plasma membrane transporters at this time, but it is highly selective for DA neurons in the striatum (Pereira *et al.*, 2016). FFN200 was developed as a more polar derivative of FFN511 to improve selectivity for dopaminergic neurons. It is also the first FFN to label dopaminergic neurons in both culture and the acute brain slice. Indeed, the signal to background for this FFN is sufficient to be used effectively in postnatal neuronal cultures and is the first such compound for this use: this may be because cultured neuron synapses have fewer synaptic vesicles than those in situ in the brain.

While amphetamine caused complete destaining of FFN200 from presynaptic sites, we found that only a small fraction of these vesicle clusters, exhibited exocytosis in response to local electrical stimulation, which was independently confirmed by the endocytic tracer FM1-43. We were able to measure not only FFN200 destaining from DA synapses, but also calcium transients measured by GCaMP, and found that both silent and active sites showed identical calcium transients. Our results suggest that, in dopaminergic axons, the majority of vesicle clusters capable of accumulating neurotransmitter are functionally silent, a discovery that required optical methods.

3.8.3 FFN206, a VMAT2 substrate effective in Drosophila

FFN206 (Figure 3.1) was developed as an effective VMAT2 substrate (Km 50 ~ 1 μM, similar to DA) that could be adapted for screening drugs or other treatments that might affect transporter activity in intact cells (Hu *et al.*, 2013). FFN206 has been used for high throughput imaging screening of secretory vesicle function in mast cells (Kiyoi *et al.*, 2018). Empirical results showed that this probe provides an outstanding signal in dopamine axons in the whole brain preparation of the fly, and has been used to elucidate mechanisms of amphetamine mediated release of DA (Freyberg *et al.*, 2016) and effects of depolarization on synaptic vesicle loading of substrate (Aguilar *et al.*, 2017).

3.8.4 FFN102, a pH sensitive DAT and VMAT2 substrate that provides a dual optical and electrochemical probe

In an effort to establish more selective and pH-sensitive FFNs for the DA system, we developed FFN102 (Figure 3.1) as a more polar coumarin-based dual substrate for VMAT and DAT: it is selective for DAT, and is not a good substrate for NET or SERT. In contrast to FFN511, in the striatal slice, nearly all of the FFN102 signal colocalizes with TH-GFP.

In addition to its properties as dual transporter substrate, FFN102 was designed as a pH sensor (Lee *et al.*, 2010), with the phenol group in the 7-position of the coumarin nucleus providing a pKa value of 6.2: this is ideal for comparing secretory vesicles, which typically have a pH of ~5 and the neutral extracellular media. The absorption spectrum of FFN102 exhibits pH dependence, exhibiting a maximum at ~ 340 nm at acidic pH and 370 nm at higher values, corresponding to the protonated and non-protonated forms, and thus FFN102 provides a ratiometric tool for estimating pH. This feature afforded a means to demonstrate that there is a wide range of pH values in vesicles, even in a single cell (Lee *et al.*, 2010).

As FFN102's emission at 451 nm is far more efficient at neutral pH with excitation at 370 nm (in the striatum, we use 2 photon excitation at 760 nm), the pH dependent change in absorption also provided a fluorescence increase during exocytosis due to the change of pH between the acidic lumen of synaptic vesicles and the neutral synapse and extrasynaptic milieu (Rodriguez *et al.*, 2013).

While FFN511 and FFN200 destain release sites as they undergo exocytosis, in contrast the fluorescence signal of FFN102 at the appropriate wavelength is greatly enhanced when the probe is released from the acidic interior of the synaptic vesicle to the neutral extracelluar solution, which drives a rapid increase in fluorescent signal, and we thus call this probe a paradigmatic "flashing FFNs". This property is very useful for regions with low monoamine signal that are insufficient for recording by electrochemistry, and we recently characterized release from substantia nigra axons in the globus pallidus externa, a region of low DA innervation from axons of passage, by measuring the flash following FFN102 release (Meszaros *et al.*, 2018).

Yet another property of FFN102 is that the OH group on its ring can be electrochemically oxidized at reasonable potentials, which allows it to be recorded by amperometry (Liu *et al.*, 2017): a recent study from Manon Guille-Collignon's laboratory has used simultaneous amperometry and optical recording of FFN102 release to measure quantal neurosecretion (Liu *et al.*, 2018). The oxidation current, like the fluorescence, is dependent on the pH and this could provide another means to measure pH changes associated with exocytosis and neurotransmission.

FFN102 together with amperometric measurements in striatal slices, was used to discover that methylphenidate, a frequently prescribed psychostimulant to treat ADHD, as well as cocaine, inhibit synaptic DA release, an unexpected effect of these drugs (Federici *et al.*, 2014).

FFN102 has been used successfully for measuring exocytosis from DA neurons in mouse retina (Ho *et al.*, 2015).

3.8.5 FFN270, a dual VMAT2-NET substrate

The first FFN to act as both a VMAT2 and a specific NET substrate is FFN270 (Figure 3.1) (Dunn *et al.*, 2018). The probe is taken up along with NE into

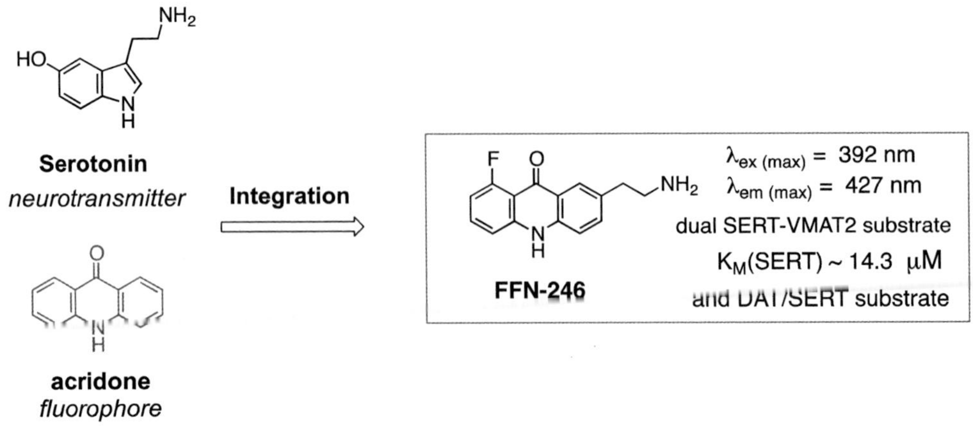

Figure 3.2 A prototype of serotonergic FFNs was developed based on the acridone fluorophore.

synaptic vesicles within NE axonal varicosities and enables measurement of synaptic vesicle content release by destaining during exocytosis. This compound was also the first to be used in the mouse brain *in vivo*, which is enabled by a cranial window over the cortex. The data suggest that NE axons may also have silent sites, although all of the compound is effectively released with peripheral injection of amphetamine. FFN270 has been used to demonstrate the release of NE from Merkel cells in the skin that covey the sense of touch (Hoffman *et al.*, 2018).

3.8.6 Towards SERT selective FFNs

Through optimization of the acridone core, we developed FFN246 (Figure 3.2) as a substrate of both SERT and VMAT2 (Henke *et al.*, 2018). FFN246 is effective for labeling serotonergic neurons in mouse brain tissue but has not yet been adapted for successful destaining by electrical stimuli, and we are continuing efforts to develop an effective approach.

3.9 Concluding remarks

The future design of FFNs should continue to enable a wide variety of uses, with a broad palette of properties that enable research in many cell types, including those with different transporters and receptors, and for additional neurotransmitters.

The introduction of pH dependent FFNs further suggests that a range of parameters related to neurosecretion could be measured. The work by Liu *et al.* on FFN102 showing that the compound can also be measured by amperometric detection points toward additional analytical techniques that could be coupled with fluorescent detection. We hope that this continues to be a fruitful area for chemists and biologists to advance novel ideas that will provide new insights towards nervous system development, function, learning and behavior, and disorders of these systems.

Acknowledgments

We thank the members of our labs and collaborators for the extraordinary ingenuity, dedication and effort that this research has and continues to require. Supported by NIDA (R01DA07418, Sulzer), the G. Harold & Y. Mathers Charitable Foundation (Sames), NIMH (R01MH086545, Sames; and R01MH108186, Sames and Sulzer), the JPB Foundation (Sulzer), McKnight Foundation (Sulzer and Sames), and Columbia University (Sames and Sulzer).

References

Agnati LF, Fuxe K, Zoli M, Ozini I, Toffano G, Ferraguti F (1986) A correlation analysis of the regional distribution of central enkephalin and beta-endorphin immunoreactive terminals and of opiate receptors in adult and old male rats. Evidence for the existence of two main types of communication in the central nervous system: the volume transmission and the wiring transmission. *Acta Physiol Scand*, 128: 201–207.

Aguilar JI *et al.* (2017) Neuronal depolarization drives increased dopamine synaptic vesicle loading via VGLUT. *Neuron*, 95: 1074–1088.e1077.

Baumgarten HG, Lachenmayer L (1972) 5,7-dihydroxytryptamine: improvement in chemical lesioning of indoleamine neurons in the mammalian brain. *Zeitschrift fur Zellforschung und mikroskopische Anatomie* (Vienna, Austria : 1948) 135: 399–414.

Baux G, Tauc L (1983) Carbachol can be released at a cholinergic ganglionic synapse as a false transmitter. *Proc Natl Acad Sci USA*, 80: 5126–5128.

Beltran B, Carrillo R, Martin T, Martin VS, Machado JD, Borges R (2011) Fluorescent beta-blockers as tools to study presynaptic mechanisms of neurosecretion. *Pharmaceuticals* (Basel, Switzerland) 4: 713–725.

Bernstein AI, Stout KA, Miller GW (2012) A fluorescent-based assay for live cell, spatially resolved assessment of vesicular monoamine transporter 2-mediated neurotransmitter transport. *J Neurosci Methods*, 209: 357–366.

Bertler A, Carlsson A, Rosengren E (1958) A method for the fluorimetric determination of adrenaline and noradrenaline in tissues. *Acta Physiol Scand*, 44:273–292.

Bjorklund A, Baumgarten HG, Nobin A (1974) Chemical lesioning of central monoamine axons by means of 5,6-dihydroxytryptamine and 5,7-dihydroxytryptamine. *Adv BiochemPsychopharmacol*, 10: 13–33.

Cardozo DL (1993) Midbrain dopaminergic neurons from postnatal rat in long-term primary culture. *Neuroscience*, 56: 409–421.

Carlsson A, Waldeck B (1958) A fluorimetric method for the determination of dopamine (3-hydroxytyramine). *Acta Physiol Scand*, 44: 293–298.

Carlsson A, Falck B, Hillarp NA (1962) Cellular localization of brain monoamines. *Acta Physiologica Scandinavica Supplementum*, 56: 1–28.

Cheng KT (2004) 1R-[(11)C]Phenylephrine. In: *Molecular Imaging and Contrast Agent Database (MICAD)*. Bethesda (MD): National Center for Biotechnology Information (US).

Colgan TJ, McLachlin CM, Cotterchio M, Howlett R, Seidenfeld AM, Mai VM (2004) Results of the implementation of liquid-based cytology-SurePath in the Ontario screening program. *Cancer*, 102: 362–367.

Dacey DM (1988) Dopamine-accumulating retinal neurons revealed by *in vitro* fluorescence display a unique morphology. *Science*, 240: 1196–1198.

Dulcis D, Jamshidi P, Leutgeb S, Spitzer NC (2013) Neurotransmitter switching in the adult brain regulates behavior. *Science*, 340: 449–453.

Dunn M, Henke A, Clark SD, Kovalyova Y, Kempadoo KA, Karpowicz RJ, Jr., Kandel ER, Sulzer D, Sames D (2018) Design of a norepinpephrine optical tracer for iaging individual norandrenergic synapses and their activity *in vivo*. *Nat Commun*, 9: 2838.

Falck B (1962) Observations on the possibilities of the cellular localization of monoamines by a fluorescent method. *Acta Physiol Scand*, 197: 1–25.

Falck B, Hillarp NA (1959) On the cellular localization of catechol amines in the brain. *Acta Anat* (Basel), 38: 277–279.

Falck B, Hillarp NA, Thieme G, Torp A (1982) Fluorescence of catechol amines and related compounds condensed with formaldehyde. *Brain Res Bull*, 9: xi–xv.

Federici M, Latagliata EC, Ledonne A, Rizzo FR, Feligioni M, Sulzer D, Dunn M, Sames D, Gu H, Nistico R, Puglisi-Allegra S, Mercuri NB (2014) Paradoxical abatement of striatal dopaminergic transmission by cocaine and methylphenidate. *J Biol Chem*, 289: 264–274.

Fleck MW, Barrionuevo G, Palmer AM (2001) Release of D,L-threo-beta-hydroxyaspartate as a false transmitter from excitatory amino acid-releasing nerve terminals. *Neurochem Int*, 39: 75–81.

Ford CP, Phillips PE, Williams JT (2009) The time course of dopamine transmission in the ventral tegmental area. *J Neurosci*, 29: 13344–13352.

Foster AC, Chen J, Runyan S, Dinh T, Venadas S, Ehring GR, Li YX, Staubli U (2015) 3-Aminoglutarate is a "silent" false transmitter for glutamate neurons. *Neuropharmacology*, 97: 436–446.

Freyberg Z *et al.* (2016) Mechanisms of amphetamine action illuminated through optical monitoring of dopamine synaptic vesicles in Drosophila brain. *Nat Commun*, 7: 10652.

Grabner CP, Zenisek D (2013) Amperometric resolution of a prespike stammer and evoked phases of fast release from retinal bipolar cells. *J Neurosci*, 33: 8144–8158.

Gros Y, Schuldiner S (2010) Directed evolution reveals hidden properties of VMAT, a neurotransmitter transporter. *J Biol Chem*, 285: 5076–5084.

Gubernator NG, Zhang H, Staal RG, Mosharov EV, Pereira DB, Yue M, Balsanek V, Vadola PA, Mukherjee B, Edwards RH, Sulzer D, Sames D (2009) Fluorescent false neurotransmitters visualize dopamine release from individual presynaptic terminals. *Science*, 324: 1441–1444.

Henke A, Kovalyova Y, Dunn M, Dreier D, Gubernator NG, Dincheva I, Hwu C, Sebej P, Ansorge MS, Sulzer D, Sames D (2018) Toward serotonin fluorescent false neurotransmitters: development of fluorescent dual serotonin and vesicular monoamine transporter substrates for visualizing serotonin neurons. *ACS Chem Neurosci.*

Hillarp NA, Fuxe K, Dahlstrom A (1966) Demonstration and mapping of central neurons containing dopamine, noradrenaline, and 5-hydroxytryptamine and their reactions to psychopharmaca. *Pharmacol Rev*, 18: 727–741.

Ho T, Jobling AI, Greferath U, Chuang T, Ramesh A, Fletcher EL, Vessey KA (2015) Vesicular expression and release of ATP from dopaminergic neurons of the mouse retina and midbrain. *Front Cell Neurosci*, 9: 389.

Hoffman BU, Baba Y, Griffith TN, Mosharov EV, Woo SH, Roybal DD, Karsenty G, Patapoulian A, Sulzer D, Lumpkin EA (2018) Merkel cells activate sensory neural pathways throught adrenergic synapses. *Neuron*, 100: 1401–1413.

Hu G, Henke A, Karpowicz RJ, Jr., Sonders MS, Farrimond F, Edwards R, Sulzer D, Sames D (2013) New fluorescent substrate enables quantitative and high-throughput examination of vesicular monoamine transporter 2 (VMAT2). *ACS Chem Biol.* 8: 1947–1954.

Karpowicz RJ, Jr., Dunn M, Sulzer D, Sames D (2013) APP^+, a fluorescent analogue of the neurotoxin MPP^+, is a marker of catecholamine neurons in brain tissue, but not a fluorescent false neurotransmitter. *ACS Chem Neurosci.* 4: 858–869.

Kim KT, Koh DS, Hille B (2000) Loading of oxidizable transmitters into secretory vesicles permits carbon-fiber amperometry. *J Neurosci*, 20: Rc101.

Kiyoi T, Liu S, Sahid MNA, Shudou M, Maeyama K, Mogi M (2018) High-throughput screening system for dynamic monitoring of exocytotic vesicle trafficking in mast cells. *PLoS One*. 13: e0198785.

Koh DS, Moody MW, Nguyen TD, Hille B (2000) Regulation of exocytosis by protein kinases and Ca(2+) in pancreatic duct epithelial cells. The *J Gen Physiol*, 116:5 07–520.

Kopin IJ (1968) False adrenergic transmitters. *Annu Rev Pharmacol*, 8: 377–394.

Kopin IJ, Fischer JE, Musacchio J, Horst WD (1964) Evidence for a false neurochemical transmitter as a mechanism for the hypotensive effect of monamine oxidase inhibitors. *Proc Natl Acad Sci USA*, 52: 716–721.

Lau T, Proissl V, Ziegler J, Schloss P (2015) Visualization of neurotransmitter uptake and release in serotonergic neurons. *J Neurosci Methods*, 241: 10–17.

Lee D, Creed M, Jung K, Stefanelli T, Wendler DJ, Oh WC, Mignocchi NL, Luscher C, Kwon HB (2017) Temporally precise labeling and control of neuromodulatory circuits in the mammalian brain. *Nat Methods*, 14: 495–503.

Lee M, Gubernator NG, Sulzer D, Sames D (2010) Development of pH-responsive fluorescent false neurotransmitters. *J Am Chem Soc*, 132: 8828–8830.

Li H, Waites CL, Staal RG, Dobryy Y, Park J, Sulzer DL, Edwards RH. (2006) Sorting of vesicular monoamine transporter 2 to the regulated secretory pathway confers the somatodendritic exocytosis of monoamines. *Neuron*, 48: 619–633.

Lindvall O, Bjorklund A (1974) The glyoxylic acid fluorescence histochemical method: a detailed account of the methodology for the visualization of central catecholamine neurons. *Histochemistry*, 39: 97–127.

Liu X, Hu L, Pan N, Grimaud L, Labbe E, Buriez O, Delacotte J, Lemaitre F, Guille-Collignon M (2018) Coupling electrochemistry and TIRF-microscopy with the fluorescent false neurotransmitter FFN102 supports the fluorescence signals during single vesicle exocytosis detection. *Biophys Chem*, 235: 48–55.

Liu X, Savy A, Maurin S, Grimaud L, Darchen F, Quinton D, Labbe E, Buriez O, Delacotte J, Lemaitre F, Guille-Collignon M (2017) A dual functional electroactive and fluorescent probe for coupled measurements of vesicular exocytosis with high spatial and temporal resolution. *Angew Chem Int Ed Engl*, 56: 2366–2370.

Liu Y, Roghani A, Edwards RH (1992a) Gene transfer of a reserpine-sensitive mechanism of resistance to MPP$^+$. *Proc Natl Acad Sci USA*, 89: 9074–9078.

Liu Y, Peter D, Roghani A, Schuldiner S, Prive GG, Eisenberg D, Brecha N, Edwards RH (1992b) A cDNA that suppresses MPP$^+$ toxicity encodes a vesicular amine transporter. *Cell*, 70: 539–551.

Maiti S, Shear JB, Williams RM, Zipfel WR, Webb WW (1997) Measuring serotonin distribution in live cells with three-photon excitation. *Science*, 275: 530–532.

Marcott PF, Mamaligas AA, Ford CP (2014) Phasic dopamine release drives rapid activation of striatal D2-receptors. *Neuron*, 84: 164–176.

Meszaros J, Cheung T, Erler MM, Kang UJ, Sames D, Kellendonk C, Sulzer D (2018) Evoked transients of pH-sensitive fluorescent false neurotransmitter reveal dopamine hot spots in the globus pallidus. *Elite* 19: e42383.

Muller A, Joseph V, Slesinger PA, Kleinfeld D (2014) Cell-based reporters reveal *in vivo* dynamics of dopamine and norepinephrine release in murine cortex. *Nat Methods*, 11: 1245–1252.

Newton MW, Crosland RD, Jenden DJ (1985) *In vivo* metabolism of a cholinergic false precursor after dietary administration to rats. *J Pharmacol Exp Ther*, 235: 157–161.

Owesson-White C, Belle AM, Herr NR, Peele JL, Gowrishankar P, Carelli RM, Wightman RM (2016) Cue-evoked dopamine release rapidly modulates D2 neurons in the nucleus Aaccumbens during motivated behavior. *J Neurosci*, 36: 6011–6021.

Oz M, Libby T, Kivell B, Jaligam V, Ramamoorthy S, Shippenberg TS (2010) Real-time, spatially resolved analysis of serotonin transporter activity and regulation using the fluorescent substrate, ASP$^+$. *J Neurochem*, 114: 1019–1029.

Patriarchi T, Cho JR, Merten K, Howe MW, Marley A, Xiong WH, Folk RW, Broussard GJ, Liang R, Jang MJ, Zhong H, Dombeck D, von Zastrow M, Nimmerjahn A, Gradinaru V, Williams JT, Tian L (2018) Ultrafast neuronal imaging of dopamine dynamics with designed genetically encoded sensors. *Science*, 360: eaat 4422.

Pereira DB, Schmitz Y, Meszaros J, Merchant P, Hu G, Li S, Henke A, Lizardi-Ortiz JE, Karpowicz RJ, Jr., Morgenstern TJ, Sonders MS, Kanter E, Rodriguez PC, Mosharov EV, Sames D, Sulzer D (2016) Fluorescent false neurotransmitter reveals functionally silent dopamine vesicle clusters in the striatum. *Nat Neurosci*, 19: 578–586.

Peter D, Jiminez J, Liu Y, Kim J, Edwards R (1994) The chromaffin granule and synaptic vesicle amine transporters differ in substrate recognition and sensitivity to inhibitors. *J Biol Chem*, 269: 7231–7237.

Pothos EN, Krantz D, Edwards R, Sulzer D (1999) Expression of VMAT2 in hippocampal neurons as a tool for exploring the plasticity of quantal transmission. *Soc Neurosci Abstr.*

Pothos EN, Larsen KE, Krantz DE, Liu Y, Haycock JW, Setlik W, Gershon MD, Edwards RH, Sulzer D (2000) Synaptic vesicle transporter expression regulates vesicle phenotype and quantal size. *J Neurosci* 20: 7297–7306.

Raghupathi R, Jessup CF, Lumsden AL, Keating DJ (2016) Fusion pore size limits 5-HT release from single enterochromaffin cell vesicles. *J Cell Physiol*, 231: 1593–1600.

Rodriguez PC, Pereira DB, Borgkvist A, Wong MY, Barnard C, Sonders MS, Zhang H, Sames D, Sulzer D (2013) Fluorescent dopamine tracer resolves individual dopaminergic synapses and their activity in the brain. *Proc Natl Acad Sci USA*, 110: 870–875.

Romano N, Yip SH, Hodson DJ, Guillou A, Parnaudeau S, Kirk S, Tronche F, Bonnefont X, Le Tissier P, Bunn SJ, Grattan DR, Mollard P, Martin AO (2013) Plasticity of hypothalamic dopamine neurons during lactation results in dissociation of electrical activity and release. *J Neurosci*, 33: 4424–4433.

Schwartz JW, Blakely RD, DeFelice LJ (2003) Binding and transport in norepinephrine transporters. Real-time, spatially resolved analysis in single cells using a fluorescent substrate. *J Biol Chem*, 278: 9768–9777.

Silva NL, Mariani AP, Harrison NL, Barker JL (1988) 5,7-Dihydroxytryptamine identifies living dopaminergic neurons in mesencephalic cultures. *Proc Natl Acad Sci USA*, 85: 7346–7350.

Steensen BH, Nedergaard S, Ostergaard K, Lambert JD (1995) Electrophysiological characterization of dopaminergic and non-dopaminergic neurones in organotypic slice cultures of the rat ventral mesencephalon. *Exp Brain Res*, 106: 205–214.

Steyer JA, Horstmann H, Almers W (1997) Transport, docking and exocytosis of single secretory granules in live chromaffin cells. *Nature*, 388: 474–478.

Sulzer D, Pothos EN (2000) Regulation of quantal size by presynaptic mechanisms. *Rev Neurosci*, 11: 159–212.

Sulzer D, Cragg SJ, Rice ME (2016) Striatal dopamine neurotransmission: regulation of release and uptake. *Basal Ganglia*, 6: 123–148.

Sulzer D, Sonders MS, Poulsen NW, Galli A (2005) Mechanisms of neurotransmitter release by amphetamines: a review. *Prog Neurobiol*, 75: 406–433.

Tucker KR, Block ER, Levitan ES (2015) Action potentials and amphetamine release antipsychotic drug from dopamine neuron synaptic VMAT vesicles. *Proc Natl Acad Sci USA*, 112: E4485–4494.

Van der Kloot W, Naves LA (1996) Accounting for the shapes and size distributions of miniature endplate currents. Biophys J 70: 2175–2184.

Williams RM, Webb WW (2000) Single granule pH cycling in antigen-induced mast cell secretion. *J Cell Sci*, 113 Pt 21: 3839–3850.

Williams RM, Shear JB, Zipfel WR, Maiti S, Webb WW (1999) Mucosal mast cell secretion processes imaged using three-photon microscopy of 5-hydroxytryptamine autofluorescence. *Biophys J*, 76: 1835–1846.

Wu Z, Foster AC, Staubli U, Wu X, Sun C, Tang X, Li YX, Chen G (2015) Effects of 3-aminoglutarate, a "silent" false transmitter for glutamate neurons, on synaptic transmission and epileptiform activity. *Neuropharmacology*, 97: 95–103.

Yelin R, Schuldiner S (1995) The pharmacological profile of the vesicular monoamine transporter resembles that of multidrug transporters. *FEBS Lett*, 377: 201–207.

CHAPTER 4

FUNCTIONAL CONNECTOME MAPPING

Daniel Eskenazi, Nao Chuhma, Susana Mingote, Samira Ztaou and Stephen Rayport*

Department of Psychiatry, Columbia University and Department of Molecular Therapeutics, New York State Psychiatric Institute, USA

*sgr1@columbia.edu

4.1 Introduction

There has been great interest in identifying brain circuits mediating specific behaviors in order to gain insight into neurological and psychiatric disorders and thereby identify new treatments (Gordon, 2016). While insight into circuit abnormalities underlying specific psychiatric disorders has been limited, rare single mutations that give rise to psychiatric symptoms have been identified, pointing to potential circuit abnormalities (Mitchell *et al.*, 2011). Connecting circuits to research domain criteria (Insel, 2014) has seen significant progress (Carcone and Ruocco, 2017). What defines a circuit and how can alterations in a circuit be queried?

Circuits are defined by specific connections. Connections vary between brain areas in strength, reciprocity, and organization — they may be topographical, hierarchical or not. While structural studies of the connections between brain regions are essential to understanding the nervous system, knowing the wiring diagram is not enough: *anatomical* connectivity does not predict *functional* connectivity (Brezina and Weiss, 1997; Angstadt *et al.*, 2001; Nusbaum *et al.*, 2001). Not only does anatomical evidence for the presence or absence of particular connections correlate poorly with the strength of connections, anatomical data do not resolve the differential actions of multiple transmitters released by the same neuron, emphasizing the importance of direct determinations of functional synaptic connectivity. Furthermore, a static representation of synaptic connectivity makes it easy to overlook heterogeneity in the efficacy, type and timing of connections that have flexible roles and undergo neuroplastic changes in mediating different behaviors (Rockland, 2015).

One way to address this is through **functional connectome mapping**, the systematic querying of all the connections made by an identified population of neurons with all types of identified follower cells (Chuhma *et al.*, 2011). This method uses recordings of direct synaptic connections between an identified presynaptic population and identified postsynaptic neurons to reveal the incidence and strength of synaptic connections. Additionally, if multiple neurotransmitters are present, which is more the rule than the exception in mammalian nervous systems (Granger *et al.*, 2017), the strength of the signal mediated by each transmitter can be determined.

Historically, part of the challenge in mapping connectivity depends on the complexity of the system (Figure 4.1) and the limits of available technology. Simpler organisms such as the sea slug *Aplysia californica* (Kandel *et al.*, 1967) and the nematode *Caenorhabditis elegans* (White *et al.*, 1976) have served as model organisms for studying neuronal connections, made possible by their relatively limited number of neurons that can be identified from animal to animal, and have near identical connectivity (Kandel, 2001). Optogenetics has made it possible to examine connections in more complex organisms, with orders of magnitude more neurons and more variation between individuals, such as the fruit fly *Drosophila melanogaster* (Lima and Miesenböck, 2005) and the mouse *Mus musculus* (Boyden *et al.*, 2005).

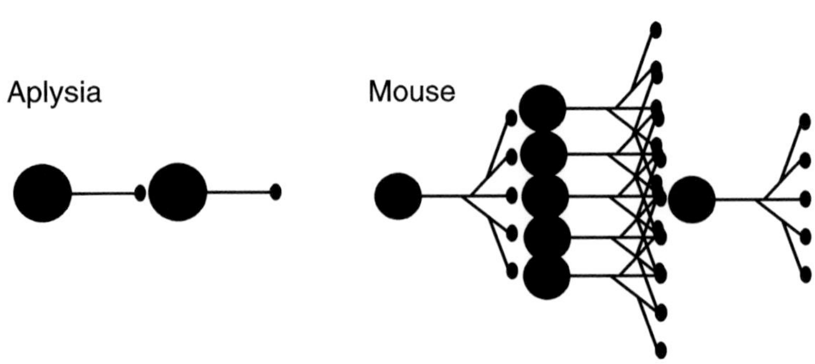

Figure 4.1 Neuronal connectivity. In the nervous system of simple invertebrates such as *Aplysia californica*, individual cells make direct connections with significant behavioral effects. In mammals such as the mouse, individual neurons make many, more distributed connections, that rarely have detectable behavioral effects.

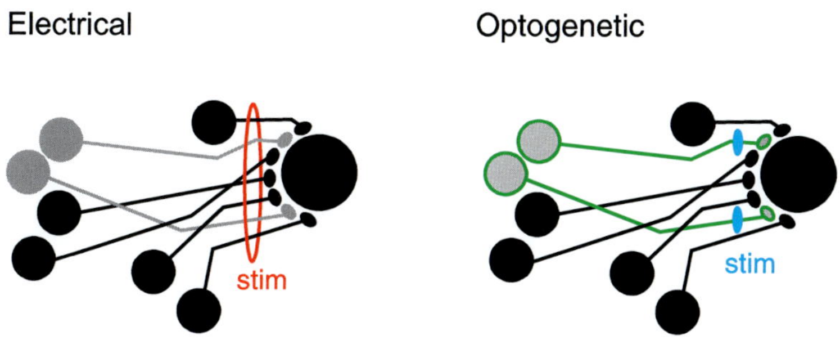

Figure 4.2 Electrical *versus* optogenetic stimulation. Electrical field stimulation (red oval) activates all nearby axons, making selective stimulation of a population of neurons (light gray) impossible due to intermingled processes from other neurons. In contrast, optogenetic stimulation utilizes selective expression of a photosensitive channel (e.g. ChR2) within a target population (green) so that photostimulation (blue ovals) activates just those neurons, even if intermingled with other neurons. Photostimulation is usually delivered in a wider field, but shown here only where it is effective, that is in stimulating the processes of neurons expressing ChR2.

While electrical stimulation of the brain may be focal, it nonetheless activates a mixed population of neurons and their processes (Figure 4.2, left). In contrast, optogenetics uses photoactivated or light-gated cation-selective membrane channels such as channelrhodopsin-2 (ChR2) (Nagel *et al.*, 2003), expressed in specific neuronal subpopulations, to enable selective cell stimulation (Figure 4.2, right). ChR2-expressing mammalian neurons can be stimulated to query circuit function both *in vitro* (Boyden *et al.*, 2005) and *in vivo* (Arenkiel *et al.*, 2007). This has been applied to circuit mapping (Petreanu *et al.*, 2007; Wang *et al.*, 2007) and photostimulation-assisted identification of neuronal populations (Lima *et al.*, 2009). Functional connectome mapping is a further refinement, in querying all synaptic connections made by a defined neuron population with identified follower neurons.

The first functional connectome map was made with the restricted expression of ChR2 in striatal spiny projection neurons (SPNs) in transgenic mice (Chuhma *et al.*, 2011). This revealed that SPNs project exclusively to substantia nigra (SN) gamma-aminobutyric acid (GABA) neurons, and not to dopamine neurons. The results argued against subsequent anatomical evidence for connections of dopamine D1 receptor-expressing SPNs to dopamine neurons (Watabe-Uchida *et al.*, 2012), which were not borne out in a further optogenetic study, using viral labeling, examining nucleus accumbens (NAc) SPN projections to the ventral tegmental area (VTA) (Xia *et al.*, 2011). Subsequent optogenetic studies have revealed that some NAc SPNs make functional connections to dopamine neurons (Bocklisch *et al.*, 2013). Differences in the results likely have to do with the transgenically labeled SPN subpopulation, and sparser labeling of NAc SPNs. Indeed, SPNs in distinct subdomains of the NAc shell do and do not project directly to dopamine neurons (Yang *et al.*, 2018). Functional connectome mapping has

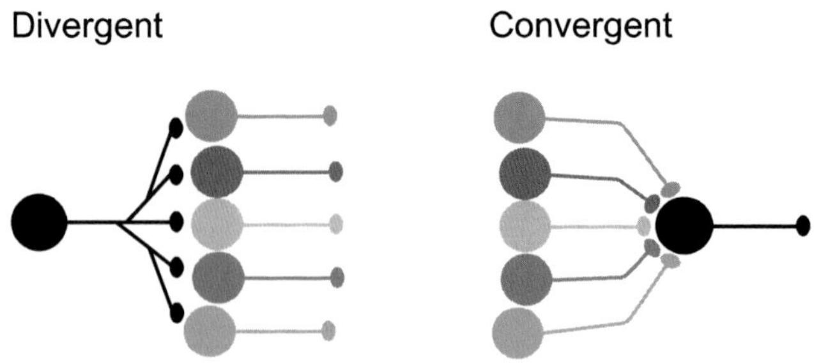

Figure 4.3 Breaking down the complexity. Most mammalian neurons make divergent connections, in which one presynaptic neuron innervates multiple postsynaptic neurons. Multiple neurons innervate a single postsynaptic neuron, making convergent connections. Both patterns pose challenges for evaluating functional connectivity in the mammalian CNS.

continued in the mapping of dopamine neuron glutamate connections to forebrain targets (Mingote *et al.*, 2015).

Dopamine neurons, as well as the other major monoaminergic cell groups, the noradrenergic and serotonergic neurons, form the most divergent cell populations in the CNS (Figure 4.3, left). Dopamine neurons reside in the ventral midbrain, in two main subpopulations in the VTA and SN. VTA neurons project principally to the ventral striatum and cortex, while SN neurons project to the dorsal striatum (dStr) (Swanson, 1982; Björklund and Dunnett, 2007; Yetnikoff *et al.*, 2014). The full projections of individual dopamine neurons were first revealed with a viral vector driving expression of a membrane-bound fluorescent protein injected in SN neurons, confirmed by immunostaining for tyrosine hydroxylase (TH), the rate-limiting enzyme in dopamine biosynthesis (Matsuda *et al.*, 2009). 3D reconstructions revealed large arborizations in the striatum, spanning striosome and matrix, compartments; some neurons projected preferentially to one compartment or the other. A single dopamine neuron axon spanned about 3% of the volume of the striatum. A study of VTA dopamine neurons distinguished various subtypes based on their projection targets (i.e. mesocorticolimbic, mesocortical, mesolimbic and mesostriatal) and showed that some innervated more than one forebrain structure (Aransay *et al.*, 2015). VTA dopamine neurons that project to the NAc target either the core or shell subregions and do not project elsewhere (Rodríguez-López *et al.*, 2017).

Dopamine neurons are subject to multiple convergent inputs that contribute to their diversity. They receive afferents from prefrontal cortex, striatum and amygdala (Geisler *et al.*, 2007; Jhou *et al.*, 2009; Watabe-Uchida *et al.*, 2012; Beier *et al.*, 2015). Gene profiling has enabled their subdivision into subpopulations based on coordinated gene expression, which correlates with their projection targets (Poulin *et al.*, 2018). Subpopulations of dopamine neurons, in turn, converge on individual striatal target neurons, and the sum total of their inputs can be measured, in mapping their functional connectome.

4.2 Functional connectome mapping

Population-selective stimulation of presynaptic terminals and identification of postsynaptic cells, where recordings are performed, are two essential components of functional connectome mapping. The strategy involves comprehensive expression of ChR2 in an identified population of neurons — here dopamine neurons — and recording responses to photostimulation in identified follower cells (Figure 4.4). Dopamine neurons, as well as the other monoaminergic neurons, are particularly well suited to a functional connectome strategy, as they are grouped in small, discrete brain regions and can be genetically targeted using their neurotransmitter transporter (Zhuang *et al.*, 2005), enabling both viral and transgenic ChR2 expression. In the striatum, principal neurons are identified based on cell soma size and confirmed by recorded electrophysiological parameters. Spiny projection neurons make up 95% of striatal neurons and are typically silent. Cholinergic interneurons (ChIs) comprise only about 1% of striatal neurons, are sparsely distributed larger neurons and are spontaneously active. Dopamine neuron inputs are activated by wide-field photostimulation and responses measured, either under current clamp to assess the impact on postsynaptic neuron activity, or under voltage clamp with pharmacological isolation to measure the incidence and strength of neurotransmitter-specific synaptic signals.

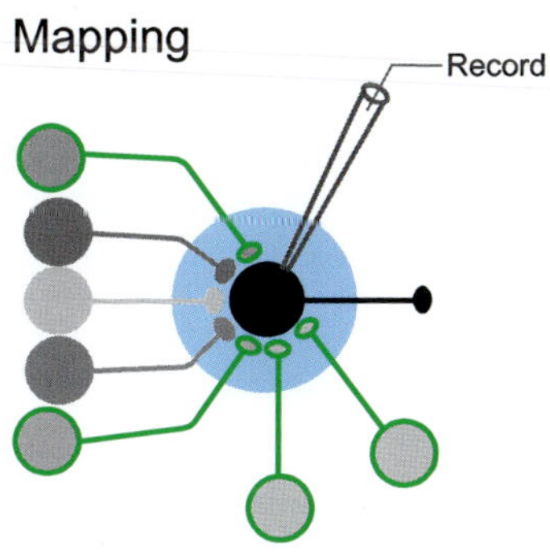

Figure 4.4 Functional connectome mapping. A genetically defined subpopulation of neurons expressing ChR2 (with green membranes) that converge on to a neuron of interest (black) are selectively activated in concert by optogenetic stimulation (blue circle of light). Only the terminals of the neurons expressing ChR2 will be stimulated, even if the presynaptic cell bodies are not clustered (e.g. two cells at bottom). Individual postsynaptic neurons are recorded, under current- or voltage clamp, to determine the incidence/strength of synaptic connections, and the impact on activity. Pharmacological isolation cocktails are used to measure synaptic currents mediated by different neurotransmitters.

Several studies using viral transfections of dopamine neurons have revealed parts of the functional connectome map of glutamate cotransmission (Stuber *et al.*, 2010; Tecuapetla *et al.*, 2010) and GABA cotransmission (Tritsch *et al.*, 2012). The functional connectome strategy has been applied to examining dopamine neuron glutamatergic connections to the principal neurons in forebrain regions, in the striatum and beyond (Mingote *et al.*, 2015), and dopamine neuron connections to ChIs and SPNs within different subregions of the striatum (Chuhma *et al.*, 2014; Mingote *et al.*, 2015; Chuhma *et al.*, 2018). These studies underscore the importance of functional connectome mapping in revealing the heterogeneity of dopamine neuron connections. While this chapter provides an overview of the rationale for and basic concepts of functional connectome mapping, more detailed methods are available (Mingote *et al.*, 2017a).

4.2.1 Genetic targeting of neurons for cell-type selective expression

A major challenge in the analysis of brain circuits has been identifying specific neurons *in vivo,* due to the heterogeneity of neurons within a given brain region. In brain slices, some neurons may be distinguished by their appearance, membrane properties, intrinsic activity, and spike shape. In the SN, dopamine neurons clustered in the pars compacta can be distinguished by their position, size and spike shape, but not so easily the smaller dopamine neurons in the VTA. Similarly, ChIs in the dStr can be distinguished by their size and tonic activity, but not so easily the smaller ChIs in the NAc (Phelps and Vaughn, 1986). Pharmacological measures can be used, such as the inhibitory response of dopamine neurons to dopamine receptor D2 agonists. Genetic markers, on the other hand, allow for targeting neurons by selective gene expression, driven by transcription factors and other developmentally regulated markers that confer neuronal identity. These genetic markers (typically cell-specific promoters) can be used to drive expression of proteins to distinguish specific neurons within a brain region, measure intracellular responses such as calcium levels or manipulate the neurons in question via membrane bound channels or receptors. Genetic strategies involve a **promoter sequence** only activated in a specific subset of neurons and a **coding region** to drive expression of a protein that accomplishes one of the goals above. These genes can then either be added into animal germlines as **transgenes** or into viral genomes used as **vectors**.

Progress has been spurred by the generation of a repertoire of mice expressing recombinant enzymes derived from bacteriophages to allow for combinatorial breeding strategies (Gong *et al.*, 2007). *Cre recombinase* recognizes specific genetic base-pair sequences, called *loxP* sites that flank a gene of interest. Based on the orientation of the *loxP* sites, Cre can either excise the gene or invert it, to either inactivate or activate it, respectively. Thus, for example, a mouse can be made carrying a transgene with the promoter region of a cell-specific marker, such as the dopamine transporter (DAT; *Slc6a3*) and the coding region for the gene for *Cre*, known as a DAT-cre mouse (Zhuang *et al.*, 2005; Bäckman *et al.*, 2006). That mouse can then be bred to a mouse carrying a transgene with a universally expressed promoter and a coding region for a protein such as ChR2 whose expression is blocked by a transcriptional STOP cassette flanked by loxP sites. In cells expressing Cre, the STOP cassette is excised allowing expression; thus, ChR2 will only be expressed in DAT+ neurons, that is dopamine neurons. Alternately, a double-inverted open (DiO) reading frame or FLEX strategy enables Cre-dependent activation or inactivation (Atasoy *et al.*, 2008; Tsai *et al.*, 2009). Using more than one set of recombinatory enzymes / recognition sites (e.g. Cre-lox, Flp-FRT, and Dre-rox) neuronal populations can be targeted based on more than one condition. *Intronic recombinase sites enabling combinatorial targeting* (INTRSECT) allows targeting of cells with single vectors using multiple-feature Boolean logic (Fenno *et al.*, 2014). For example, driving ChR2 expression from both DAT and vesicular glutamate transporter 2 (VGLUT2) promotors selectively targets dopamine-glutamate neurons (Poulin *et al.*, 2018).

Viral vectors can be used to introduce ChR2 into dopamine neurons using DAT-cre mice, allowing for anatomic as well as neuron-subtype-specificity, and bypassing transient developmental expression. Several viral vectors may be used, differing in their cellular affinity (e.g. herpes-simplex virus, HSV, has tropism for neurons over glial or endothelial cells), genetic payload (e.g. 5 kilobase pairs for adeno-associated virus, AAH, *versus* 8–10 kb for HSV), and duration of expression (e.g. days to weeks), all of which must be factored into the specific project at hand (Edry *et al.*, 2011). Viruses can also carry promoters specific to single neuronal subtypes to drive genes of interest in wild type animals, without the need for transgenic animals (Ferguson *et al.*, 2011), and used in a combinatorial fashion to drive ChR2 expression in mice as well as non-human primates (Stauffer *et al.*, 2016). Advantages to using viral vectors include their wide availability, relative ease of production, ability to be combined and high expression, with multiple alleles per cell. However, expression varies from animal-to-animal due to differences in the localization of stereotaxic injections, variable viral spread, and transfection efficiency (e.g. expressed copy number). Viral transfections with ChR2 in DAT-cre mice have opened up functional connectivity studies in dopamine neurons (Stuber *et al.*, 2010; Tecuapetla *et al.*, 2010; Tritsch *et al.*, 2012; Chuhma *et al.*, 2014). While these factors generally average out across animals, reliability and comprehensiveness of expression are better achieved using a transgenic strategy.

By reducing variability, transgenic expression is important for functional connectome mapping. This became feasible with optogenetic mice (Madisen *et al.*,

2012). The key issues involve lower ChR2 expression, from just one or two alleles, and determining the specificity of ChR2 expression in the neurons of interest. DAT-IRES-Cre/+; ROSA26-flox-STOP-CAG-ChR2-YFP (Ai32) homozygous mice (with two ChR2 alleles; $Ai32^{DAT}$ mice) have sufficient expression to confer reliable photoexcitation, making both single and train photostimulation feasible, and quantitative functional mapping possible (Chuhma *et al.*, 2018). Ectopic expression due to transient developmental DAT expression is not a confound, as the DAT-IRES-cre promoter is highly selective for dopamine neurons. The presumptive dopamine specific promoter TH has been shown in some studies to express ectopically in non-DAergic neurons; thus, the selectivity of new constructs must be individually assessed (Lammel *et al.*, 2015; Pupe and Wallén-Mackenzie, 2015; Stuber *et al.*, 2015). Similarly, for targeting serotonergic neurons, the transient expression of the serotonin transporter (Narboux-Nême *et al.*, 2008) confounds transgenic approaches. Crucially, the sparsely distributed, TH-expressing, putative DAergic interneurons seen in the striatum do not express DAT (Ibáñez-Sandoval *et al.*, 2010), and do not express ChR2-EYFP. Thus, $Ai32^{DAT}$ mice, can be used for the specific photoactivation of dopamine neuron terminals in the striatum, and so enable dopamine neuron functional connectome mapping.

4.2.2 Defining presynaptic neurons

Monoaminergic neurons are particularly suited to the functional connectomes approach owing to the availability of specific genetic markers for the various neuronal subtypes, the grouping of their cell bodies in the midbrain. The DAT promoter is highly specific for dopamine neurons. Several viral transfection studies in DAT-IRES-cre mice have shown fidelity of ChR2 expression, with comprehensiveness approaching 95% of dopamine neurons (Tsai *et al.*, 2009; Cohen *et al.*, 2012; Tritsch *et al.*, 2012; Chaudhury *et al.*, 2013; Chuhma *et al.*, 2014). While transient ectopic developmental Cre expression in $Ai32^{DAT}$ mice might reduce specificity of ChR2 expression, a recent comparison of viral and transgenic expression of ChR2 in DAT-IRES-cre mice found near identical numbers of connections to ChIs (Straub *et al.*, 2014). These data argue for the specificity of ChR2 expression in dopamine neurons in $Ai32^{DAT}$ mice.

Two other strategies may be used to restrict ChR2 expression to dopamine neurons in adult mice. DAT-PF-tTA (JAX 027178); tetO-ChR2 (RIKEN Bioresource Center) mice enable doxycycline-induced ChR2 expression in adulthood, obviating transient developmental expression. However, the endogenous DAT allele is compromised, and so the mice are DAT heterozygous. Alternately, DAT-creERT2 mice (JAX 016583); Ai32 mice enable tamoxifen-dependent ChR2 expression (Papathanou *et al.*, 2018). The transgenic strategies offer reproducible and comprehensive expression of ChR2 in DAT-expressing neurons.

4.2.3 Identifying postsynaptic neurons

Identifying postsynaptic cells recipient to optogenetically activated inputs and the measurement of responses in postsynaptic cells within and between brain regions is the next key element for functional connectome mapping. The divergence of projections from a defined subpopulation of neurons may have varying effects based on presynaptic differences, such as the neurotransmitter or transmitters released, and postsynaptic differences in receptor expression. In *ex vivo* brain slices, individual target cells can be identified based on morphology, soma size and recorded electrophysiological parameters, some of which can be assessed in loose patch configuration, such as whether the cell is spontaneously active. Cells can be back-labeled with fluorescent tracers such as retrobeads, which for instance enabled the first successful identification of dopamine neurons in cell culture (Rayport *et al.*, 1992). A less tedious method, and one that also allows for *in vivo* monitoring and analysis is using transgenic reporters, as became widely available through the Gene Expression Nervous System Atlas (GENSAT) project (Schmidt *et al.*, 2013).

In establishing the dopamine neuron functional connectome in the striatum, Ai32^{DAT} mice can be bred with mice containing a third transgenic reporter allele for each of the four principal cell types: direct- and indirect-pathway SPNs, ChIs and fast-spiking interneurons (FSIs). SPNs constitute 95% of striatal neurons and are segregated into two subpopulations: direct pathway SPNs (dSPNs) that project to the SN and preferentially express the D1 receptor or indirect pathway SPNs (iSPNs) that project to the Globus Pallidus and preferentially express the D2 receptor. While SPNs are easily found, dSPNs and iSPNS cannot be distinguished by appearance or electrophysiological properties, so Ai32^{DAT} mice may be crossed with Tg(Drd1a-tdTomato) or Tg(Drd2-EGFP) mice to label dSPNs red or iSPNs green, respectively. These two subpopulations are almost completely segregated in the dStr and NAc core (Shuen *et al.*, 2008; Gangarossa *et al.*, 2013b, 2013a), so lack of D1-tdTomato or D2-GFP can be used to detect the other subpopulation (Valjent *et al.*, 2009). Both D1-tdTomato; Ai32^{DAT} mice and D2-EGFP; Ai32^{DAT} mice are needed for NAc shell studies due to greater D1-D2 overlap.

ChIs are identifiable by their larger size (20–50 μm in diameter) and confirmed by their tonic activity and hyperpolarization-activated current. Post recording, immunostaining for choline acetyltransferase (ChAT, the rate-limiting enzyme in acetylcholine synthesis) confirms ChI identification using these criteria (Chuhma *et al.*, 2014). As ChIs comprise one to two percent of striatal neurons, they take effort to find; breeding in the ChAT-GFP can facilitate their identification. While the striatum has several classes of GABA interneurons (Kreitzer, 2009; Tepper *et al.*, 2010), the principal class of GABA are FSIs, which express parvalbumin (PV). FSIs have a unique electrophysiological signature, but with soma diameters similar to SPNs, and comprising less than one percent of striatal neurons, they are difficult to find, so the PV-GFP allele is important for their identification.

4.2.4 Recording direct synaptic connections

The three main functional connectome measures are neurotransmitter mediation, and the incidence and strength of connections. To determine the neurotransmitter mediation of a given response, methods include pharmacological blockade of postsynaptic receptors, knockout of synthetic enzymes such as glutaminase (the glutamate recycling enzyme) or knockout of vesicular transporters, such as VGLUT2 (which fills synaptic vesicles and enables glutamate release). The incidence of a connection can be determined by the presence of a significant postsynaptic response, one that exceeds a criterion such as an amplitude two standard deviations greater than the baseline noise (Chuhma *et al.*, 2014; Mingote *et al.*, 2015; Figure 4.5). The strength of a connection can be determined by direct measurement of current under voltage-clamp or by using voltage- or calcium-sensitive dyes.

While synaptic responses under voltage clamp evoked with single-pulse photostimulation provide measures of synaptic strength, the circuit function of particular synaptic connections is determined by the impact on the firing of postsynaptic neurons. The next step in functional connectome mapping is thus examination of firing modulation due to dopamine neuron input, under current clamp. Phasic firing of dopamine neurons can be mimicked by train photostimulation (five pulses at 20 Hz with 5 msec pulse duration). For ChIs, the only tonically active striatal neurons, the effect of this stimulation can be seen in changes in spontaneous firing. For SPNs and FSIs (which are silent in the slice), depolarizing current can be injected (for a few seconds) to elicit regular firing during which train photostimulation is delivered. The firing ratio is then calculated as the ratio of firing frequency during train photostimulation to the preceding baseline firing frequency.

While fast, short-latency, reliable responses are likely to be monosynaptic, that is a direct connection, there is the possibility that an intervening neuron is activated, engendering a polysynaptic response. This may be addressed using tetrodotoxin (TTX) to block voltage-gated sodium channels, thus eliminating the propagation of action potentials mediating a polysynaptic response, together with 4-aminopyridine (4-AP) to block potassium channels, enabling ChR2-induced activation of presynaptic calcium channels and transmitter release (Petreanu *et al.*, 2009). In the striatum, there is little chance of polysynaptic responses, since SPNs release GABA, and are inhibitory. In the NAc medial shell (m-shell), dopamine neuron glutamate cotransmission can drive ChIs to fire, and ChIs are capable of glutamate cotransmission, mediated by VGluT3 (Higley *et al.*, 2011). However, concerted activation of ChIs would be required to drive dopamine neuron terminals (Cachope *et al.*,

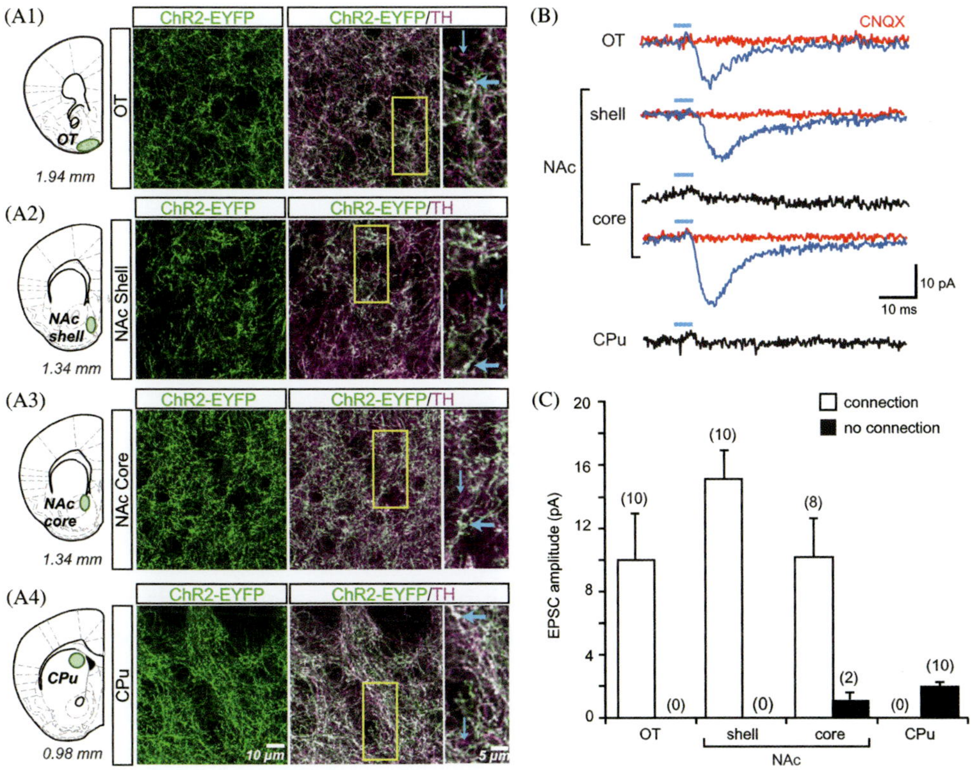

Figure 4.5 ChR2-EYFP expression in dopamine neuron axons and photostimulated glutamatergic responses in the striatum. (A) Confocal immunofluorescence images of ChR2-EYFP+ and TH+ axons in the olfactory tubercle (OT) (A1), NAc shell (A2), NAc core (A3), and dStr (A4) are shown. Schematics (left) show the coronal slices, with distance from bregma indicated below each section, that were used for immunostaining and *in vitro* EPSC recordings; sites of image acquisition and recording are outlined in green. The three columns of images (right) show ChR2-EYFP+ axons (green), the merge of ChR2-EYFP+ (green) with TH+ (magenta) to reveal ChR2-EYFP+/TH+ (white) axons and expanded images of the regions of interest (yellow rectangle, middle column). Thick blue arrows indicate ChR2-EYFP+/TH+ axons; thin blue arrows indicate ChR2-EYFP−/TH+ axons. No ChR2-EYFP+/TH− axons were seen in the striatum. (B) Recordings were made from SPNs in the four sites in the striatum. EPSCs shown are the average of 10 consecutive traces. Traces with significant light-evoked responses are shown in blue; those without responses are shown in black. Blue bars above traces indicate photostimulation (5 ms pulses, at 0.1 Hz). CNQX (40 μm) completely blocked light-evoked EPSCs (red traces). (C) Summary of the amplitudes of recorded light-evoked EPSCs in different regions of the striatum. The graph shows the average amplitudes in SPNs with EPSCs (connection; white bars) and without (no connection; black bars). The number of cells recorded is shown in parentheses. (Reproduced with permission from Mingote *et al.*, 2015).

2012; Threlfell *et al.*, 2012). Dopamine neuron glutamate inputs to SPNs in the NAc have been shown to be monosynaptic by TTX and 4-AP (Tecuapetla *et al.*, 2010). Connections to both SPNs, ChIs and FSIs all are short latency and reliable, consistent with monosynapticity (Chuhma *et al.*, 2014).

4.3 Modulation of the functional connectome map

With quantitative measures of the strength of identified synaptic connections, it becomes possible to evaluate the relative importance of synaptic effects both regionally and temporally. Following drug exposure perturbations in the functional connectome map offer a roadmap for elucidating drug actions in the brain. Drug-induced synaptic plasticity has been studied extensively (Lüscher, 2013; Ikemoto and Bonci, 2014), but no studies have yet revealed the sequence of changes from molecular actions at synapses to the impact on circuitry and then on behavior. Drug-induced plasticity should affect the functional connectome map, and this should be readily quantifiable.

Indeed, there is a remarkable degree of plasticity in the dopamine neuron functional connectome map following a single dose of amphetamine, pointing specifically to dopamine neuron connections to ChIs in the m-shell (Chuhma *et al.*, 2014). About two hours after a single low dose of amphetamine (2 mg/kg), there is a significant reduction in dopamine neuron driven glutamate-mediated bursts in m-shell ChIs, while dopamine neuron driven pauses in dStr ChIs are unaffected. High dose amphetamine (16 mg/kg) reduces both dopamine neuron-driven bursts in m-shell ChIs and pauses in dStr ChIs (Figure 4.6). In contrast, glutamate cotransmission to m-shell SPNs is not affected, after either low or high doses of amphetamine. Acute amphetamine generally causes enhancement of glutamate synaptic connections (Faleiro *et al.*, 2004; Lüscher and Malenka, 2011; Underhill *et al.*, 2014) and dopamine release (Sulzer *et al.*, 2005; Natarajan and Yamamoto, 2011; Sulzer, 2011). However, the opposite is seen for dopamine neuron cotransmission: a reduction in dopamine neuron glutamate cotransmission following low-dose amphetamine, and a reduction in fast dopamine transmission following high-dose.

Amphetamine effects are time- and dose-dependent (Covey *et al.*, 2014; Siciliano *et al.*, 2014), and differ in striatal subregions (Avelar *et al.*, 2013). The initial functional connectome observations were made at a single time point, after behavioral stimulation had subsided. The effect could be different acutely, during the behavioral response, or 24 hours later, which is when most studies of single dose effects have been done. Mapping synaptic transmission, transmitter release, and postsynaptic receptors after amphetamine administration are likely to reveal regionally and temporally heterogeneous modulatory effects on dopamine neuron synapses and sites of modulation.

4.4 Novel tools for functional connectome studies

Recently developed tools may extend functional connectome mapping. dLight1 is an intensity-based genetically encoded dopamine indicator that allows optical

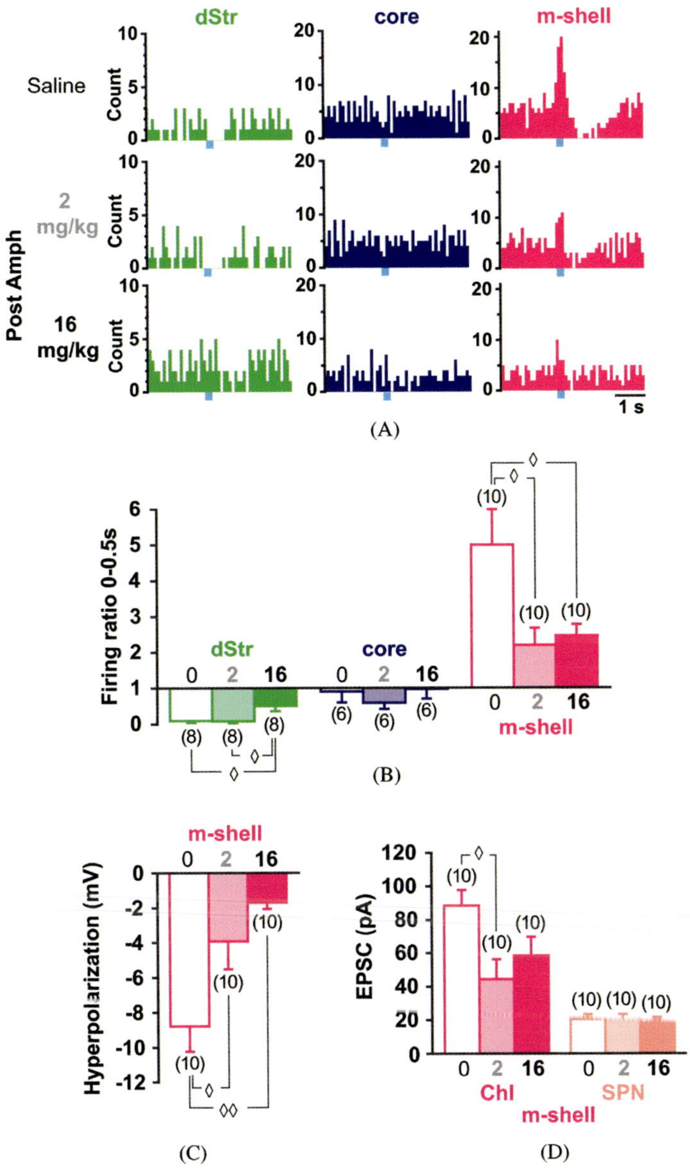

Figure 4.6 Amphetamine-induced plasticity is regionally heterogeneous. (A) Peristimulus histograms of firing in the three regions in mice previously treated with saline (top), low-dose (middle), and high-dose (bottom) amphetamine. (B) Firing ratio measurements following saline (open bars), low-dose (light-shaded bars) or high-dose (dark-shaded bars) amphetamine show that low-dose attenuates ChI bursts selectively in the m-shell, while high-dose attenuates ChI bursts in the dStr. (C) The postburst hyperpolarization in the m-shell shows a progressive dose-dependent reduction. (D) EPSCs evoked by single-pulse photostimulation (at 0.1 Hz) in m-shell ChIs were attenuated following amphetamine, whereas there was no effect in SPNs. ◊ and ◊◊ indicate $p < 0.05$ and $p < 0.01$, respectively, by one-way ANOVA with Scheffe's post hoc test. Data are presented as mean ± SEM. (Reproduced with permission from Chuhma *et al.*, 2014).

recording of dopamine dynamics with high spatiotemporal resolution in behaving mice (Patriarchi *et al.*, 2018). A genetically-modified D1 receptor, dLight1 activation is proportional to dopamine at physiological concentrations, activated by the D1-specific agonist dihydrexidine, blocked by the D1-specific antagonist SCH-23390, and not functionally coupled to adenylate cyclase (thus it does not drive intracellular increases in cAMP), nor does it show beta-arrestin mediated internalization. Fluorescence emitted from the dLight1 protein *in vivo* can be recorded using fiber photometry (Akerboom *et al.*, 2012; Chen *et al.*, 2013). dLight1 has been used in NAc during appetitive Pavlovian conditioning and demonstrated changes in dopamine concentration at subsecond timescales consistent with a reward prediction error, corroborating results from fast-scan cyclic voltammetry (FSCV). dLight1 has also been used *in vivo* to record from motor cortex layer II/III neurons during a visuomotor association task demonstrating dopamine transients during rewarded trials. dLight1, unlike FSCV, distinguishes dopamine from norepinephrine (Kovach *et al.*, 1984; Heien *et al.*, 2003). Thus, dLight1 has the potential to serve as a way to measure postsynaptic neuron activity in a genetically-identified subset of neurons optically, and simultaneously.

Genetically encoded GPCR-activation-based-DA ($GRAB_{DA}$) sensors enable measurements of dopamine dynamics in freely behaving animals with high spatiotemporal precision (Sun *et al.*, 2018). $GRAB_{DA}$ is a modification of the D2 dopamine receptor, is blocked by the D2-specific antagonist haldoperidol but does not engage the primary signaling pathways downstream of D2R (e.g. negative cAMP coupling, beta-arrestin internalization). However, one caveat is that it is unclear whether D2R coupling to G protein-coupled inwardly-rectifying potassium channels (GIRK) were affected, which could lead to potassium-current-mediated hyperpolarization. Another limitation is that DA release by norepinephrine, though at a ~10-fold lower EC_{50}. This technique lends itself to *in vivo* monitoring and has been used in conjunction with fiber photometry in mice to record DA release in dStr during a Pavlovian conditioning task and in the NAc during sexual activity. This revealed time-locked dopamine increases that were not detected with previous methods such as microdialysis or FSCV.

In order to assess dopamine signaling on a briefer timescale than GPCR-derived sensors, a dopamine ligand-gated chloride channel LGC-53 derived from *Caenorhabditis elegans* can be used as a high-fidelity sensor of synaptic dopamine release that allows for observations of the quantal basis of release, paired pulse facilitation/depression and synchronous *versus* asynchronous transmitter release (Kress *et al.*, 2014). LGC-53 has been tested in *Xenopus* oocytes, HEK293 cells and mouse striatal slices. Similar to $GRAB_{DA}$, LGC-53 is most robustly activated by dopamine, but is also activated by norepinephrine and to a lesser extent epinephrine. Although LGC-53 shows dopamine agonist selectivity, it is sensitive to GABA-R antagonists. Nonetheless, LGC-53's temporal dynamics have proved helpful in distinguishing mono- from di-synaptic responses. In a similar vein, other groups have used viral overexpression of G-protein inwardly rectifying potassium channels (GIRK2; $K_{ir}32$) in target neurons in combination with electrical and optogenetic stimulation of dopamine terminals to study inhibitory postsynaptic currents (Marcott *et al.*, 2014).

4.5 Beyond the functional connectome: Synapse to circuit to behavior

The ultimate goal of functional connectome studies is to provide a bridge between *ex vivo* synaptic studies and *in vivo* studies that incorporate awake behaving animals, thereby connecting synapse to circuit to behavior. For example, one main finding from functional connectome mapping of dopamine neurons is that they make the strongest glutamate connections in the m-shell of the NAc (Figure 4.7) (Mingote *et al.*, 2015). Furthermore, when dopamine neurons are optogenetically driven at burst firing frequencies akin to their endogenous phasic firing, they drive synchronized burst-pause sequences in ChIs (Chuhma *et al.*, 2014). ChIs activity has been theorized to play a role in encoding salience (Lim *et al.*, 2014), which is altered in disorders such as schizophrenia and addiction. Taken together, these findings from functional connectome mapping studies suggest that alterations to dopamine neuron glutamate cotransmission in m-shell ChIs may modulate salience processing.

In fact, this hypothesis was directly tested by selectively modulating dopamine neuron glutamate cotransmission by reducing levels of the glutamate-recycling enzyme glutaminase (*Gls1*) in dopamine neurons using the DAT promoter (Mingote *et al.*, 2017). As expected, mice with DAT-driven conditional heterozygous reduction of *Gls1* had reductions in glutamate cotransmission, at phasic firing frequencies with a resultant attenuation of m-shell ChI responses. Though the mice had intact responses to acute amphetamine, they showed attenuated sensitization to amphetamine. Sensitization is manifest in increasingly robust responses to the same dose of amphetamine (Robinson and Berridge, 1993), and considered to be a key behavioral component in addiction (Berridge and Robinson, 2016). Additionally, mice with a DAT-driven reduction in *Gls1* expression demonstrated potentiated latent inhibition (LI). LI is a test of how pre-exposure to a conditioned stimulus, an audio tone for example, can interfere with freezing behavior to subsequent pairing of that tone and an unconditioned aversive stimulus, say an electrical foot shock, compared to mice that did not receive previous exposure. That the DAT-driven *Gls1* reduction demonstrated a more robust ability to refrain from freezing is suggestive

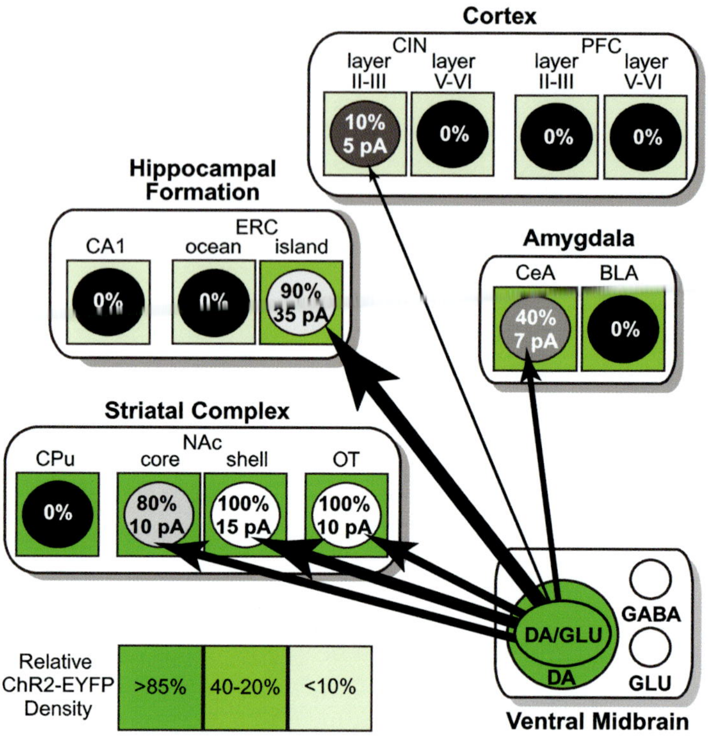

Figure 4.7 Dopamine neuron glutamatergic functional connectome in the forebrain. In the ventral midbrain, the large green circle represents dopamine neurons expressing ChR2-EYFP; the inner green ellipse (DA/GLU) represents the subpopulation of dopamine (DA) neurons that expresses ChR2-EYFP and coreleases glutamate (Glu). Within projection target areas, subregions contain putative principal neurons, diagrammed as circles. Shades of green surrounding the circles indicate the density of ChR-EYFP+ axons (scale on the bottom left); ChR2-EYFP density values are expressed as the percentage of the average density in the NAc shell (100%). Connections to the different projection areas are shown as black arrows, with relative strengths indicated by the thickness of the arrows and the incidence of connections indicated by gray shading of the circles (100% white to 0% black). In the circles, the percentage of connections found is above and average strength of the connections below. (Reproduced with permission from Mingote *et al.*, 2015).

of a psychosis-resilient phenotype (Young *et al.*, 1993, 2005). Functional connectome mapping has thus revealed dopamine neuron glutamate cotransmission as a potential therapeutic target for schizophrenia and addiction.

Functional connectome mapping offers a way to connect findings from synapse to circuit to behavior to further the understanding of complex neuronal systems, with the potential for identifying novel interventions for the treatment of neuropsychiatric disorders. The above tools when used systematically in defined dopaminoceptive neurons will help map out the functional connectome across brain regions. Being able to evaluate ever more precise circuits has the potential to reveal how interconnectivity contributes to global brain function and dysfunction, especially as it underlies neuropsychiatric disorders. This too will certainly help identify potential targets and refine treatments for disorders that continue to be refractory to currently available options, especially gene- and cell-based therapies in humans (Piguet *et al.*, 2017; Deverman *et al.*, 2018).

Acknowledgments

We thank Vlad Velicu for his thoughtful comments. DE is supported by T32 MH018870. Functional connectome mapping in our laboratory has been supported by R01 DA038966, R01 MH087758, P50 MH086404, and NARSAD.

References

Akerboom J *et al.* (2012) Optimization of a GCaMP calcium indicator for neural activity imaging. *J Neurosci*, 32: 13819–13840.

Angstadt JD, Donmoyer JE, Stretton AO (2001) The number of morphological synapses between neurons does not predict the strength of their physiological synaptic interactions: a study of dendrites in the nematode Ascaris suum. *J Comp Neurol*, 432: 512–527.

Aransay A, Rodríguez-López C, García-Amado M, Clascá F, Prensa L (2015) Long-range projection neurons of the mouse ventral tegmental area: a single-cell axon tracing analysis. *Front Neuroanat*, 9: 59.

Arenkiel BR, Peca J, Davison IG, Feliciano C, Deisseroth K, Augustine GJ, Ehlers MD, Feng G (2007) *In vivo* light-induced activation of neural circuitry in transgenic mice expressing channelrhodopsin-2. *Neuron*, 54: 205–218.

Atasoy D, Aponte Y, Su HH, Sternson SM (2008) A FLEX switch targets Channelrhodopsin-2 to multiple cell types for imaging and long-range circuit mapping. *J Neurosci*, 28: 7025–7030.

Avelar AJ, Juliano SA, Garris PA (2013) Amphetamine augments vesicular dopamine release in the dorsal and ventral striatum through different mechanisms. *J Neurochem*, 125: 373–385.

Bäckman CM, Malik N, Zhang Y, Shan L, Grinberg A, Hoffer BJ, Westphal H, Tomac AC (2006) Characterization of a mouse strain expressing Cre recombinase from the 3' untranslated region of the dopamine transporter locus. *Genesis*, 44: 383–390.

Beier KT, Steinberg EE, DeLoach KE, Xie S, Miyamichi K, Schwarz L, Gao XJ, Kremer EJ, Malenka RC, Luo L (2015) Circuit architecture of VTA dopamine neurons revealed by systematic input-output mapping. *Cell*, 162: 622–634.

Berridge KC, Robinson TE (2016) Liking, wanting, and the incentive-sensitization theory of addiction. *Am Psychol*, 71: 670–679.

Björklund A, Dunnett SB (2007) Dopamine neuron systems in the brain: an update. *Trends Neurosci*, 30: 194–202.

Bocklisch C, Pascoli V, Wong JCY, House DRC, Yvon C, de Roo M, Tan KR, Lüscher C (2013) Cocaine disinhibits dopamine neurons by potentiation of GABA transmission in the ventral tegmental area. *Science*, 341: 1521–1525.

Boyden ES, Zhang F, Bamberg E, Nagel G, Deisseroth K (2005) Millisecond-timescale, genetically targeted optical control of neural activity. *Nat Neurosci*, 8: 1263–1268.

Brezina V, Weiss KR (1997) Analyzing the functional consequences of transmitter complexity. *Trends Neurosci*, 20: 538–543.

Cachope R, Mateo Y, Mathur BN, Irving J, Wang H-L, Morales M, Lovinger DM, Cheer JF (2012) Selective activation of cholinergic interneurons enhances accumbal phasic dopamine release: setting the tone for reward processing. *Cell Rep*, 2: 33–41.

Carcone D, Ruocco AC (2017) Six years of research on the national institute of mental health's research domain criteria (RDoC) initiative: A systematic review. *Front Cell Neurosci*, 11: 46.

Chaudhury D *et al.* (2013) Rapid regulation of depression-related behaviours by control of midbrain dopamine neurons. *Nature*, 493: 532–536.

Chen T-W, Wardill TJ, Sun Y, Pulver SR, Renninger SL, Baohan A, Schreiter ER, Kerr RA, Orger MB, Jayaraman V, Looger LL, Svoboda K, Kim DS (2013) Ultrasensitive fluorescent proteins for imaging neuronal activity. *Nature*, 499: 295–300.

Chuhma N, Mingote S, Moore H, Rayport S (2014) Dopamine neurons control striatal cholinergic neurons via regionally heterogeneous dopamine and glutamate signaling. *Neuron*, 81: 901–912.

Chuhma N, Mingote S, Yetnikoff L, Kalmbach A, Ma T, Ztaou S, Sienna AC, Tepler S, Poulin JF, Ansorge M, Awatramani R, Kang UJ, Rayport S (2018) Dopamine neuron glutamate cotransmission evokes a delayed excitation in lateral dorsal striatal cholinergic interneurons. *eLife*, 7: e39786.

Chuhma N, Tanaka KF, Hen R, Rayport S (2011) Functional connectome of the striatal medium spiny neuron. *J Neurosci* 31: 1183–1192.

Cohen JY, Haesler S, Vong L, Lowell BB, Uchida N (2012) Neuron-type-specific signals for reward and punishment in the ventral tegmental area. *Nature*, 482: 85–88.

Covey DP, Roitman MF, Garris PA (2014) Illicit dopamine transients: reconciling actions of abused drugs. *Trends Neurosci* 37: 200–210.

Deverman BE, Ravina BM, Bankiewicz KS, Paul SM, Sah DWY (2018) Gene therapy for neurological disorders: progress and prospects. *Nat Rev Drug Discov*, 17: 641–659.

Edry E, Lamprecht R, Wagner S, Rosenblum K (2011) Virally mediated gene manipulation in the adult CNS. *Front Mol Neurosci*, 4: 57.

Faleiro LJ, Jones S, Kauer JA (2004) Rapid synaptic plasticity of glutamatergic synapses on dopamine neurons in the ventral tegmental area in response to acute amphetamine injection. *Neuropsychopharmacology*, 29: 2115–2125.

Fenno LE *et al.* (2014) Targeting cells with single vectors using multiple-feature Boolean logic. *Nat Methods*, 11: 763–772.

Ferguson SM, Eskenazi D, Ishikawa M, Wanat MJ, Phillips PEM, Dong Y, Roth BL, Neumaier JF (2011) Transient neuronal inhibition reveals opposing roles of indirect and direct pathways in sensitization. *Nat Neurosci*, 14: 22–24.

Gangarossa G, Espallergues J, de Kerchove d'Exaerde A, El Mestikawy S, Gerfen CR, Hervé D, Girault J-A, Valjent E (2013a) Distribution and compartmental organization of GABAergic medium-sized spiny neurons in the mouse nucleus accumbens. *Front Neural Circuits*, 7: 22.

Gangarossa G, Espallergues J, Mailly P, De Bundel D, de Kerchove d'Exaerde A, Hervé D, Girault J-A, Valjent E, Krieger P (2013b) Spatial distribution of D1R- and D2R-expressing medium-sized spiny neurons differs along the rostro-caudal axis of the mouse dorsal striatum. *Front Neural Circuits*, 7: 124.

Geisler S, Derst C, Veh RW, Zahm DS (2007) Glutamatergic afferents of the ventral tegmental area in the rat. *J Neurosci*, 27: 5730–5743.

Gong S, Doughty M, Harbaugh CR, Cummins A, Hatten ME, Heintz N, Gerfen CR (2007) Targeting Cre recombinase to specific neuron populations with bacterial artificial chromosome constructs. *J Neurosci*, 27: 9817–9823.

Gordon JA (2016) On being a circuit psychiatrist. *Nat Neurosci*, 19: 1385–1386.

Granger AJ, Wallace ML, Sabatini BL (2017) Multi-transmitter neurons in the mammalian central nervous system. *Curr Opin Neurobiol*, 45: 85–91.

Heien MLAV, Phillips PEM, Stuber GD, Seipel AT, Wightman RM (2003) Overoxidation of carbon-fiber microelectrodes enhances dopamine adsorption and increases sensitivity. *Analyst*, 128: 1413–1419.

Higley MJ, Gittis AH, Oldenburg IA, Balthasar N, Seal RP, Edwards RH, Lowell BB, Kreitzer AC, Sabatini BL (2011) Cholinergic interneurons mediate fast VGluT3-dependent glutamatergic transmission in the striatum. *PLoS ONE*, 6: e19155.

Ibáñez-Sandoval O, Tecuapetla F, Unal B, Shah F, Koós T, Tepper JM (2010) Electrophysiological and morphological characteristics and synaptic connectivity of tyrosine hydroxylase-expressing neurons in adult mouse striatum. *J Neurosci*, 30: 6999–7016.

Ikemoto S, Bonci A (2014) Neurocircuitry of drug reward. *Neuropharmacology*, 76 Pt B: 329–341.

Insel TR (2014) The NIMH Research Domain Criteria (RDoC) Project: precision medicine for psychiatry. *Am J Psychiatry*, 171: 395–397.

Jhou TC, Fields HL, Baxter MG, Saper CB, Holland PC (2009) The rostromedial tegmental nucleus (RMTg), a GABAergic afferent to midbrain dopamine neurons, encodes aversive stimuli and inhibits motor responses. *Neuron*, 61: 786–800.

Kandel ER (2001) The molecular biology of memory storage: a dialog between genes and synapses. *Biosci Rep*, 21: 565–611.

Kandel ER, Frazier WT, Waziri R, Coggeshall RE (1967) Direct and common connections among identified neurons in Aplysia. *J Neurophysiol*, 30: 1352–1376.

Kovach PM, Ewing AG, Wilson RL, Wightman RM (1984) *In vitro* comparison of the selectivity of electrodes for *in vivo* electrochemistry. *J Neurosci Methods*, 10: 215–227.

Kreitzer AC (2009) Physiology and pharmacology of striatal neurons. *Annu Rev Neurosci*, 32: 127–147.

Kress GJ, Shu H-J, Yu A, Taylor A, Benz A, Harmon S, Mennerick S (2014) Fast phasic release properties of dopamine studied with a channel biosensor. *J Neurosci*, 34: 11792–11802.

Lammel S, Steinberg EE, Földy C, Wall NR, Beier K, Luo L, Malenka RC (2015) Diversity of transgenic mouse models for selective targeting of midbrain dopamine neurons. *Neuron*, 85:429–438.

Lim SAO, Kang UJ, McGehee DS (2014) Striatal cholinergic interneuron regulation and circuit effects. *Front Synaptic Neurosci*, 6: 22.

Lima SQ, Hromádka T, Znamenskiy P, Zador AM (2009) PINP: a new method of tagging neuronal populations for identification during *in vivo* electrophysiological recording. *PLoS One*, 4: e6099.

Lima SQ, Miesenböck G (2005) Remote control of behavior through genetically targeted photostimulation of neurons. *Cell*, 121: 141–152.

Lüscher C (2013) Drug-evoked synaptic plasticity causing addictive behavior. *J Neurosci*, 33: 17641–17646.

Lüscher C, Malenka RC (2011) Drug-evoked synaptic plasticity in addiction: from molecular changes to circuit remodeling. *Neuron*, 69: 650–663.

Madisen L *et al.* (2012) A toolbox of Cre-dependent optogenetic transgenic mice for light-induced activation and silencing. *Nat Neurosci*, 15: 793–802.

Marcott PF, Mamaligas AA, Ford CP (2014) Phasic dopamine release drives rapid activation of striatal D2-receptors. *Neuron*, 84: 164–176.

Matsuda W, Furuta T, Nakamura KC, Hioki H, Fujiyama F, Arai R, Kaneko T (2009) Single nigrostriatal dopaminergic neurons form widely spread and highly dense axonal arborizations in the neostriatum. *J Neurosci*, 29: 444–453.

Mingote S, Chuhma N, Kalmbach A, Thomsen GM, Wang Y, Mihali A, Sferrazza C, Zucker-Scharff I, Siena A-C, Welch MG, Lizardi-Ortiz J, Sulzer D, Moore H, Gaisler-Salomon I, Rayport S (2017) Dopamine neuron dependent behaviors mediated by glutamate cotransmission. *Elife*, 6.

Mingote S, Chuhma N, Kusnoor SV, Field B, Deutch AY, Rayport S (2015) Functional connectome analysis of dopamine neuron glutamatergic connections in forebrain regions. *J Neurosci*, 35: 16259–16271.

Mingote S, Chuhma N, Rayport S (2017a) Optogenetic mapping of synaptic connections in mouse brain slices to define the functional connectome of identified neuronal populations. *Bio-protocol*, 7(1): e2090.

Mitchell KJ, Huang ZJ, Moghaddam B, Sawa A (2011) Following the genes: a framework for animal modeling of psychiatric disorders. *BMC Biol*, 9: 76.

Nagel G, Szellas T, Huhn W, Kateriya S, Adeishvili N, Berthold P, Ollig D, Hegemann P, Bamberg E (2003) Channelrhodopsin-2, a directly light-gated cation-selective membrane channel. *Proc Natl Acad Sci USA*, 100: 13940–13945.

Narboux-Nême N, Pavone LM, Avallone L, Zhuang X, Gaspar P (2008) Serotonin transporter transgenic (SERTcre) mouse line reveals developmental targets of serotonin specific reuptake inhibitors (SSRIs). *Neuropharmacology*, 55: 994–1005.

Natarajan R, Yamamoto BK (2011) The basal ganglia as a substrate for the multiple actions of amphetamines. *Basal Ganglia*, 1: 49–57.

Nusbaum MP, Blitz DM, Swensen AM, Wood D, Marder E (2001) The roles of co-transmission in neural network modulation. *Trends Neurosci*, 24: 146–154.

Papathanou M, Creed M, Dorst MC, Bimpisidis Z, Dumas S, Pettersson H, Bellone C, Silberberg G, Lüscher C, Wallén-Mackenzie Å (2018) Targeting VGLUT2 in mature dopamine neurons decreases mesoaccumbal glutamatergic transmission and identifies a role for glutamate co-release in synaptic plasticity by increasing baseline AMPA/NMDA ratio. *Front Neural Circuits*, 12: 64.

Patriarchi T, Cho JR, Merten K, Howe MW, Marley A, Xiong W-H, Folk RW, Broussard GJ, Liang R, Jang MJ, Zhong H, Dombeck D, von Zastrow M, Nimmerjahn A, Gradinaru V, Williams JT, Tian L (2018) Ultrafast neuronal imaging of dopamine dynamics with designed genetically encoded sensors. *Science*, 360, eaat 4422

Petreanu L, Huber D, Sobczyk A, Svoboda K (2007) Channelrhodopsin-2-assisted circuit mapping of long-range callosal projections. *Nat Neurosci*, 10: 663–668.

Petreanu L, Mao T, Sternson SM, Svoboda K (2009) The subcellular organization of neocortical excitatory connections. *Nature*, 457: 1142–1145.

Phelps PE, Vaughn JE (1986) Immunocytochemical localization of choline acetyltransferase in rat ventral striatum: a light and electron microscopic study. *J Neurocytol*, 15: 595–617.

Piguet F, Alves S, Cartier N (2017) Clinical gene therapy for neurodegenerative diseases: past, present, and future. *Hum Gene Ther*, 28: 988–1003.

Poulin J-F, Caronia G, Hofer C, Cui Q, Helm B, Ramakrishnan C, Chan CS, Dombeck DA, Deisseroth K, Awatramani R (2018) Mapping projections of molecularly defined dopamine neuron subtypes using intersectional genetic approaches. *Nat Neurosci*, 21: 1260–1271.

Pupe S, Wallén-Mackenzie Å (2015) Cre-driven optogenetics in the heterogeneous genetic panorama of the VTA. *Trends Neurosci*, 38: 375–386.

Rayport S, Sulzer D, Shi WX, Sawasdikosol S, Monaco J, Batson D, Rajendran G (1992) Identified postnatal mesolimbic dopamine neurons in culture: morphology and electrophysiology. *J Neurosci*, 12: 4264–4280.

Robinson TE, Berridge KC (1993) The neural basis of drug craving: an incentive-sensitization theory of addiction. *Brain Res Brain Res Rev*, 18: 247–291.

Rockland KS (2015) About connections. *Front Neuroanat*, 9: 61.

Rodríguez-López C, Clascá F, Prensa L (2017) The mesoaccumbens pathway: A retrograde labeling and single-cell axon tracing analysis in the mouse. *Front Neuroanat*, 11: 25.

Schmidt EF, Kus L, Gong S, Heintz N (2013) BAC transgenic mice and the GENSAT database of engineered mouse strains. *Cold Spring Harb Protoc*, 2013.

Shuen JA, Chen M, Gloss B, Calakos N (2008) Drd1a-tdTomato BAC transgenic mice for simultaneous visualization of medium spiny neurons in the direct and indirect pathways of the basal ganglia. *J Neurosci*, 28: 2681–2685.

Siciliano CA, Calipari ES, Ferris MJ, Jones SR (2014) Biphasic mechanisms of amphetamine action at the dopamine terminal. *J Neurosci*, 34: 5575–5582.

Stauffer WR, Lak A, Yang A, Borel M, Paulsen O, Boyden ES, Schultz W (2016) Dopamine neuron-specific optogenetic stimulation in Rhesus Macaques. *Cell*, 166: 1564–1571.e6.

Straub C, Tritsch NX, Hagan NA, Gu C, Sabatini BL (2014) Multiphasic modulation of cholinergic interneurons by nigrostriatal afferents. *J Neurosci*, 34: 8557–8569.

Stuber GD, Hnasko TS, Britt JP, Edwards RH, Bonci A (2010) Dopaminergic terminals in the nucleus accumbens but not the dorsal striatum corelease glutamate. *J Neurosci*, 30: 8229–8233.

Stuber GD, Stamatakis AM, Kantak PA (2015) Considerations when using cre-driver rodent lines for studying ventral tegmental area circuitry. *Neuron*, 85: 439–445.

Sulzer D (2011) How addictive drugs disrupt presynaptic dopamine neurotransmission. *Neuron*, 69: 628–649.

Sulzer D, Sonders MS, Poulsen NW, Galli A (2005) Mechanisms of neurotransmitter release by amphetamines: a review. *Prog Neurobiol*, 75: 406–433.

Sun F *et al.* (2018) A genetically encoded fluorescent sensor enables rapid and specific detection of dopamine in flies, fish, and mice. *Cell*, 174: 481–496.e19.

Swanson LW (1982) The projections of the ventral tegmental area and adjacent regions: a combined fluorescent retrograde tracer and immunofluorescence study in the rat. *Brain Res Bull*, 9: 321–353.

Tecuapetla F, Patel JC, Xenias H, English D, Tadros I, Shah F, Berlin J, Deisseroth K, Rice ME, Tepper JM, Koos T (2010) Glutamatergic signaling by mesolimbic dopamine neurons in the nucleus accumbens. *J Neurosci*, 30: 7105–7110.

Tepper JM, Tecuapetla F, Koós T, Ibáñez-Sandoval O (2010) Heterogeneity and diversity of striatal GABAergic interneurons. *Front Neuroanat*, 4: 150.

Threlfell S, Lalic T, Platt NJ, Jennings KA, Deisseroth K, Cragg SJ (2012) Striatal dopamine release is triggered by synchronized activity in cholinergic interneurons. *Neuron*, 75: 58–64.

Tritsch NX, Ding JB, Sabatini BL (2012) Dopaminergic neurons inhibit striatal output through non-canonical release of GABA. *Nature*, 490: 262–266.

Tsai H-C, Zhang F, Adamantidis A, Stuber GD, Bonci A, de Lecea L, Deisseroth K (2009) Phasic firing in dopaminergic neurons is sufficient for behavioral conditioning. *Science*, 324: 1080–1084.

Underhill SM, Wheeler DS, Li M, Watts SD, Ingram SL, Amara SG (2014) Amphetamine modulates excitatory neurotransmission through endocytosis of the glutamate transporter EAAT3 in dopamine neurons. *Neuron*, 83: 404–416.

Valjent E, Bertran-Gonzalez J, Hervé D, Fisone G, Girault J-A (2009) Looking BAC at striatal signaling: cell-specific analysis in new transgenic mice. *Trends Neurosci*, 32: 538–547.

Wang H, Peca J, Matsuzaki M, Matsuzaki K, Noguchi J, Qiu L, Wang D, Zhang F, Boyden E, Deisseroth K, Kasai H, Hall WC, Feng G, Augustine GJ (2007) High-speed mapping of synaptic connectivity using photostimulation in Channelrhodopsin-2 transgenic mice. *Proc Natl Acad Sci USA*, 104: 8143–8148.

Watabe-Uchida M, Zhu L, Ogawa SK, Vamanrao A, Uchida N (2012) Whole-brain mapping of direct inputs to midbrain dopamine neurons. *Neuron*, 74: 858–873.

White JG, Southgate E, Thomson JN, Brenner S (1976) The structure of the ventral nerve cord of Caenorhabditis elegans. *Philos Trans R Soc Lond B Biol Sci*, 275: 327–348.

Xia Y, Driscoll JR, Wilbrecht L, Margolis EB, Fields HL, Hjelmstad GO (2011) Nucleus accumbens medium spiny neurons target non-dopaminergic neurons in the ventral tegmental area. *J Neurosci*, 31: 7811–7816.

Yang H, de Jong JW, Tak Y, Peck J, Bateup HS, Lammel S (2018) Nucleus accumbens subnuclei regulate motivated behavior via direct inhibition and disinhibition of VTA dopamine subpopulations. *Neuron*, 97: 434–449.e4.

Yetnikoff L, Lavezzi HN, Reichard RA, Zahm DS (2014) An update on the connections of the ventral mesencephalic dopaminergic complex. *Neuroscience*, 282: 23–48.

Young AM, Joseph MH, Gray JA (1993) Latent inhibition of conditioned dopamine release in rat nucleus accumbens. *Neuroscience*, 54: 5–9.

Young AMJ, Moran PM, Joseph MH (2005) The role of dopamine in conditioning and latent inhibition: what, when, where and how? *Neurosci Biobehav Rev*, 29: 963–976.

Zhuang X, Masson J, Gingrich JA, Rayport S, Hen R (2005) Targeted gene expression in dopamine and serotonin neurons of the mouse brain. *J Neurosci Methods*, 143: 27–32.

CHAPTER 5

OPTOGENETICS

Erik Naylor, Daniel V. Aillon and David A. Johnson*
Pinnacle Technology, Inc., USA
*davej@pinnaclet.com

5.1 Introduction

Optogenetics is a powerful technology harnessing "the combination of genetic and optical methods to achieve gain or loss of function of well-defined events in specific cells of living tissue" (Deisseroth, 2011). Optogenetic techniques excite or inhibit neuronal activity through the directed application of specific wavelengths of light to membrane-bound receptors. Taking advantage of modern advances in this technique, researchers have reported groundbreaking discoveries in many areas of neurobiology such as learning (Wolff *et al.*, 2014), memory (Goshen *et al.*, 2011; Beyeler *et al.*, 2014), behavior (Vogelstein *et al.*, 2014) reward (Adamantidis *et al.*, 2011; Danjo *et al.*, 2014), vision (Glickfeld *et al.*, 2013), neural circuitry (Kohara *et al.*, 2014), sleep (Adamantidis *et al.*, 2007; Tsunematsu *et al.*, 2011; Kim *et al.*, 2012; Carter *et al.*, 2013; Tsunematsu *et al.*, 2014; Williams *et al.*, 2014), and seizure control (Kokaia, 2011; Katzel *et al.*, 2014).

Optogenetic techniques rely on the insertion of cell membrane receptors (opsins) genetically altered to activate when exposed to a specific wavelength of light. A large catalog of opsins has been created to either excite or inhibit neuronal action potentials or activate intracellular signaling depending on the specific wavelength of light to be applied. Recently developed opsins are capable of conditionally activating complex intracellular processes (Spoida *et al.*, 2014) and providing active feedback of cellular activity through fiber photometry (Gunaydin *et al.*, 2014).

Precise, localized firing patterns generated by optogenetic stimulation are effective tools for manipulating neuronal networks. However, to better explore physiological changes evoked through optogenetics, it is many times necessary to employ additional monitoring methodologies (Kim *et al.*, 2017; Rivnay *et al.*, 2017). Measurement of downstream physiological effects through simultaneous recording of multiple parameters in conjunction with optogenetic stimulation provides a valuable insight into real-time physiological changes brought about by the optogenetic manipulation. Recording parameters, such as EEG and enzyme based biosensor activity, requires additional electronics (amplifiers, filters and potentiostats) as well as the combination of optical and electrical cabling in a lightweight form-factor suitable for use with a rodent model.

A discussion of optogenetics has been previously presented in Volume 1 of this series (Kosillo *et al.*, 2015). In this chapter, we will expand and update the previously presented information then, in context, discuss a relatively straightforward mouse experiment using optogenetic signaling to stimulate the hippocampus. The resulting seizure propagation is measured by cortical EEG, while local glutamate changes are measured with biosensors. Design choices and technical details of the experiment will be discussed.

5.1.1 Choice of optogenetic construct

Use of a specific optogenetic model is probably the most important decision surrounding an experiment and options should be carefully weighed based on the expected experimental outcomes. Light responding opsins are not native to mammalian organisms and therefore need to be artificially introduced. Broadly, there are two primary methods for accomplishing this — viral vectors or genetically modified breeding lines. When using wild-type animals, an artificial viral vector must be used to introduce the opsin into animals that have not previously been genetically modified to respond to light. Viruses containing genomic transcripts for the opsin of interest along with a specific promoter for gene expression are injected into the target tissue or brain region. Adeno-associated virus (AAV) is a popular choice for its ability to infect both nondividing and dividing cells and for its lack of apparent pathogenicity in rodents (Daya *et al.*, 2008). Once introduced, viral mechanisms allow target genetic material to transverse into the cell, enabling production of membrane–bound light sensitive opsins (Davidson *et al.*, 2003). This methodology excels at localized positioning of the opsin expression. However, expression level, opsin duration and specificity of expression are mediated by a combination of the specific virus vector, promoter, and individual physiology and therefore can vary greatly among animals (Klein *et al.*, 2002).

Creation of transgenic animal lines containing specific opsins relies on the deliberate insertion of the opsin of interest into the genome of a founder line. This is most commonly accomplished through microinjection of DNA into a fertilized egg and implantation of the modified egg into a recipient female animal. The resulting offspring are then evaluated for expression of the intended opsin typically through use of a colocalized fluorescent tag such as green fluorescent protein (GFP). Once opsin expression is verified, progeny are then bred to establish an ongoing transgenic line (Feng *et al.*, 2004). The principal advantage to this technique is that, once established, generational offspring exhibit consistent, predictable expression of opsins. Verification of initial opsin expression as well as variability in genetic insertions can significantly increase the time required to develop a viable line.

5.1.2 Opsin choices and availability

The first light-activated cation channels to be employed in model organisms were channelrhodopsin-1 (ChR1) and channelrhodopsin-2 (ChR2). Originally isolated from *Chlamydomonas reinhardtii,* these proteins were harnessed to work as optically activated regulators for mammalian action potential firing and had begun to be incorporated into neuronal physiology (Deisseroth, 2015). Since those early experiments, there has been a widespread adoption of optogenetics in the neuroscience community resulting in a vast array of available opsins wherein ChR2 is, by far, the most commonly employed. ChR2 is a 7-transmembrane-helix that, when activated by light around 470 nm wavelength, opens as a nonselective cation channel allowing the entry of Na^+ ions resulting in neuronal depolarization (Nagel *et al.*, 2003). Stimulation by sufficient blue light intensity in the 440–490 nm range results in the selective firing of neurons expressing this protein. Because of this utility, many channelrhodopsin variants have been developed. Just in the last 10 years channelrhodopsins have been modified to exhibit higher amplitude and slower discharge currents (Nagel *et al.*, 2005; Lin *et al.*, 2009; Wang *et al.*, 2009; Wen *et al.*, 2010; Berndt *et al.*, 2011). Variants of ChR2 have also been created that are faster or slower to turn off after illumination such as the ChETA, ChR2(D156A), and the SFO mutants (Berndt *et al.*, 2009; Bamann *et al.*, 2010; Gunaydin *et al.*, 2010). Color shifting has been introduced as well, with the advent of channelrhodopsins with peak cation gating at 580–650 nm (VChR1, MChR1, and C1V1) (Zhang *et al.*, 2008; Chow *et al.*, 2010; Govorunova *et al.*, 2011; Yizhar *et al.*, 2011; Klapoetke, 2014). Enhanced calcium permeability has also been introduced in the CatCh variant resulting in the increased light sensitivity and repolarization time (Kleinlogel *et al.*, 2011). Most recently, the newest channelrhodopsin variants (dACRs and nACRs and GtACR1) (Kim *et al.*, 2018) have been designed with specific anion-conducting properties to support their use as an optogenetic tool to inhibit neuronal firing with light (Kato *et al.*, 2018). Given the rapid pace of the methodology used to create these variants there may be little slowing to the pace at which they are created (Chow *et al.*, 2011).

In contrast to neuronal depolarization caused by most channelrhodopsins, halorhodopsins result in hyperpolarization via inward driven chloride pumps when exposed to green–yellow light (550–590 nm). Early halorhodopsins were obtained from the archaea *Natronobacterium pharaonis* (NpHR) (Han *et al.*, 2007; Zhang *et al.*, 2007) although many other organisms have also been identified to express halorhodopsins capable of mediating hyperpolarization in neurons (Chow *et al.*, 2011). In general, halorhodopsins demonstrate reduced expression in neurons compared to channelrhodopsins and typically rely on additional trafficking sequences, such as potassium channels, to functionally express at high levels in mammalian cells (Ma *et al.*, 2001; Gradinaru *et al.*, 2008; Zhao *et al.*, 2008; Gradinaru *et al.*, 2010). Early halorhodopsins also tended to have a slower recovery time following extended illumination, lasting for many minutes following complete photocurrent saturation (Hegemann *et al.*, 1985; Bamberg *et al.*, 1993; Chow *et al.*, 2011).

Enhancements to halorhodopsin (eNpHR2) have resulted in increased membrane targeting (Gradinaru *et al.*, 2008) and, more importantly, substantially enhanced inhibitory capacity even in the presence of illumination as low as 3.5 mW/mm^2 (Liu *et al.*, 2010). Although not as extensively modified as channelrhodopsin, the improvements in halorhodopsin have made it possible to successfully use it in combination with ChR2. Indeed, several papers describe successful depolarization/hyperpolarization of the cell membrane using this pairing (Liu *et al.*, 2010; Madisen *et al.*, 2012; Golden *et al.*, 2016; Sadler *et al.*, 2017).

Other opsins have been introduced to the toolkit over the years. Light-driven proton pumps have been developed to remove positive charged ions from the inside of the neuron, effectively enabling hyperpolarization of the membrane when illuminated with green–yellow (550–600 nm) light. Archaerhodopsin–3 (Arch) from *H. sodomense* (Chow *et al.*, 2010) and the more recently identified archaerhodopsin isolated from the TP009 strain (ArchT) (Han *et al.*, 2011) can be used to enable large-scale silencing of brain regions. Furthermore, G-protein coupled receptors (GPCR) have been adapted to drive coupled signaling such as calcium release or K^+ ion channel activation (Li *et al.*, 2005; Ye *et al.*, 2011). These particular light-driven opsins have been used to control reinforcing behavior (Airan *et al.*, 2009) and to silence neural activity in the spinal cord (Li *et al.*, 2005; Ye *et al.*, 2011). Taken together, these opsins represent a vast array of cellular functions that can be potentially controlled simply through the administration of light pulses. Further expansion of this technology will push researchers to delve into more and more complex neuronal functions that can be directly studied using this method.

Although these opsins have been described in the literature, obtaining any specific one for use can present some hurdles. Many constructs can be ordered directly from Stanford (*web.stanford.edu/group/dlab/optogenetics/request_dna*) or the UNC vector core (*www.med.unc.edu/genetherapy/vectorcore/*). Typically, these can be directly introduced into wild–type animals using microinjection techniques followed by an *in vivo* expression period of 3–5 weeks. However, depending on the intended usage, cost may be an issue for entities other than research institutes. It is also possible to order transgenic animals directly from some suppliers such as The Jackson Labs (*www.jax.org*) or Horizon Discovery (*www.horizondiscovery.com*). Depending on the supply, transgenic animals may be available for shipment immediately, eliminating the *in vivo* expression period necessary for virus constructs. In some cases, breeding pairs may also be obtained allowing the researcher to maintain their own breeding colony. As the availability of opsin constructs continues to expand, researchers should continually explore multiple options to find those that best fit their research and budget needs.

5.1.3 Light delivery

Numerous vendors provide benchtop and wireless optogenetic light delivery sources and components. The light delivery method depends on the intended application and available resources. In general, the absorption spectra of typical opsins are relatively broad (Klapoetke *et al.*, 2014). The irradiance (mW/mm^2) required to trigger the opsin is often assumed to be 10 mW/mm^2 (Boyden *et al.*, 2005), although the actual intensity required for activation is very opsin dependent and lower irradiance is often sufficient (Aravanis *et al.*, 2007). It is worth noting that the units used in the literature and vendor offerings are inconsistent. Often, illumination intensity (mW) is provided whereas, for a given experimental setup, the irradiance at the point of opsin activation is the important parameter. This irradiance depends on the intensity (power) of the source, coupling mechanism, number of couplings (for example, laser to fiber, or optical commutator to fiber) in the delivery system, wavelength, location dependent absorption and scattering (Al-Juboori *et al.*, 2013; Jacques, 2013), and the divergence of the light in the area of interest. In many cases, fiber optics are used for light delivery but surface illumination is also common using two-photon (fixed or scanning) (Yang *et al.*, 2010; Oron *et al.*, 2012) and red shifted opsins (Smith *et al.*, 2014; Senova *et al.*, 2017). Structured or patterned illumination (Bovetti *et al.*, 2015) is also commonly used. Lasers provide powerful, monochromatic, and coherent light that is available in a wide range of wavelengths and couples easily to optical fibers; however, laser safety and random intensity fluctuations (speckle) must be addressed. Lasers tend to be benchtop units coupled via optical connections and commutators. Light emitting diodes (LEDs) are available in a wide range of wavelengths. While LEDs are not as monochromatic as lasers, they do provide more than sufficient light in the wavelengths necessary for optogenetic experimentation. LEDs are not coherent, so speckle is not an issue and they can be purchased either as benchtop units, to be coupled via fiber optics and optical commutators, or they can be mounted at or near the opsin activation region. For this reason, LEDs are normally the choice for wireless measurements on freely moving animals. However, it is more difficult to couple LED light to fiber optics because the power output of LEDs is much lower than that of lasers. Furthermore, when LEDs are placed in direct contact with tissue, localized heating must be controlled. In fact, heating (Arias–Gil, 2016) and phototoxicity (Delbeke, 2017) must be considered in any optogenetics experiment.

5.2 Experimental description

Transgenic C57 mice [B6.Cg–Tg(Thy1–COP4/EYFP)9Gfng/J] acquired from Jackson Laboratory (Bar Harbor, ME) were employed as the optogenetic model (n = 6). This mouse strain has been engineered to express the ChR2 protein tied to a Thy1 promoter. Histological examination reveals ChR2–YFP expression in hippocampal CA1 and CA3 pyramidal neurons, nuclei of the thalamus, midbrain, and brainstem as well as layer 5 cortical neurons (Arenkiel *et al.*, 2007). It was our goal to induce seizure-like activity in these animals though optogenetic kindling within the hippocampal region and record both the resulting seizures as well as extracellular glutamate concentration in real time via an enzyme-mediated biosensor (Osawa *et al.*, 2013).

5.2.1 EEG Recording

Discrimination of brain electrical state activity relies on the accurate determination of electroencephalograph (EEG) signals from a freely moving animal. A full explanation of the function and physiology is beyond the scope of this chapter but has been described elsewhere (Naylor *et al.*, 2011; 2012; 2015). Briefly, conductive electrodes are used to record an electrical potential between two points on the brain. This voltage differential is then plotted over time and can be used to describe the overall electrical activity of the brain around and between the two measurement points. Cortical EEG activity within the brain (recorded in this experiment using screws placed directly on the cortex) yields a wealth of information regarding gross neuronal activity. Depth recording using more closely spaced electrodes (recorded in this experiment by using two twisted pair wires positioned at the end of the optical stimulation cannula) is ideal for recording localized neuronal firing in response to a stimulus, in this case an optical pulse.

Both the frequency and the amplitude of the recorded EEG signal are important measures that need to be considered when analyzing this type of data. Overall EEG frequency is essential for determination of sleep/wake state and seizure-like activity. Physiologically, EEG frequency indicates the degree of overall neuronal activity and the amount to which that firing is occurring in a synchronized manner. Normal, waking cortical activity will contain an abundance of higher frequencies with components in the Alpha (8–12 Hz), Beta (13–30 Hz) and Gamma (> 30 Hz) ranges. It should be noted that there are no strict definitions for these band ranges and they will tend to vary with respect to the upper and lower bounds depending on the laboratory and technique. Non-REM sleep EEG, by contrast will have a much greater degree of synchronized neuronal activity and contains a majority of lower frequency components in the Delta (0.5–4 Hz) and Theta (4–7 Hz) ranges. Frequency for seizure-like EEG activity is more variable but typically displays sharp, defined frequency peaks and repeated activity in a specific frequency range lasting the entirety of the seizure event representing short-bursts of high-energy, synchronized neuronal activity.

In addition to frequency, EEG signal amplitude is also an important component for analysis of EEG data. The amplitude of the EEG signal indicates the number of neurons that are firing at any particular frequency. The neurophysiologist Donald Hebb put it best when he noted, "neurons that fire together wire together" when referring to neuronal models of associative learning. Low amplitude signals indicate a disparity of neurons firing at any given frequency, while higher amplitudes indicate many neurons firing in synchrony at a given frequency. Potential differences ranging in size from 50–150 μV arise as a result of normal activity over a variety of frequencies. Larger amplitudes (200–350 μV) are noted during non-REM sleep when many neurons are firing together at lower frequencies. Very large amplitude signals (>400 μV) can be observed during seizure events as more and more neurons are recruited to fire at a given seizure frequency.

5.2.2 Hardware requirements

The hardware required for this recording is detailed in Figure 5.1A. EEG and biosensor data were filtered and amplified using an 8401–HS DCAS system while control of the optogenetic fiber firing was accomplished using an 8480 Optogenetic Interface Module (Pinnacle Technology, Inc., Lawrence, KS). Signals were routed through a 10-pin, low torque swivel and lightweight tether. The head-mounted LED and fiber optic coupling eliminated the need for an optical swivel, allowing for all electrical connections. EEG and biosensor data were collected via a head-mounted preamplifier that allowed for connections to the optical fiber, biosensor and provided 100X gain for all three EEG channels. Preamplification of EEG

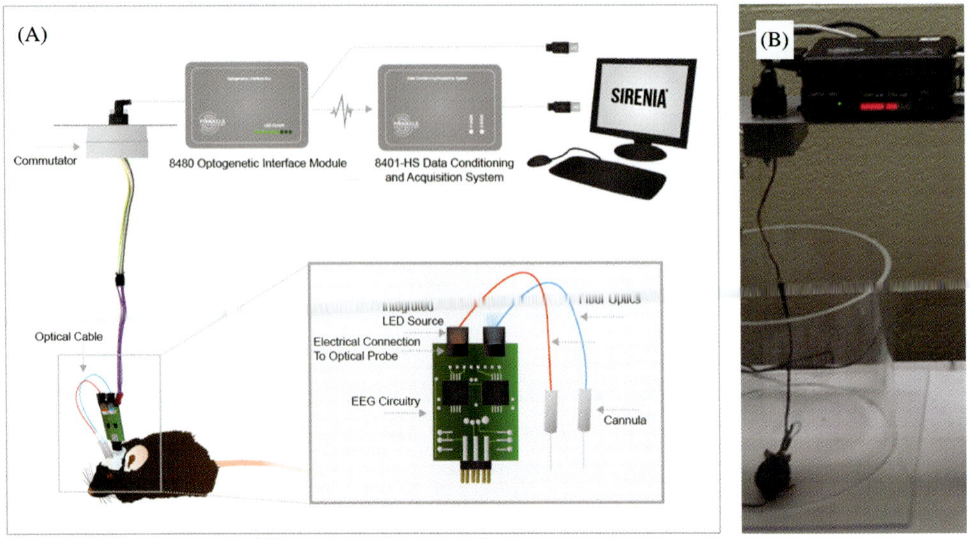

Figure 5.1 Complete optogenetics + EEG + biosensor recording system. (A) System diagram incorporating a lightweight head stage, swivel, recording amplifiers and optogenetics control circuits with a computer interface. (B) The complete system shown mounted above a standard mouse cage.

signals was essential for the elimination of most cable–associated movement artifact. Other than the preamplifier and tether, all electronic components were mounted directly above the cage (Figure 5.1B).

5.2.3 Surgical implantation and recording

Mice were implanted with a cannula (BASi, West Lafayette, IN) and depth electrode in the hippocampus (A/P –2.6, M/L +2.5, D/V –2.5) along with two cortical electrodes (Pinnacle Technology, Inc., Lawrence, KS). The cortical leads were designed to measure general cortical electrical activity while the depth electrode served to record neuronal electrical activity at a location close to the site of optical stimulation. This configuration is diagrammed in Figure 5.2. In the instances when glutamate activity was also recorded, a second cannula was placed in the prefrontal cortex (A/P +1.6, M/L +2.5, D/V –2.0) during the initial surgery (Figures 5.2A and 5.2B). All procedures undertaken on animals were previously approved by the University of Kansas IACUC. Animals were allowed at least seven days to recover from surgical implantation.

Following recovery from surgery, a blue (445 nm) LED butt-coupled to a fiber probe was inserted into the guide cannula targeting the hippocampus. The cannula-based implantation localized light exposure to the dorsal hippocampus as shown in Figure 5.2C. Light intensity was measured to be 100 mW/mm^2 or

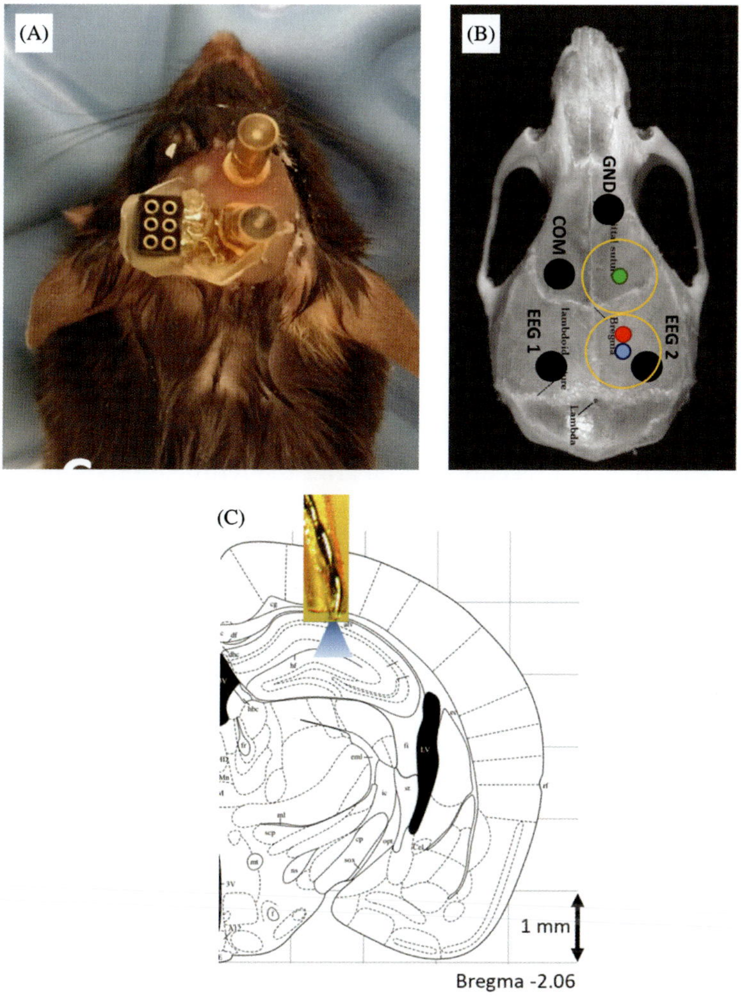

Figure 5.2 (A) 6-pin EEG connector + cannulas for seizure induction/recording. (B) Diagram of optical probe (blue circle), depth electrode (red circle) and biosensor (green circle) placement. Yellow circles represent cannula outer diameter and black circles represent EEG electrode lead placement. (C) Diagram of fiber optic probe placement and estimated light projection within hippocampus. (B) and (C) adapted from The Mouse Brain in Stereotaxic Coordinates, 2nd ed. by Paxinos and Franklin.

greater and delivered at 20 Hz with a 10% duty cycle. To kindle seizure activity, each animal was stimulated for 20 seconds six times a day. Stimulations were delivered once every 30 minutes over a 3-hour period for periods lasting from 2–7 days in a manner similar to previously described seizure kindling protocols (Osawa *et al.*, 2013; Khoshkhoo *et al.*, 2017). In two of the animals, a glutamate biosensor was inserted into the guide cannula once seizure activity was observable. Biosensors rely on the measurement of enzymatically produced H_2O_2 to communicate near real-time information regarding concentration change of a specific analyte (Wilson *et al.*, 2000, 2008). *In vivo* biosensors provide precise and accurate reporting of analyte concentration changes within a freely-moving animal with response time of approximately one second (Hu *et al.*, 1997a, 1997b). This rapid sampling allows neurochemical events occurring on a fast timescale to be tied to observable physiology. In this case the biosensor allows for the continuous measurement of glutamate concentration over the course of the induced seizure events.

5.3 Findings

The application of optogenetic pulses resulted in predictable, measurable seizure-like effects persisting and sustaining even after the termination of the optogenetic pulses (Figure 5.3A). Repeatable increases in glutamate concentration were observed to correspond with induced seizure events (Figure 5.3B) demonstrating robustness of the seizure induction. In 11 observable cases a measurable, abrupt decline in extracellular glutamate activity was seen corresponding with the termination of the seizure event (Figures 5.3A and 5.3B). The etiology of this drop is unknown however, it may be related to the increased generation of glutamate as a source of neuronal energy resulting from high levels of stimulation during the seizure event. When this stimulation is terminated, a drop-off is observed because there is no further need for this source of energy. Such a mechanism is consistent with the Astrocyte-Neuron-Lactate Shuttle (ANLS) (Pellerin *et al.*, 2012). The length of the event, lasting approximately 2–4 seconds, is much too short to be detected with traditional microdialysis techniques but was easily discerned within the one second resolution of biosensor monitoring.

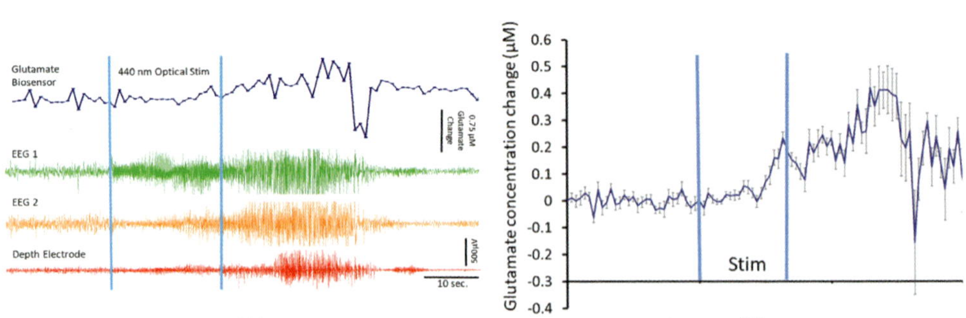

(A) (B)

Figure 5.3 (A) Simultaneous cortical EEG, depth electrode and glutamate biosensor recording during seizure induction via optical fiber probe (blue lines) (B) Average glutamate change in response to multiple, optically induced seizure events (n = 11). Optical stimulation was applied during the period between the blue lines. Bars represent S.E.M.

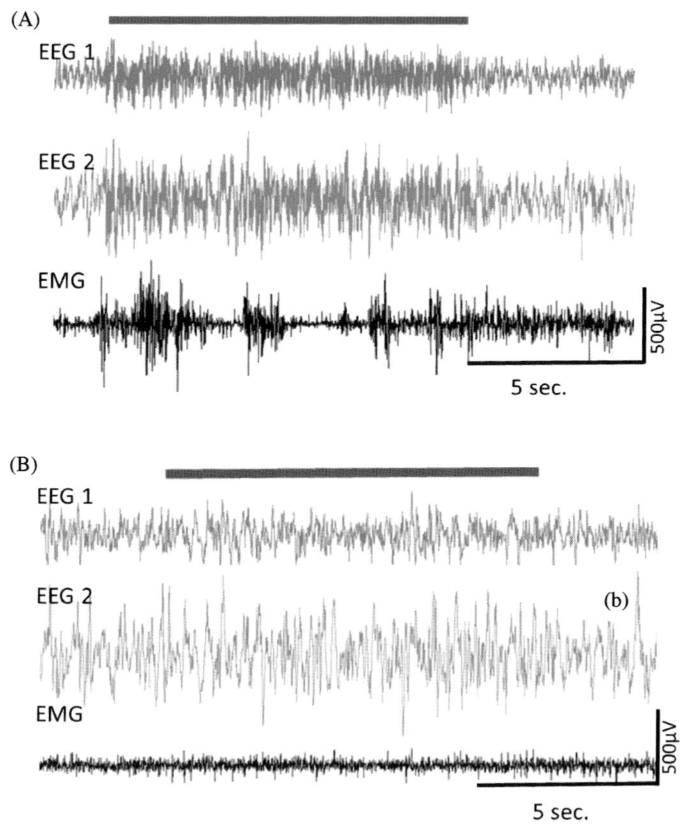

Figure 5.4 Effect on EEG during high frequency (40 Hz) stimulation with blue light (A) and repeated in the same animal using red light (B). Additional high-frequency EEG stimulation reflecting greater neuronal activation is visible under blue light stimulation and not observed when stimulation is shifted to the higher frequency (red) range.

To control for other experimental variables commonly associated with opsin experiments, such as localized expression, cannula placement, temperature, and opsin specificity, the blue (445 nm) LED was exchanged for a red (660 nm) LED within the same cannula. Stimulation by blue light resulted in changes to the EEG signal corresponding with the stimulus frequency (Figure 5.4A). Stimulation by red light pulses of identical frequency, duration, and duty cycle resulted in no observable changes in the EEG waveforms (Figure 5.4B). The ability to switch the light source from an opsin-activating (blue) wavelength to an opsin-indifferent (red) wavelength, allows for the direct observation of the effects of opsin activity while controlling for animal-specific effects such as surgical implantation differences and individual brain morphology as well as potential physical confounds such as Becquerel effects and probe temperature. The ability to switch wavelengths also provides confidence in the functionality of the photobiology as well as the instrumentation.

5.4 Summary

Precise, targeted optogenetic stimulation at 20 Hz for 20 seconds delivered via a fiber optic cable butt-coupled to a high-intensity LED provided sufficient power to activate ChR2 activity in commercial transgenic mice. Stimulation effects had sufficient power to generate kindled seizure events and reliably reproduce large-scale seizure activity observable well beyond the termination of the initial optogenetic pulse train. Coupling of EEG recording allowed precise visualization of the seizure events while addition of a biosensor provided details of the extracellular glutamate concentration within the prefrontal cortex during seizure events.

In our model animals, optogenetic expression was driven by the Thy1 promoter. By design, Thy1 linked expression in this model results in ChR2 expression in multiple, unrelated areas of the brain. Delivery of broad-based light to the top of the head, via an LED mounted directly on the skull (Iwai *et al.*, 2011), or placed next to the outer cortex (Huber *et al.*, 2008) would potentially cause activation of target areas beyond those intended. It was therefore necessary to carefully plan the method of light delivery such that sufficient blue light was delivered to our target area without the introduction of light to additional, non-target areas that may also respond to optogenetic stimulation. Implanting a cannula for specific guidance of the optical fiber optic probe is the most effective technique while the direct attachment of the fiber probe to the LED emitter ensured that the light source remained in close proximity.

The coupling method was also highly important as light loss due to direct coupling of optical components can range from 50–90% depending on the materials and method used. For our experiments, it was necessary to emit at least 100 mW/mm^2 out of a 200 μm tip to ensure sufficient opsin activation. During probe construction, light meter readings were continually observed to optimize placement of the optical fiber for maximum light transmission. This method resulted in a capture efficiency of 60% of the total light produced within that diameter on the LED die.

Overall these experiments demonstrate the advantage of using localized optogenetic stimulation via LEDs coupled with multiple physiological recording methodologies to gain a more complete picture of seizure physiology.

At a time when pharmaceutical companies have severely curtailed or altogether ceased their internal neuroscience drug discovery efforts due to repeated failures and lack of new drug candidates, optogenetic tools offer new discovery pathways. Targeting neuronal circuits with unprecedented specificity in pathway activation, cellular-specific capability and compatibility with cutting edge electrophysiology and electrochemical detection of neurotransmitters all open up new potential for research. As our knowledge of neuronal circuitry deepens, new disease models can be created. With a better understanding of diseases on a circuit level, new therapeutic targets can be discovered, and compounds can be more efficiently screened.

References

Adamantidis AR, Tsai HC, Boutrel B, Zhang F, Stuber GD, Budygin EA, Tourino C, Bonci A, Deisseroth K, de Lecea L (2011) Optogenetic interrogation of dopaminergic modulation of the multiple phases of reward-seeking behavior. *J Neurosci*, 31: 10829–10835.

Adamantidis AR, Zhang F, Aravanis AM, Deisseroth K, de Lecea L (2007) Neural substrates of awakening probed with optogenetic control of hypocretin neurons. *Nature*, 450: 420–424.

Airan RD, Thompson KR, Fenno LE, Bernstein H, Deisseroth K (2009) Temporally precise *in vivo* control of intracellular signalling. *Nature*, 458: 1025–1029.

Al-Juboori SI, Dondzillo A, Stubblefield EA, Felsen G, Lei TC, Klug A (2013) Light scattering properties vary across different regions of the adult mouse brain. *PLoS One*, 8: e67626.

Aravanis AM, Wang LP, Zhang F, Meltzer LA, Mogri MZ, Schneider MB, Deisseroth K (2007) An optical neural interface: *in vivo* control of rodent motor cortex with integrated fiberoptic and optogenetic technology. *J Neural Eng*, 4: S143–156.

Arenkiel BR, Peca J, Davison IG, Feliciano C, Deisseroth K, Augustine JG, Ehlers MD, Feng G (2007) *In vivo* light-induced activation of neural circuitry in transgenic mice expressing channelrhodopsin-2. *Neuron*, 54: 205–218.

Bamann C, Gueta R, Kleinlogel S, Nagel G, Bamberg E (2010) Structural guidance of the photocycle of channelrhodopsin-2 by an interhelical hydrogen bond. *Biochemistry*, 49: 267–278.

Bamberg E, Tittor J, Oesterhelt D (1993) Light-driven proton or chloride pumping by halorhodopsin. *Proc Natl Acad Sci USA*, 90: 639–643.

Berndt A, Schoenenberger P, Mattis J, Tye KM, Deisseroth K, Hegemann P, Oertner TG (2011) High-efficiency channelrhodopsins for fast neuronal stimulation at low light levels. *Proc Natl Acad Sci USA*, 108: 7595–7600.

Berndt A, Yizhar O, Gunaydin LA, Hegemann P, Deisseroth K (2009) Bi-stable neural state switches. *Nat Neurosci*, 12: 229–234.

Beyeler A, Eckhardt CA, Tye KM (2014) Deciphering memory function with optogenetics. *Prog Mol Biol Transl Sci*, 122: 341–390.

Bovetti S, Fellin T (2015) Optical dissection of brain circuits with patterned illumination through the phase modulation of light. *J Neurosci Methods*, 241: 66–77.

Boyden ES, Zhang F, Bamberg E, Nagel G, Deisseroth K (2005) Millisecond-timescale, genetically targeted optical control of neural activity. *Nat Neurosci*, 8: 1263–1268.

Carter ME, de Lecea L, Adamantidis A (2013) Functional wiring of hypocretin and LC-NE neurons: implications for arousal. *Front Behav Neurosci*, 7: 43.

Chow BY, Chuong AS, Klapoetke NC, Boyden ES (2011) Synthetic physiology strategies for adapting tools from nature for genetically targeted control of fast biological processes. *Methods Enzymol*, 497: 425–443.

Chow BY, Han X, Dobry AS, Qian X, Chuong AS, Li M, Henninger MA, Belfort GM, Lin Y, Monahan PE, Boyden ES (2010) High-performance genetically targetable optical neural silencing by light-driven proton pumps. *Nature*, 463: 98–102.

Danjo T, Yoshimi K, Funabiki K, Yawata S, Nakanishi S (2014) Aversive behavior induced by optogenetic inactivation of ventral tegmental area dopamine neurons is mediated by dopamine D2 receptors in the nucleus accumbens. *Proc Natl Acad Sci USA*, 111: 6455–6460.

Davidson BL, Breakefield XO (2003) Viral vectors for gene delivery to the nervous system. *Nat Rev Neurosci*, 4: 353–364.

Daya S, Berns KI (2008) Gene therapy using adeno-associated virus vectors. *Clin Microbiol Rev*, 21: 583–593.

Deisseroth K (2011) Optogenetics. *Nat Methods*, 8: 26–29.

Deisseroth K (2015) Optogenetics: 10 years of microbial opsins in neuroscience. *Nat Neurosci*, 18: 1213–1225.

Feng G, Lu J, Gross J (2004) Generation of transgenic mice. *Methods Mol Med*, 99: 255–267.

Glickfeld LL, Histed MH, Maunsell JH (2013) Mouse primary visual cortex is used to detect both orientation and contrast changes. *J Neurosci*, 33: 19416–19422.

Golden SA, Heshmati M, Flanigan M, Christoffel DJ, Guise K, Pfau ML, Aleyasin H, Menard C, Zhang H, Hodes GE, Bregman D, Khibnik L, Tai J, Rebusi N, Krawitz B, Chaudhury D, Walsh JJ, Han MH, Shapiro ML, Russo SJ (2016) Basal forebrain projections to the lateral habenula modulate aggression reward. *Nature*, 534: 688–692.

Goshen I, Brodsky M, Prakash R, Wallace J, Gradinaru V, Ramakrishnan C, Deisseroth K (2011) Dynamics of retrieval strategies for remote memories. *Cell*, 147: 678–689.

Govorunova EG, Spudich EN, Lane CE, Sineshchekov OA, Spudich JL (2011) New channelrhodopsin with a red-shifted spectrum and rapid kinetics from Mesostigma viride. *MBio*, 2: e00115–00111.

Gradinaru V, Thompson KR, Deisseroth K (2008) eNpHR: a Natronomonas halorhodopsin enhanced for optogenetic applications. *Brain Cell Biol*, 36: 129–139.

Gradinaru V, Zhang F, Ramakrishnan C, Mattis J, Prakash R, Diester I, Goshen I, Thompson KR, Deisseroth K (2010) Molecular and cellular approaches for diversifying and extending optogenetics. *Cell*, 141: 154–165.

Gunaydin LA, Grosenick L, Finkelstein JC, Kauvar IV, Fenno LE, Adhikari A, Lammel S, Mirzabekov JJ, Airan RD, Zalocusky KA, Tye KM, Anikeeva P, Malenka RC, Deisseroth K (2014) Natural neural projection dynamics underlying social behavior. *Cell*, 157: 1535–1551.

Gunaydin LA, Yizhar O, Berndt A, Sohal VS, Deisseroth K, Hegemann P (2010) Ultrafast optogenetic control. *Nat Neurosci*, 13: 387–392.

Han X, Boyden ES (2007) Multiple-color optical activation, silencing, and desynchronization of neural activity, with single-spike temporal resolution. *PLoS One*, 2: e299.

Han X, Chow BY, Zhou H, Klapoetke NC, Chuong A, Rajimehr R, Yang A, Baratta MV, Winkle J, Desimone R, Boyden ES (2011) A high-light sensitivity optical neural silencer: development and application to optogenetic control of non-human primate cortex. *Front Syst Neurosci*, 5: 18.

Hegemann P, Oesterbelt D, Steiner M (1985) The photocycle of the chloride pump halorhodopsin. I: Azide-catalyzed deprotonation of the chromophore is a side reaction of photocycle intermediates inactivating the pump. *EMBO J*, 4: 2347–2350.

Hu Y, Wilson GS (1997a) Rapid changes in local extracellular rat brain glucose observed with an *in vivo* glucose sensor. *J Neurochem*, 68: 1745–1752.

Hu Y, Wilson GS (1997b) A temporary local energy pool coupled to neuronal activity: fluctuations of extracellular lactate levels in rat brain monitored with rapid-response enzyme-based sensor. *J Neurochem*, 69: 1484–1490.

Huber D, Petreanu L, Ghitani N, Ranade S, Hromadka T, Mainen Z, Svoboda K (2008) Sparse optical microstimulation in barrel cortex drives learned behaviour in freely moving mice. *Nature*, 451: 61–64.

Iwai Y, Honda S, Ozeki H, Hashimoto M, Hirase H (2011) A simple head-mountable LED device for chronic stimulation of optogenetic molecules in freely moving mice. *Neurosci Res*, 70: 124–127.

Jacques SL (2013) Optical properties of biological tissues: a review. *Phys Med Biol*, 58(11): R37–61.

Kato HE, Kim YS, Paggi JM, Evans KE, Allen WE, Richardson C, Inoue K, Ito S, Ramakrishnan C, Fenno LE, Yamashita K, Hilger D, Lee SY, Berndt A, Shen K, Kandori H, Dror RO, Kobilka BK, Deisseroth K (2018) Structural mechanisms of selectivity and gating in anion channelrhodopsins. *Nature*, 561(7723): 349–354.

Katzel D, Nicholson E, Schorge S, Walker MC, Kullmann DM (2014) Chemical-genetic attenuation of focal neocortical seizures. *Nat Commun*, 5: 3847.

Khoshkhoo S, Vogt D, Sohal VS (2017) Dynamic, cell-type-specific roles for GABAergic interneurons in a mouse model of optogenetically inducible seizures. *Neuron*, 93: 291–298.

Kim A, Latchoumane C, Lee S, Kim GB, Cheong E, Augustine GJ, Shin HS (2012) Optogenetically induced sleep spindle rhythms alter sleep architectures in mice. *Proc Natl Acad Sci USA*, 109: 20673–20678.

Kim CK, Adhikari A, Deisseroth K (2017) Integration of optogenetics with complementary methodologies in systems neuroscience. *Nat Rev Neurosci*, 18: 222–235.

Kim YS, Kato HE, Yamashita K, Ito S, Inoue K, Ramakrishnan C, Fenno LE, Evans KE, Paggi JM, Dror RO, Kandori H, Kobilka BK, Deisseroth K (2018) Crystal structure of the natural anion-conducting channelrhodopsin GtACR1. *Nature*, 561(7723): 343–348.

Klapoetke NC, Murata Y, Kim SS, Pulver SR, Birdsey-Benson A, Cho YK, Morimoto TK, Chuong AS, Carpenter EJ, Tian Z, Wang J, Xie Y, Yan Z, Zhang Y, Chow BY, Surek B, Melkonian M, Jayaraman V, Constantine-Paton M, Wong GK, Boyden ES (2014) Independent optical excitation of distinct neural populations. *Nat Methods*, 11: 338–346.

Klein RL, Hamby ME, Gong Y, Hirko AC, Wang S, Hughes JA, King MA, Meyer EM (2002) Dose and promoter effects of adeno-associated viral vector for green fluorescent protein expression in the rat brain. *Exp Neurol*, 176: 66–74.

Kleinlogel S, Feldbauer K, Dempski RE, Fotis H, Wood PG, Bamann C, Bamberg E (2011) Ultra light-sensitive and fast neuronal activation with the Ca^{2+}-permeable channelrhodopsin CatCh. *Nat Neurosci*, 14: 513–518.

Kohara K, Pignatelli M, Rivest AJ, Jung HY, Kitamura T, Suh J, Frank D, Kajikawa K, Mise N, Obata Y, Wickersham IR, Tonegawa S (2014) Cell type-specific genetic and optogenetic tools reveal hippocampal CA2 circuits. *Nat Neurosci*, 17: 269–279.

Kokaia M (2011) Light-activated channels in acute seizures. *Epilepsia*, 52 Suppl 8: 16–18.

Kosillo P, Brimblecombe KR, Threlfell S, Cragg SJ (2015) Coupling voltammetry with optogenetics to reveal axonal control of dopamine transmission by striatal acetylcholine. In *Compendium of In Vivo Monitoring in Real-Time Molecular Neuroscience: Volume 1: Fundamentals and Applications*. Wilson GS, Michael AC, Eds., World Scientific Publishing, pp. 201–223.

Li X, Gutierrez DV, Hanson MG, Han J, Mark MD, Chiel H, Hegemann P, Landmesser LT, Herlitze S (2005) Fast noninvasive activation and inhibition of neural and network activity by vertebrate rhodopsin and green algae channelrhodopsin. *Proc Natl Acad Sci USA*, 102: 17816–17821.

Lin JY, Lin MZ, Steinbach P, Tsien RY (2009) Characterization of engineered channelrhodopsin variants with improved properties and kinetics. *Biophys J*, 96: 1803–1814.

Liu X, Tonegawa S (2010) Optogenetics 3.0. *Cell*, 141: 22–24.

Ma D, Zerangue N, Lin YF, Collins A, Yu M, Jan YN, Jan LY (2001) Role of ER export signals in controlling surface potassium channel numbers. *Science*, 291: 316–319.

Madisen L, Mao T, Koch H, Zhuo JM, Berenyi A, Fujisawa S, Hsu YW, Garcia AJ 3rd, Gu X, Zanella S, Kidney J, Gu H, Mao Y, Hooks BM, Boyden ES, Buzsaki G, Ramirez JM, Jones AR, Svoboda K, Han X, Turner EE, Zeng H (2012) A toolbox of Cre-dependent optogenetic transgenic mice for light-induced activation and silencing. *Nat Neurosci*, 15: 793–802.

Nagel G, Brauner M, Liewald JF, Adeishvili N, Bamberg E, Gottschalk A (2005) Light activation of channelrhodopsin-2 in excitable cells of Caenorhabditis elegans triggers rapid behavioral responses. *Curr Biol*, 15: 2279–2284.

Nagel G, Szellas T, Huhn W, Kateriya S, Adeishvili N, Berthold P, Ollig D, Hegemann P, Bamberg E (2003) Channelrhodopsin-2, a directly light-gated cation-selective membrane channel. *Proc Natl Acad Sci USA*, 100: 13940–13945.

Naylor E, Aillon DV, Barrett BS, Wilson GS, Johnson DA, Harmon HP, Gabbert S, Petillo PA (2012) Lactate as a biomarker for sleep. *Sleep*, 35: 1209–1222.

Naylor E, Aillon DV, Gabbert S, Harmon H, Johnson DA, Wilson GS, Petillo PA (2011) Simultaneous real-time measurement of EEG/EMG and L-glutamate in mice: A biosensor study of neuronal activity during sleep. *J Electroanal Chem*, 656: 106–113.

Naylor E, Petillo PA (2015) Using biosensors to probe fundamental questions of sleep. In *Compendium of In Vivo Monitoring in Real-Time Molecular Neuroscience: Volume 1: Fundamentals and Applications*. Wilson GS, Michael AC, Eds., World Scientific Publishing,pp 1–26.

Oron D, Papagiakoumou E, Anselmi F, Emiliani V (2012) Two-photon optogenetics. *Prog Brain Res*, 196: 119–143.

Osawa S, Iwasaki M, Hosaka R, Matsuzaka Y, Tomita H, Ishizuka T, Sugano E, Okumura E, Yawo H, Nakasato N, Tominaga T, Mushiake H (2013) Optogenetically induced seizure and the longitudinal hippocampal network dynamics. *PLoS One*, 8: e60928.

Pellerin, L., P.J. Magistretti (2012) Sweet Sixteen of ANLS. *J Cereb Blood Flow Metab*, 32(7): 1152–1166

Rivnay J, Wang H, Fenno L, Deisseroth K, Malliaras GG (2017) Next-generation probes, particles, and proteins for neural interfacing. *Sci Adv*, 3: e1601649.

Sadler KE, McQuaid NA, Cox AC, Behun MN, Trouten AM, Kolber BJ (2017) Divergent functions of the left and right central amygdala in visceral nociception. *Pain*, 158: 747–759.

Senova S, Scisniak I, Chiang CC, Doignon I, Palfi S, Chaillet A, Martin C, Pain F (2017) Erratum: Experimental assessment of the safety and potential efficacy of high irradiance photostimulation of brain tissues. *Sci Rep*, 7: 46199.

Smith IT, Smith SL (2014) Getting it through your thick skull. *Nat Neurosci*, 17: 1018–1019.

Spoida K, Masseck OA, Deneris ES, Herlitze S (2014) Gq/5-HT2c receptor signals activate a local GABAergic inhibitory feedback circuit to modulate serotonergic firing and anxiety in mice. *Proc Natl Acad Sci USA*, 111: 6479–6484.

Tsunematsu T, Kilduff TS, Boyden ES, Takahashi S, Tominaga M, Yamanaka A (2011) Acute optogenetic silencing of orexin/hypocretin neurons induces slow-wave sleep in mice. *J Neurosci*, 31: 10529–10539.

Tsunematsu T, Ueno T, Tabuchi S, Inutsuka A, Tanaka KF, Hasuwa H, Kilduff TS, Terao A, Yamanaka A (2014) Optogenetic manipulation of activity and temporally controlled cell-specific ablation reveal a role for MCH neurons in sleep/wake regulation. *J Neurosci*, 34: 6896–6909.

Vogelstein JT, Park Y, Ohyama T, Kerr RA, Truman JW, Priebe CE, Zlatic M (2014) Discovery of brainwide neural-behavioral maps via multiscale unsupervised structure learning. *Science*, 344(6182): 386–392.

Wang H, Sugiyama Y, Hikima T, Sugano E, Tomita H, Takahashi T, Ishizuka T, Yawo H (2009) Molecular determinants differentiating photocurrent properties of two channelrhodopsins from chlamydomonas. *J Biol Chem*, 284: 5685–5696.

Wen L, Wang H, Tanimoto S, Egawa R, Matsuzaka Y, Mushiake H, Ishizuka T, Yawo H (2010) Opto-current-clamp actuation of cortical neurons using a strategically designed channelrhodopsin. *PLoS One*, 5: e12893.

Williams RH, Chee MJ, Kroeger D, Ferrari LL, Maratos-Flier E, Scammell TE, Arrigoni, E (2014) Optogenetic-mediated release of histamine reveals distal and autoregulatory mechanisms for controlling arousal. *J Neurosci*, 34: 6023–6029.

Wilson GS, Hu Y (2000) Enzyme-based biosensors for *in vivo* measurements. *Chem Rev*, 100: 2693–2704.

Wilson GS, Johnson MA (2008) *In-vivo* electrochemistry: what can we learn about living systems? *Chem Rev*, 108: 2462–2481.

Wolff SB, Grundemann J, Tovote P, Krabbe S, Jacobson GA, Muller C, Herry C, Ehrlich I, Friedrich RW, Letzkus JJ, Luthi A (2014) Amygdala interneuron subtypes control fear learning through disinhibition. *Nature*, 509: 453–458.

Yang XF, Schmidt BF, Rode DL, Rothman SM (2010) Optical suppression of experimental seizures in rat brain slices. *Epilepsia*, 51: 127–135.

Ye H, Daoud-El Baba M, Peng RW, Fussenegger M (2011) A synthetic optogenetic transcription device enhances blood-glucose homeostasis in mice. *Science*, 332: 1565–1568.

Yizhar O, Fenno LE, Prigge M, Schneider F, Davidson TJ, O'Shea DJ, Sohal VS, Goshen I, Finkelstein J, JT Paz, Stehfest K, Fudim R, Ramakrishnan C, Huguenard JR, Hegemann P, Deisseroth K (2011) Neocortical excitation/inhibition balance in information processing and social dysfunction. *Nature*, 477: 171–178.

Zhang F, Prigge M, Beyriere F, Tsunoda SP, Mattis J, Yizhar O, Hegemann P, Deisseroth K (2008) Red-shifted optogenetic excitation: a tool for fast neural control derived from *Volvox carteri*. *Nat Neurosci*, 11: 631–633.

Zhang F, Wang LP, Brauner M, Liewald JF, Kay K, Watzke N, Wood PG, Bamberg E, Nagel G, Gottschalk A, Deisseroth K (2007) Multimodal fast optical interrogation of neural circuitry. *Nature*, 446: 633–639.

Zhao S, Cunha C, Zhang F, Liu Q, Gloss B, Deisseroth K, Augustine GJ, Feng G (2008) Improved expression of halorhodopsin for light-induced silencing of neuronal activity. *Brain Cell Biol*, 36: 141–154.

CHAPTER 6

ELECTROCHEMICAL IMAGING OF EXOCYTOTIC FUSION EVENTS USING ELECTROCHEMICAL DETECTOR ARRAYS

Shailendra S. Rathore*, Meng Huang*, Ying Zhao*,†, Qinghua Fang*,† and Manfred Lindau*,†,‡

*School of Applied & Engineering Physics, Cornell University, NY, USA

†Laboratory for Nanoscale Cell Biology, Max-Planck-Institute for Biophysical Chemistry, Göttingen, Germany

‡ML95@cornell.edu

6.1 Introduction

In chromaffin cells, single vesicle fusion events are mostly studied by amperometric recordings using carbon fiber microelectrodes (CFMs) (Wightman *et al.*, 1991; Chow *et al.*, 1992) or by fluorescence techniques that detect release of fluorophores accumulating in acidic compartments (Steyer *et al.*, 1997), fluorescently labeled peptides or proteins (Kaether *et al.*, 1997; Gandasi *et al.*, 2015), fluorescent membrane markers such as FM dyes (Betz *et al.*, 1992; Henkel and Betz, 1995; Gaffield and Betz, 2006), fluorescent false neurotransmitters (Gubernator *et al.*, 2009), or by detecting vesicular pH changes using luminal pH sensitive indicators, such as pHluorins (Miesenbock *et al.*, 1998). It is generally assumed that these different techniques monitor the same events, fusion of the same dense core vesicles. Monitoring fluorescence changes and amperometric signals from the same vesicle is not simultaneously possible using CFMs because the vesicles from which release is detected amperometrically are located beneath the CFM tip where their fluorescence change is very difficult, if not impossible, to image.

For high resolution, fluorescence imaging of release events at the cell surface that is in contact with the coverslip is required because Total Internal Reflection Fluorescence (TIRF) excitation is used (Axelrod *et al.*, 1983; Steyer *et al.*, 1997), which minimizes background fluorescence and improves the signal-to-noise ratio. To detect release events towards the coverslip amperometrically, two approaches have been used. One uses surface patterned transparent electrodes, such that fluorescence is excited through the electrode (Sun and Gillis, 2006; Kisler *et al.*, 2012). Such transparent electrode materials include indium tin oxide (ITO), nitrogen-doped diamond-like carbon (DLC) deposited on top of ITO, or very thin (12–17 nm) gold films (Kisler *et al.*, 2012). While this makes it possible to observe a fluorescence change related to an amperometric spike, it remains uncertain if the two events are indeed originating from the same vesicle, because in these experiments amperometric events were detected at much higher frequency than fluorescence events. A second approach uses an ElectroChemical Detector (ECD) array of surface patterned microelectrodes that are not transparent but have a transparent space between them such that the cell sits in this space between the electrodes directly on the coverslip (Dias *et al.*, 2002; Hafez *et al.*, 2005). This approach has the advantage that the time-resolved amperometric currents measured by the individual electrodes detect different fractions of the released molecules, depending on their distance from the release site, and therefore allow determination of the time course as well as localization of quantal events.

6.2 Electrochemical imaging using ECD arrays

6.2.1 ECD array design

The original ECD array design shown in Figure 6.1 (Dias *et al.*, 2002) consists of 4 Pt electrodes surface patterned on a standard microscope coverslip No. 1 (150–165 μm thick). The ECD area consists of a window that is opened in an insulating layer exposing the tips of 4 Pt conductors that are otherwise insulated. The active electrodes are formed by these exposed tips which are ~3–5 μm wide and ~150 nm thick. Each of these four electrodes is at the end of a surface patterned insulated Pt conductor that has at its other end a contact pad connecting to one channel of a four-channel current amplifier. In the center between the electrodes an approximately 5 μm diameter area is present where the cell will be placed such that its bottom surface makes direct contact with the glass coverslip.

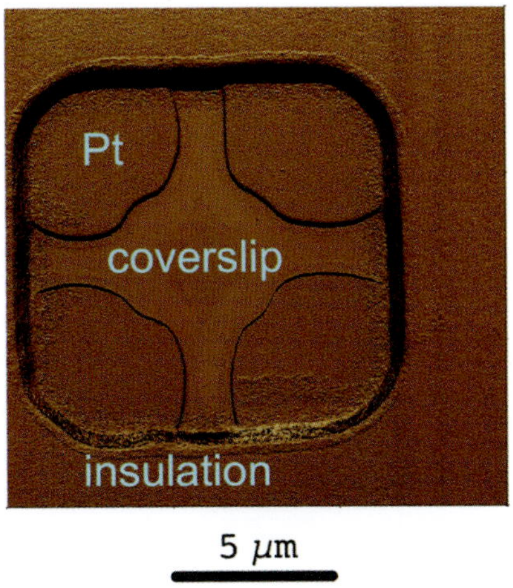

Figure 6.1 AFM image of four electrode electrochemical detector array. Dimensions are in μm (modified from Dias *et al.* (2002)).

6.2.2 ECD array fabrication

At present, our lab fabricates two different types of ECD arrays. In the first type the Pt conductors are insulated with an SiO_2 layer (Berberian *et al.*, 2009), a robust material for regular use that is easy to clean but cells have to be positioned on the individual ECD array. In the second type the Pt conductors are insulated with a thick SU-8 photoresist layer forming microwells for cell trapping (Liu *et al.*, 2011; Huang *et al.*, 2018) directly over the ECD arrays. The entire sequence of steps of the ECD array fabrication process is shown in Figure 6.2.

For more efficient fabrication the ECD arrays are now fabricated on 4-inch 0.175 mm thick (#1.5) D263 glass wafers rather than 1-inch coverslips. After cleaning with acetone, isopropanol (IPA) and de-ionized (DI) water, the glass wafer is primed with hexamethyldisilazane (HMDS) and then spin-coated with AZ nLOF 2020 negative photoresist (PR) at 3000 rpm for 30 s to gain a 2 μm thick uniform layer. The wafer is soft baked on a hotplate at 110°C for 1 min. A photomask with the ECD patterns is used for pattern transfer onto the wafer. The wafer is then aligned with the photomask and exposed with 85 mJ/cm^2 using the near-UV light source (365–405 nm) of the contact aligner. The post exposure bake is again at 110°C for 1 min to form the cross-link structure in the photoresist and remove gas bubbles generated during the exposure. The wafer is developed with AZ 726 MIF developer for 70s, rinsed with DI water and blown dry with a N_2 gun.

Before metal deposition, the wafer goes through an O_2 plasma descum process with 70 Standard Cubic Centimeter per Minute (sccm) O_2 and 50 W power at 80°C for 5 min. A sputtering system is used to first deposit 10 nm Ti and then 150 nm Pt. Alternatively, an electron beam evaporator may be used to deposit 2 nm Ti

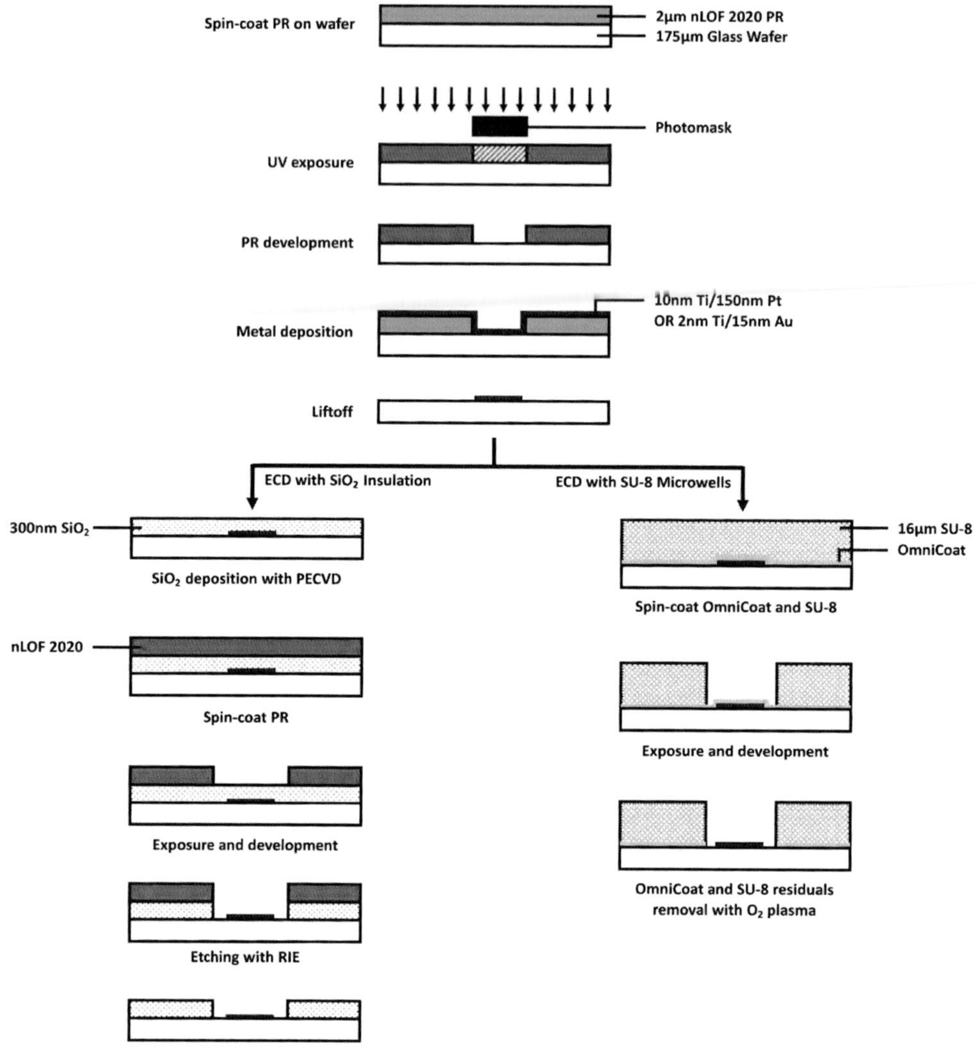

Figure 6.2 ECD fabrication flow diagram, for description of steps see text.

and 15 nm Au, forming a transparent Au electrode. The deposited wafer is then placed into MICROPOSIT remover 1165 overnight for liftoff. On the second day, the wafer is taken out from the beaker and rinsed with DI water and isopropyl alcohol. It is then inspected under the microscope to confirm clean liftoff.

6.2.2.1 ECD array with 300 nm SiO_2 insulation

To fabricate the ECD with regular SiO_2 insulation, the entire surface of the glass wafer is deposited with 300 nm SiO_2 using plasma-enhanced chemical vapor deposition. It is then primed with HMDS and spin-coated with AZ nLOF 2020 negative photoresist at 3000 rpm for 30s and soft-baked at 110°C for 1 min. The wafer is aligned with the patterns on the photomask and then exposed with 85 mJ/ cm^2 to open the windows over the ECD electrode array for etching. It is again post baked at 110°C for 1 min and developed with AZ 726 MIF developer for 70 s. The descum process is performed with the same settings described above. A final hard bake is performed at 110°C for 5 min. A reactive ion etching (RIE) system is used to etch the exposed SiO_2 350 nm in depth. The anticipated etching depth exceeds the deposited SiO_2 layer thickness by 50 nm in order to ensure removal of all SiO_2 making sure that the electrodes are exposed to the environment. The wafer is placed into MICROPOSIT remover 1165 overnight to remove the remaining photoresist. The SiO_2 insulation layer is a protection for the beneath metal connection.

6.2.2.2 ECD array with microwells in SU-8 insulation

To fabricate the ECD array with SU-8 insulation and microwells, the glass wafer is dehydrated on a hotplate at 150°C for 30 min after liftoff. It is spin-coated with OmniCoat at 3000 rpm for 30 s and baked at 200°C for 1 min. The OmniCoat layer is a primer for SU-8 as SU-8 adheres poorly to glass or Au. The glass wafer is then spin-coated with SU-8 2025 at 4000 rpm for 1 min. The SU-8 coated wafer is baked at 65°C for 3 min and then 95°C for 5 min. The resulting SU-8 layer has a thickness of ~16 μm. A photomask with an 18 μm diameter circular opening for each ECD array is used for the microwell pattern transfer. The wafer is aligned with the photomask and exposed with the near-UV light source through a 365 nm filter with 182 mJ/cm^2 to open up the microwells to expose the electrodes. After 10 min rest followed by the post exposure bake at 65°C for 1 min and 95°C for 5 min, the glass wafer is placed inside a half-cut wafer cassette facing down and immersed in SU-8 developer for 6.5 min. During the development, mild agitation is performed with a stir bar at 50 rpm. After the development, the wafer is placed in isopropyl alcohol for 2 min, rinsed with DI water and blown dry with a N_2 gun. The wafer is hard baked on a hotplate at 200°C for 20 min to heal any cracks formed during the development. To remove any remaining SU-8 and OmniCoat residues on the ECD arrays, the glass wafer undergoes a harsh O_2 plasma etch with 500 W power, 500 sccm N_2 and 4500 sccm O_2 at 300°C. The fabricated SU-8 microwells are slightly larger than bovine chromaffin cells making them suitable for efficient trapping of these cells. The size of the microwells may be modified by changing the patterns on the photomask and SU-8 thickness for other cells. For smaller mouse chromaffin cells, a 12 μm well diameter and well depth may be appropriate.

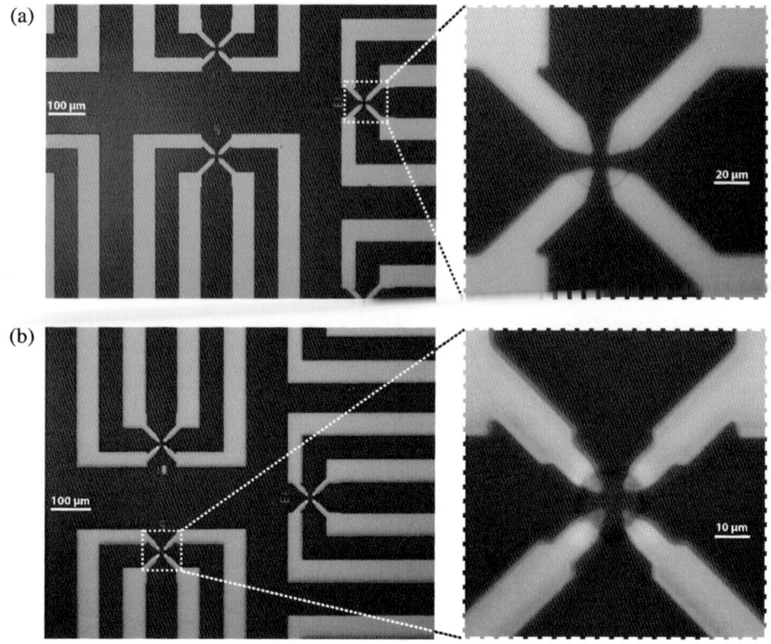

Figure 6.3 Microscope images of ECDs with SiO_2 insulation (a) and SU-8 microwells (b) acquired in epi-illumination mode (left, 5× objective; right 20× objective). Due to the thick SU-8 layer, the focal plane for the zoomed in image in (b) was set in between the SU-8 surface plane and the electrode plane to visualize the microwell and the electrodes in the same view.

6.2.2.3 Final steps of ECD array fabrication

The SiO_2 or SU-8 insulated glass wafer is subsequently primed with HMDS and spin-coated with MICROPOSIT S1813 photoresist at 4000 rpm for 30 s. The wafer is soft baked at 90°C for 1 min. The photoresist layer acts as a protection for the ECDs during cutting and handling. The wafer is then cut into nine individual 1 inch × 1 inch ECD coverslips with a dicing saw. To remove the protection photoresist, the ECD coverslips are soaked in acetone overnight before using them for experiments. Figure 6.3 shows the microscope images of ECD arrays with SiO_2 insulation (Figure 6.3a) and with SU-8 microwells (Figure 6.3b).

6.2.3 Simultaneous electrochemical and TIRF imaging

6.2.3.1 Mounting and connecting the ECD array

The coverslip containing the ECD arrays (Figure 6.4a) was air dried after washing with 70% alcohol and water and then mounted on the custom-made microscope stage designed to hold the ECD. The ECD arrays were covered

(a) (b) (c)

Figure 6.4 (a) Image of coverslip with 14 surface patterned ECD arrays. Each ECD array has four electrodes, which are connected to a total of 56 contact pads at the periphery of the coverslip. (b,c) ECD coverslip mounted on the microscope stage with one ECD array connected to the 4-channel amplifier via spring-loaded pins (yellow arrows), (b) top view, and (c) side view.

with ~200 μL bath solution containing (in mM) 140 NaCl, 5 KCl, 5 $CaCl_2$, 1 $MgCl_2$, 10 HEPES/NaOH and 20 glucose (pH 7.3) and a Ag|AgCl reference electrode was immersed into the buffer (Figure 6.4b,c). The electrodes of the chosen ECD array were connected to the 4-channel amplifier box by spring-loaded pins via the contact pads at the periphery of the ECD coverslip as shown in Figure 6.4.

6.2.3.2 Loading of fluorescent transmitter and cell handling

As an example, we describe imaging of false fluorescent transmitter (FFN511) release from chromaffin granules in combination with ECD array recording. Bovine adrenal glands were provided by a local slaughterhouse and chromaffin cells prepared as described (Parsons *et al.*, 1995). Briefly, after collecting the adrenal gland it was immediately placed in ice cold buffer A containing 118 mM NaCl, 3.3 KCl, 1 mM $NaH_2PO_4.H_2O$, 1 mM $MgSO_4.7H_2O$, 10 mM Glucose, 25 mM HEPES-NaOH pH 7.3, 2 mM L-Glutamine, 100 units-mg/mL Pen-Strep, 0.1 mg/mL Gentamycin, supplemented with MEM vitamins and MEM Amino acids according to manufacturer's instructions.

Upon reaching the lab, glands were cleaned by removing fat and other tissue and perfused through the aortal opening by connecting to a perfusion setup having carbogen flow of 3–4 bubbles/s all the time. The perfusate was allowed to drain through the several cuts made at the distal end of the gland for 1 hr at 37°C with the change of buffer three times intermittently after 20 min interval. Then glands were digested by another perfusion with enzyme (collagenase 1 mg/ml in buffer A with BSA) for 1 hr at 37°C. After complete digestion glands look fluffy, were cut and opened in a sterile dish kept inside a laminar flow hood to collect the medulla (yellowish tissue) in smallest possible clumps. All the clumps were gently homogenized with a Teflon pestle on nylon mesh strainer and re-incubated with fresh collagenase buffer in a sterile closed Erlenmeyer flask vented for carbogen flow (3–4 bubbles/s). After 1 hr incubation at 37°C cells were filtered through sterile gauze and washed twice with buffer A with BSA by centrifugation at 500 rpm for 10 min. Finally, the cell pellet was re-suspended in culture medium and filtered through a 70 μm sterile nylon cell strainer. Cells were cultured on coverslips (8 mm diameter) with a density of 1–1.5 × 10^4 cells/coverslip in a humidified incubator maintained at 37°C with 8% CO_2.

For loading with FFN511, a coverslip was taken out after 24 hrs, washed with bath solution and incubated with 7 μM FFN511 (Abcam) dissolved in bath solution for 10 min in the dark at room temperature. After incubation the cover slip was washed again with bath solution to remove free FFN511 dye from the coverslip surface.

For experiments, an 8 mm coverslip with cultured cells is placed near the edge of the ECD coverslip and a fluorescent FFN511-loaded cell identified under epifluorescence excitation. The illumination mode is then switched to bright field to avoid bleaching of the fluorophore. The selected cell is loosely patched by a glass micro-pipette filled with bath solution using a micromanipulator. By gently adjusting the micromanipulator the cell is lifted up from the glass and placed in the center area of ECD array, where the SiO_2 insulation is removed and the tips of 4 ECD electrodes are exposed. To avoid damage of the cell membrane while mechanically detaching it from the glass coverslip, we recommend selecting cells that do not attach to the coverslip firmly. We typically pick cells that exhibit slight mobility during a gentle finger-tapping on the anti-vibration table but appear morphologically healthy in the bright-field image (bright, round, without any visible granularity in the cytoplasm).

6.2.3.3 Fluorescence imaging

A series of fluorescence images of the chromaffin cell loaded with FFN511 placed on the ECD array was acquired using an EMCCD camera (Andor iXon) mounted on a Zeiss Axiovert 135 TV microscope modified for TIRF imaging and equipped with a Zeiss Plan-Fluar 1.45 N.A. 100x oil immersion objective. TIRF excitation was performed using a lamp-based (X-cite, 120PC) system (Zhao *et al.*, 2013) through a 480/40 nm excitation filter and a 505 LP dichroic with 535/50 nm emission filter. For each series recording, 3000 images were acquired with 100 ms exposure time and ~101.8 ms frame duration including the readout interval. Cells were stimulated to induce exocytosis by gently pressing the cell on the ECD array surface and release of FFN511 from a vesicle was detected as a loss of fluorescence that coincided with release of catecholamines from the vesicle as detected by the simultaneous ECD recording. To determine the correct time relation between the imaging data and the ECD recording, the camera fire output is recorded together with the four amperometry traces.

6.2.3.4 ECD array recording and analysis

Transmitter release from a single fusion event was detected by the four ECD electrodes as amperometric current spikes generated by oxidation of transmitter at electrodes. Different amplitudes of these spikes in the different electrodes result from different diffusion distances between the release site and the particular electrode with the largest spike detected by the electrode closest to the release site.

6.2.3.4.1 *Electrochemical localization of release events*

To precisely determine the relation between the location of a release site and the fraction of molecules detected by the different electrodes, the geometrical outline of the ECD array electrodes is obtained from a bright field image (Figure 6.5a), which is acquired with the same objective that is used for the fluorescence image acquisition before the fluorescence imaging starts. The geometrical outline is determined by choosing a suitable threshold followed by manual "touch-up" to correct points that were above or below the threshold within or outside the electrode areas (Figure 6.5b).

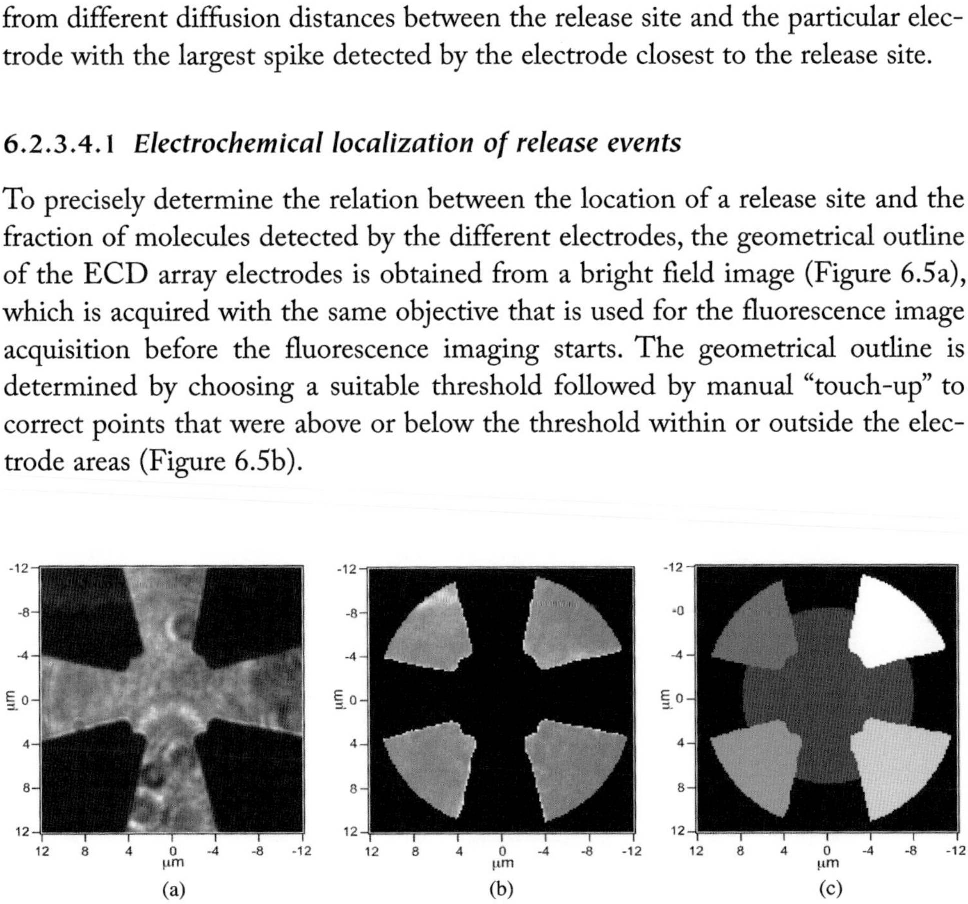

Figure 6.5 (a) Bright field ECD image used for generation of electrode charge fraction map. (b) ECD electrode outline generated by selecting the non-insulated area (where the electrodes are exposed) and background subtraction. (c) Within the gray circle (50 pixel radius of 8 μm) release from sites on a 480 nm grid are used for random walk simulations to determine the electrode charge fractions for each release site.

Once the geometrical borders of the ECD array electrodes are defined, random walk simulations (RWSs) are performed for a grid of release locations 480 nm apart (three camera pixels), covering the circular area of 8 μm radius at the center of the four electrodes of ECD (Figure 6.5c). In this simulation, the cell is represented by a semi-infinite cylinder with 10 μm diameter suspended 100 nm above the ECD array. For each release location on the grid an RWS is performed for 100,000 molecules. When a molecule touches the bottom of the cylinder or the glass it is reflected. When it touches an ECD electrode it is removed from the simulation and counted for that electrode. Molecules that did not hit an electrode and diffused outside the 10 μm circle are considered lost and removed from the simulation. In this way a map is generated of the fraction of molecules detected by the different electrodes for each simulated release site. The fractional charges for locations between the simulated grid locations are calculated by interpolation.

During the experiment, the currents from the ECD array are recorded by a 4-channel amperometry amplifier (VA-10M4; NPI Electronic), filtered at 500 Hz using the built-in two-pole Bessel filters, and sampled at 1 kHz using a 16-bit analog-to-digital converter (NIDAQ) together with the camera fire output. Individual release events produce amperometric spikes in all four electrodes of the ECD array (Figure 6.6). The amperometric spikes detected by the individual electrodes as well as the summed events adding the four recorded traces are analyzed using the quantal analysis software originally developed for carbon fiber amperometry (Mosharov and Sulzer, 2005).

Before this analysis the amperometric traces collected by the four electrodes were filtered using a Gaussian digital low pass infinite impulse response filter with 500 Hz cutoff frequency that attenuates the baseline noise without distorting the amperometric spike shape and a 60 Hz notch filter with 6 Hz width to attenuate

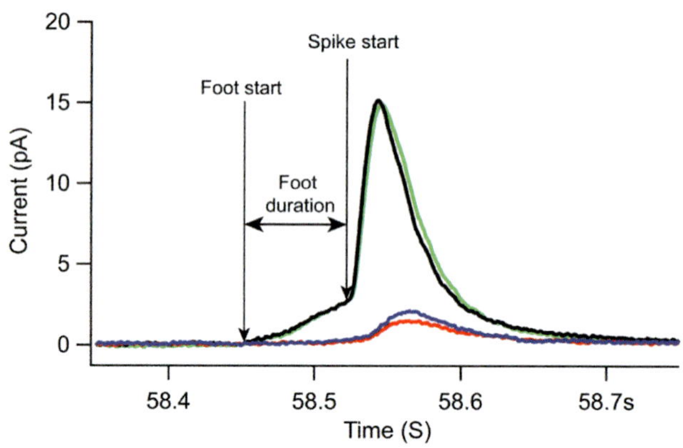

Figure 6.6 Example of amperometric spikes recorded by the ECD array for a single release event.

line frequency noise. After converting A/D converter counts to picoampere based on amplifier gain, the four traces from the four electrodes as well as the sum of the four traces were saved in IGOR binary format.

Each of the five traces was then analyzed using the quantal analysis software (Mosharov and Sulzer, 2005). In the first step the software is used to determine recommended filter settings for the quantal analysis. These recommended settings are determined using the summed trace giving typical values of 120–150 Hz for the binomial filter, ~60 Hz for the additional binomial filter and ~30 Hz for the 1st derivative filter, for details see (Mosharov and Sulzer, 2005)). These filter settings are then used for the analysis of all five traces.

For each of the five traces (four ECD traces and sum of the four traces) the parameters listed in Table 6.1 of the detected spikes are first saved in separate files and subsequently combined into one file containing the data from all five traces. During the combing all the events were grouped according to their respective time of appearance. The criteria used to group the events was that T_max of each individual trace should be between the T_Start and T_End of the sum of the four traces.

The partial quantal size values of amperometric spikes with similar spike start time that were detected by at least three electrodes are summed and the fractional quantal charge values detected by the individual electrodes are calculated. Due to diffusional delays the start times may differ slightly, which is accounted for by a tolerance of maximal 100 ms. The release locations of these events are then fitted minimizing the sum of the squared deviations between the measured fractional charges and the interpolated fractional quantal size values of the RWS map. Due to the high time resolution of amperometric recordings this electrochemical imaging (ECI) of fusion events allows for the analysis of rapid changes associated with fusion pore formation.

Table 6.1 Spike parameters calculated during quantal analysis.

Wave name	**Meaning**
T_Start	Spike Start time
T_max	Spike peak time
Peak_Q	Quantal size of trace
Foot_d	Foot duration
Peak_Imax	Amplitude of trace
T_End	Spike end time
Peak_t05	Half width of spike
Rise_time	Rise time of spike
Foot_Amp	Foot amplitude

6.2.3.5 Correlation of release events detected by the ECD array with fluorescence changes

For a precise correlation between the release sites determined from the ECD data and the fluorescence images, it is of particular importance to align the ECD geometry used to generate the RWS map (Figure 6.5) with the ECD geometry as it appears in a given fluorescence image. This may require small shifts because the sample may shift slightly between acquisition of the bright field image and beginning of the fluorescence movie and also during acquisition of the fluorescence movie. These shifts can be determined either "by eye" or by fitting a pixel shift. For this fit we use as fit parameters the x and y displacement, a variable for background and a variable intensity factor. To determine the shift between the bright field image and the start of the fluorescence movie we compare the average of the first 10 frames, improving signal-to-noise ratio over that of a single frame, to the bright field image. We then track the pixel shifts during the movie by fitting the change in each fluorescence frame relative to the initial 10-frame average. Due to the possibility that these fits get stuck in local minima, the pixel shifts must be visually inspected for accuracy.

Knowing possible sample shifts, the locations determined from the ECD recording can then be directly converted to locations in the fluorescence movie. Typically, a 2×2 to 7×7 pixel area surrounding the release site determined by the ECD is analyzed, either as a cropped movie or as an average intensity trace covering a time period of a few seconds before and after the fusion event. When many events are averaged, the 7×7 pixel area surrounding the fusion sites can provide information on the spatial localization of a fluorescence change relative to the fusion sites (Zhao *et al.*, 2013).

6.2.3.6 Relation between release of FFN511 and of catecholamines from individual vesicles

FFN511 loaded bovine chromaffin cells were placed at the center of the ECD, where the transparent glass surface of the ECD allows the fluorescence imaging of FFN511-loaded secretory vesicles in TIRF microscopy mode. Simultaneously, the ECI method using ECD arrays was used to determine the precise time and location of catecholamine release in single exocytic events. The release of the FFN511 dye from a single vesicle fusion event causes a localized loss of fluorescence (Gubernator *et al.*, 2009).

The amperometric spikes of different amplitudes were simultaneously recorded by the four ECD electrodes as a result of the oxidation of different numbers of transmitter molecules at the respective electrodes. The time of largest and earliest signal, detected by the electrode nearest to the release site, was taken as the spike start time indicating fusion pore formation. To establish the temporal association between fluorescence change and an amperometrically detected fusion event, ~20 s long (10 s before and 10 s after the event) amperometric recordings and cropped fluorescence images centered at the fusion sites determined by ECI and covering the same time period were analyzed (Figure 6.7). Unexpectedly, only 21% (62 out of 298 amperometric events had a visible fluorescence change at the fusion site associated with them. In the fluorescence movies 6 fluorescence decrease events were detected that did not have an amperometric event associated corresponding the location of the fluorescence change. These events presumably reflect undocking of vesicles moving out of the TIRF excitation zone.

To determine the precise correlation between fusion pore formation indicated by the amperometric spike start time and fluorescence loss indicating FFN511 release, 45 of the 62 amperometric spikes that were correlated in time and location with a visible loss of fluorescence event were included in the analysis. The remaining 17 events were associated with amperometric events overlapping in time such

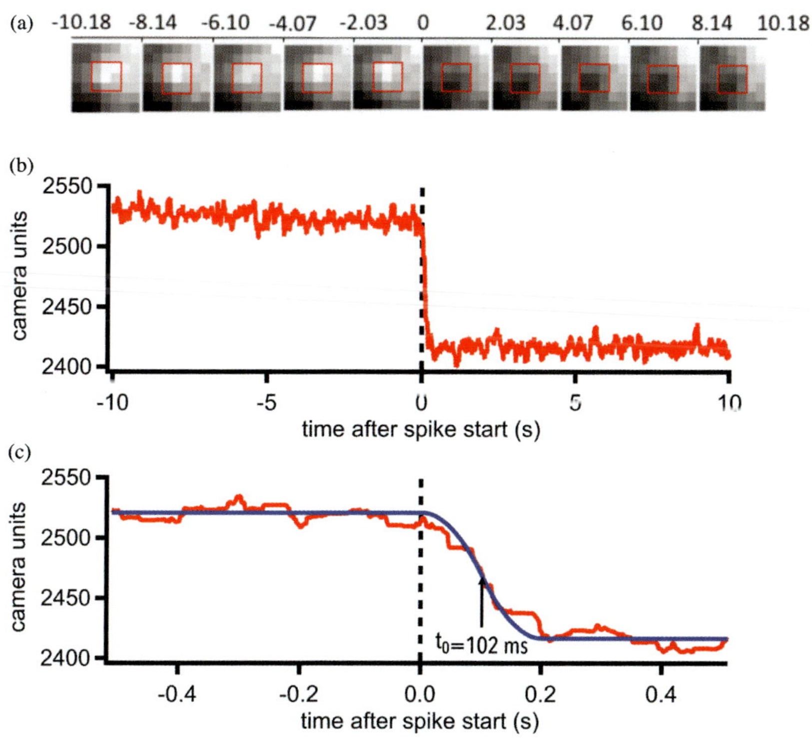

Figure 6.7 Precise temporal correlation of 45 averaged exocytic events with fluorescence change in FFN511 loaded bovine chromaffin cells. (a) Image sequence of 7 × 7-pixel region centered at the release site. Each panel shows the average of 20 consecutive frames. (b) ECOM analysis of averaged 3 × 3-pixel region (framed by red square) with the time of spike start.

that the correct spike start time was uncertain. The averaged movie of the cropped regions (Figure 6.7a) shows a clear loss of fluorescence centered at the fusion site as determined by ECI and confirms the correct localization of fusion events by ECI. The time course of fluorescence intensity in the central 3x3 pixel region of interest (ROI) (Figure 6.7a, red square) shows a rapid fluorescence change initiating at the amperometric spike start time (Figure 6.7b).

To obtain time super-resolution information that is not limited by the camera exposure time we applied the recently developed Event Correlated Microscopy (ECOM) method (Zhao *et al.*, 2013; Fang *et al.*, 2019). This method is based on conversion of the ROI intensities from the individual imaging frames to a series of data points with short dwell time, assigning the same intensity of the ROI to all data points covering the image frame duration. This conversion makes it possible to align the imaging data with sub-frame precision to the start times of the amperometric spikes. For a step change in intensity, averaging the aligned traces provides a time course that extends over two frame durations with a half time at the time of the step change. The time of the step can be determined by fitting the ECOM step response function to the imaging data. The fit is shown on expanded time scale in Figure 6.7c (black trace) revealing a delay of 102 ± 8 ms between fusion pore formation and average FFN release, presumably reflecting slow diffusion of FFN511 through the fusion pore and away from the fusion site. This diffusion process produces a gradual loss of fluorescence that is not instantaneous. The ECOM data trace in Figure 6.7c thus reflects the convolution of the time course of the fluorescence change with the ECOM step response function. The delay of ~100 ms thus reflects the average diffusion time of FFN511 from the release site. This time delay is comparable, albeit somewhat longer than the average half width of the amperometric spikes (64 ms for the respective largest spikes with shortest diffusion time to the closest electrode), which is expected because in amperometric detection the catecholamine molecules are rapidly consumed during the oxidation process.

6.3 Conclusion and future prospects

Electrochemical imaging of individual fusion events using ECD arrays provides the time and location of single fusion events with high spatial and temporal resolution while it allows simultaneous imaging of fluorescence changes at the fusion sites in epifluorescence (Hafez *et al.*, 2005) or TIRF excitation mode (Zhao *et al.*, 2013). The method reveals directly the relation between release of catecholamines and fluorescent markers. The high time resolution of amperometric recordings of 1 ms or less provides precise time information on fusion pore opening and fusion pore expansion. In contrast, fluorescence imaging is performed with much lower time resolution to obtain sufficient signal-to-noise ratio. However, the recently developed time super-resolution ECOM analysis method (Fang *et al.*, 2019) allows for determination of the average time of a fluorescence change relative to the time of the amperometric spike. As shown here, the release of FFN511 lags fusion pore formation by ~100 ms reflecting its diffusion out of the vesicle and away from the fusion site. Surprisingly, only a small fraction of catecholamine release events is accompanied by FFN511 release. Further experiments will be needed to determine if this is the case because only a small fraction of chromaffin granules are loaded with the false transmitter or if FFN511 is retained in the granules due to brief transient fusion pore openings which release only a fraction of vesicular catecholamines (reviewed in (Ren *et al.*, 2016)) and possibly undetectable amounts of FFN511.

The ECI method has also been applied to study conformational changes in the SNARE complex using a SNAP-25 based SNARE complex reporter that incorporates CFP and Venus as a Förster Resonance Energy Transfer (FRET) pair (Zhao *et al.*, 2013). These experiments revealed a rapid conformational change that precedes fusion pore formation in bovine chromaffin cells by ~90 ms. The method will enable experiments that were otherwise impossible. It will allow direct estimates of the number of molecules changing conformation in single fusion events and will enable detailed studies of molecular interactions in their relation to fusion pore formation and fusion pore expansion with unprecedented resolution.

From a technical point of view, ECI has been mostly performed with 4-electrode ECD arrays (Dias *et al.*, 2002; Hafez *et al.*, 2005; Ngatchou *et al.*, 2010; Zhao *et al.*, 2013) although 3-electrode ECDs are in principle sufficient (Berberian *et al.*, 2009). So far the optimal number and geometry of ECD arrays optimizing ECI localization precision has not been systematically investigated but such studies are currently under way.

Acknowledgments

The authors thank Professor Kevin Gillis and Dr. Xin Liu for support and advice, Joan S. Lenz for her excellent technical assistance, and Owasco Meat Co., Inc. for providing the bovine adrenal glands. ECD-fabrication was performed in Cornell Nanofabrication Facility (CNF), a member of the National Nanotechnology Coordinated Infrastructure (NNCI), which is supported by the National Science Foundation (Grant ECCS-1542081). This work was supported by the Cornell Nanobiotechnology Center (NSF), National Institutes of Health (NIH) grants R01MH095046, R43MH109212, R01GM085808 and R01GM121787, and the European Research Council (ERC) advanced grant ADG 322699.

References

Axelrod D, Thompson NL, Burghardt TP (1983) Total internal reflection fluorescent microscopy. *J Microsc*, 129: 19–28.

Berberian K, Kisler K, Fang Q, Lindau M (2009) Improved surface–patterned platinum microelectrodes for the study of exocytotic events. *Anal Chem*, 81: 8734–8740.

Betz WJ, Mao F, Bewick GS (1992) Activity–dependent fluorescent staining and destaining of living vertebrate motor nerve terminals. *J Neurosci*, 12: 363–375.

Chow RH, Rüden Lv, Neher E (1992) Delay in vesicle fusion revealed by electrochemical monitoring of single secretory events in adrenal chromaffin cells. *Nature*, 356: 60–63.

Dias AF, Dernick G, Valero V, Yong MG, James CD, Craighead HG, Lindau M (2002) An electrochemical detector array to study cell biology on the nanoscale. *Nanotechnology*, 13: 285–289.

Fang Q, Zhao Y, Lindau M (2019) Precise time super-resolution by event correlation microscopy. *Biophys. J*, in press. DOI: 10.1016/j.bpj.2019.03.034.

Gaffield MA, Betz WJ (2006) Imaging synaptic vesicle exocytosis and endocytosis with FM dyes. *Nat Protoc*, 1: 2916–2921.

Gandasi NR, Vesto K, Helou M, Yin P, Saras J, Barg S (2015) Survey of red fluorescence proteins as markers for secretory granule exocytosis. *PLoS One*, 10: e0127801.

Gubernator NG, Zhang H, Staal RG, Mosharov EV, Pereira DB, Yue M, Balsanek V, Vadola PA, Mukherjee B, Edwards RH, Sulzer D, Sames D (2009) Fluorescent false neurotransmitters visualize dopamine release from individual presynaptic terminals. *Science*, 324: 1441–1444.

Hafez I, Kisler K, Berberian K, Dernick G, Valero V, Yong MG, Craighead HG, Lindau M (2005) Electrochemical imaging of fusion pore openings by electrochemical detector arrays. *Proc Natl Acad Sci USA*, 102: 13879–13884.

Henkel AW, Betz WJ (1995) Staurosporine blocks evoked release of FM1–43 but not acetylcholine from frog motor nerve terminals. *J Neurosci*, 15: 8246–8258.

Huang M, Delacruz JB, Ruelas JC, Rathore SS, Lindau M (2018) Surface–modified CMOS IC electrochemical sensor array targeting single chromaffin cells for highly parallel amperometry measurements. *Pfluegers Arch/Eur J Physiol*, 470: 113–123.

Kaether C, Salm T, Glombik M, Almers W, Gerdes H-H (1997) Targeting of green fluorescent protein to neuroendocrine secretory granules: a new tool for real time studies of regulated protein secretion. *Eur J Cell Biol*, 74: 133–142.

Kisler K, Kim BN, Liu X, Berberian K, Fang Q, Mathai CJ, Gangopadhya S, Gillis KD, Lindau M (2012) Transparent electrode materials for simultaneous amperometric detection of exocytosis and fluorescence microscopy. *JBiomater Nanobiotechnol*, 3: 243–253.

Liu X, Barizuddin S, Shin W, Mathai CJ, Gangopadhyay S, Gillis KD (2011) Microwell device for targeting single cells to electrochemical microelectrodes for high-throughput amperometric detection of quantal exocytosis. *Anal Chem*, 83: 2445–2451.

Miesenbock G, De Angelis DA, Rothman JE (1998) Visualizing secretion and synaptic transmission with pH-sensitive green fluorescent proteins. *Nature*, 394: 192–195.

Mosharov EV, Sulzer D (2005) Analysis of exocytotic events recorded by amperometry. *Nat Methods*, 2: 651–658.

Ngatchou AN, Kisler K, Fang Q, Walter AM, Zhao Y, Bruns D, Sorensen JB, Lindau M (2010) Role of the synaptobrevin C terminus in fusion pore formation. *Proc Natl Acad Sci USA*, 107: 18463–18468.

Parsons TD, Coorssen JR, Horstmann H, Almers W (1995) Docked granules, the exocytic burst, and the need for ATP hydrolysis in endocrine cells. *Neuron*, 15: 1085–1096.

Ren L, Mellander LJ, Keighron J, Cans AS, Kurczy ME, Svir I, Oleinick A, Amatore C, Ewing AG (2016) The evidence for open and closed exocytosis as the primary release mechanism. *Q Rev Biophys*, 49: e12.

Steyer JA, Horstmann H, Almers W (1997) Transport, docking and exocytosis of single secretory granules in live chromaffin cells. *Nature*, 388: 474–478.

Sun X, Gillis KD (2006) On-chip amperometric measurement of quantal catecholamine release using transparent indium tin oxide electrodes. *Anal Chem*, 78: 2521–2525.

Wightman RM, Jankowski JA, Kennedy RT, Kawagoe DT, Schroeder TJ, Leszczyszyn DJ, Near JA, Diliberto Jr. EJ, Viveros OH (1991) Temporally resolved catecholamine spikes correspond to single vesicle release from individual chromaffin cells. *Proc Natl Acad Sci USA*, 88: 10754–10758.

Zhao Y, Fang Q, Herbst AD, Berberian KN, Almers W, Lindau M (2013) Rapid structural change in synaptosomal-associated protein 25 (SNAP25) precedes the fusion of single vesicles with the plasma membrane in live chromaffin cells. *Proc Natl Acad Sci USA*, 110: 14249–14254.

CHAPTER 7

MICRO-NANO-ELECTRODE ARRAY (MNEA) SENSORS FOR REAL-TIME, PARALLEL DETECTION OF NEUROTRANSMITTERS AND ELECTROPHYSIOLOGY IN THE RAT BRAIN

Yilin Song[*,†], Wenjing Wei[*,†], Guihua Xiao[*,†], Ziyue Li[*,†], Mixia Wang[*,†], Shengwei Xu[*,†] and Xinxia Cai[*,†,‡]

*State Key Laboratory of Transducer Technology, Institute of Electronics, Chinese Academy of Sciences, Beijing, 100190, P.R. China

† University of Chinese Academy of Sciences, Beijing, 100049, P.R. China

‡xxcai@mail.ie.ac.cn

7.1 Introduction

The brain is the most complex and elaborate structure in the human body. It is composed of approximately 100 billion neurons, each making approximately 10 thousand synaptic connections to other neurons (Bartheld *et al.*, 2016; Defelipe *et al.*, 2011). Communication between neurons is inherently both electrical and chemical in nature, involving changes in cell membrane potentials and neurotransmitter release, respectively (Kandel *et al.*, 2000). These dual-mode communication processes transmit massive amounts of neural information. First, when neurons switch from a quiet state to an exciting state, they emit action potentials (spikes) in various temporal patterns at various firing rates to encode neural information (Barnett *et al.*, 2007). Second, when the action potential propagates along axons to the synapses, neurotransmitters are released by presynaptic vesicles to activate both pre- and postsynaptic receptors (Lodish *et al.*, 2000). Some neurotransmitters are excitatory, such as glutamate (Glu), and some are inhibitory, such as gamma-aminobutyric acid (Fagg *et al.*, 1983). Others, including dopamine (DA), serotonin (5-hydroxytryptamine, 5-HT), norepinephrine (NE), acetylcholine (ACh), and so on, may be excitatory or inhibitory depending on the specific receptors they activate (Cepeda *et al.*, 2010; Hoyer *et al.*, 2002; Albuquerque *et al.*, 2009). Therefore, complex chemical information is carried by the neurotransmitters. More importantly, the dual mode of information flow from neurotransmitters and neural electrophysiology are mutually interdependent and intermodulated.

Many neurological diseases, such as epilepsy and Parkinson's disease, are disorders of neural electrophysiological activity and neurotransmitter imbalance. For example, the hippocampus of epilepsy patients exhibits abnormally high-frequency oscillations following spike firings related to seizure genesis (Urrestarazu *et al.*, 2007, 2010; Ulbert *et al.*, 2004) and greatly increased concentrations of Glu (Idil *et al.*, 2010). The basal ganglia of patients with Parkinson's disease exhibit greatly decreased concentrations of DA (Hornykiewicz *et al.*, 2002) and neural spike firing patterns related to tremor of the limbs (Follett K *et al.*, 2010). However, the above findings have been derived from independent experiments, so that the exact intermodulation mechanisms between neurotransmitter concentrations and neural electrophysiology signals remain unclear. For this reason, there is an urgent need to conduct real-time and parallel detection of the dual modes of neural communication in the brain to further investigate their dynamic relationship, which is valuable for fundamental neuroscientific research, drug screening, and evaluation of potential treatments for neurological diseases.

Throughout the history of neuroscience research, implantable sensors have promoted progress in the in vivo detection of both electrical and chemical communication. Microdialysis probes (Chaurasia *et al.*, 2007) and carbon fiber microelectrodes (Mundroff and Wightman, 2002) are recognized as classics among neurochemical detection devices. Metal microwires have been used extensively to make neuroelectrical recordings. Micromachined microelectrode arrays have been developed for neuroelectrical recording (Fekete 2015) as well as for electrochemical detection (Burmeister *et al.*, 2001): their spatiotemporal resolution is a key benefit to dual-mode, parallel detection.

The microdialysis principle was first employed in the early 1960s, when push pull cannulas, dialysis sacs, and dialytrodes were inserted into the rodent brain to study neurotransmitter release (Chaurasia *et al.*, 2007). The continuously perfused microdialysis probe was invented in 1971. It was implanted into a monkey brain for glutamate injection and long-term monitoring of neurochemicals including amino acids and glycoproteins (Delgado *et al.*, 1971). Today, microdialysis probes are commercialized and approved by the US Food and Drug Administration (FDA) and the European Union for use in human brain (Chaurasia *et al.*, 2007; Ernesto *et al.*, 2001). The probe acts in a manner similar to a capillary vessel. It consists of a shaft with a semipermeable hollow fiber membrane at the tip, which is connected to inlet and outlet tubes. Neurochemicals in the surrounding brain tissue diffuse through the semipermeable membrane into the perfusate, which is collected and subsequently analyzed by liquid chromatography, mass spectrometry, electrochemistry, fluorescence methods, and so on. Therefore, the microdialysis method offers detection of multiple neurochemicals in real time with high accuracy. But, given that the collection and analysis of chemicals in the perfusate can be time-consuming, the temporal resolution of microdialysis methods can be slower than 10 min, so it is not possible to capture neurochemical transients. Also, the spatial resolution is limited by the typical diameters (200–500 μm) and lengths (1–4 mm) of the active sampling portion of microdialysis probes (Nandi *et al.*, 2009).

The carbon fiber microelectrode (CFM) was invented and first applied to neurochemical detection *in vivo* in the 1980s. The CFM was implanted into rat striatum for the electrochemical detection of catecholamine neurotransmitters (Gonon *et al.*, 1978; 1980; 1981; Huffman *et al.*, 2008). Fabrication of CFMs involves epoxy to seal a single carbon fiber (diameter 5–7 μm) into the tip of a pulled glass capillary. Either the tip is polished to make a carbon-fiber disk microelectrode, or the fiber is cut to a desired length (typically 75–150 μm) to make a carbon-fiber cylinder microelectrode (Pike *et al.*, 2009). Electroactive neurotransmitters, such as DA and 5-HT, are oxidized at the surface of the carbon fiber under appropriate working potentials and the oxidation current recorded in real time reflects dynamic changes of the neurotransmitter concentration. Constant-potential amperometry and fast-scan cyclic voltammetry are commonly used electrochemical methods (Robinson *et al.*, 2003; Phillips and Wightman, 2003; Budai, 2010). The selectivity of the CFM is greatly improved by surface modification with the ion-selective polymer, Nafion (Brazell *et al.*, 1987). The possibility of detecting non-electroactive chemicals using enzyme modified CFMs has also been demonstrated (Lugomorales *et al.*, 2013). Due to the in situ electrochemical reaction on the carbon fiber micro-tip, CFMs offer high spatiotemporal resolution. Sub-second neurotransmitter release has been detected in the brain of both rodents and humans (Robinson *et al.*, 2003; Kenneth *et al.*, 2011).

Microelectrodes for neural electrophysiological recordings *in vivo* were developed in the 1940s (Renshaw *et al.*, 1940). Different from the macro electrodes fixed on scalp or skull for electroencephalography (EEG) recordings, the microelectrodes are implanted into the brain tissue to facilitate capturing the extracellular action potentials (neural spikes) from neuron and local field potentials (LFPs) from clusters of neurons (Seymour *et al.*, 2017). Although glass micropipettes (Li *et al.*, 1953) and carbon fiber microelectrodes (Millar, 1991) were used early on for recording neural action potentials they were relatively inconvenient for implantation. Once the ability of platinum wires to record spike activity in the cortex was discovered (Woldring *et al.*, 1950; Wolbarsht *et al.*, 1960), insulated metal wire microelectrodes became prevalent. Most of them were made from inert metals with a high Young's modulus, such as tungsten, stainless steel, or platinum-iridium alloy, coated with glass or polymer insulators leaving the tips exposed. Microwires with diameters of 10~150 μm can capture action potentials and LFPs from neurons (Cogan *et al.*, 2008). To investigate the electrical activity of multiple neurons simultaneously, micro-electrode bundles were developed in 1967 and applied in neurosurgery (Marg and Adams, 1967). Microwire arrays of more than 30 electrodes with fixed spacing were then invented, with the electrodes held in place by object-holder grids (Krüger and Bach, 1981). These arrays are commercialized and widely used for both chronic and acute (<1 day) neural recordings. In some cases, the electrodes are packaged in solidified polyethylene glycol to prevent bending and deviations during implantation (Nicolelis *et al.*, 1997; Desai *et al.*, 2010). The temporal resolution of the microelectrodes, usually less than 1 ms, is determined by the data acquisition hardware.

At the end of the 20th century, with the development of Micro-Electro-Mechanical Systems (MEMS), microelectrode arrays (MEA) fabricated on silicon substrates were invented. As the most representative of the MEAs, the 3-dimensional "Utah Array" is suitable for multichannel electrophysiological recordings in the cortex (Huber *et al.*, 1992; Rousche *et al.*, 1998) and the 2-dimensional "Michigan Array" is used for deep brain recordings (Hoogerwerf *et al.*, 1994; Kewley *et al.*, 1997). Based on lithography, thin film deposition, and micro-etching techniques, tens to hundreds of microelectrode channels may be integrated on the silicon probes. The diameter and spacing of each microelectrode are accurately controlled to less than 1 μm. The silicon shafts are sufficiently thin and rigid for easy implantation without the bending or breaking. Microelectrode sites with dimensions of less than 30 μm are considered suitable for single-unit recordings (Fekete *et al.*, 2015; Seymour *et al.*, 2017). The recently developed "Neuropixels Probe" places 384 12×12-μm microelectrode sites on a single silicon shank with the Complementary Metal-Oxide-Semiconductor (CMOS) fabrication process (Jun *et al.*, 2017). Modification of array elements with nano-materials and structures has improved their signal-to-noise ratio (Yang *et al.*, 2004). Additionally, when combined with electrochemical methods, the MEAs have also been used to detect neurotransmitters. For example, the microelectrode sites on the Michigan Array were modified with Nafion and used to detect DA in the rat striatum (Johnson *et al.*, 2006; 2008). A series of ceramic-based MEAs modified with enzymes have been specifically designed for the real-time detection of Glu and ACh *in vivo* (Burmeister *et al.*, 2000; 2008; Stephens *et al.*, 2010).

When conducting real-time, parallel detection of dual mode neural signals, it is desirable to place the electrophysiology and neurotransmitter sensors as close as possible within brain tissue. This is expected to allow capture of electrical and chemical activities from the same neurons, making it possible to investigate the possible correlations between the dual modes of communication. Since action potentials and neurotransmitter release both occur on the millisecond scale, the response of the sensors must be fast enough to detect transient signals. High throughput multi-channel detection is also desirable, as it allows for the simultaneous detection of different neurons and several types of neurotransmitters within a localized neural environment. When increasing the number of detection channels it is important to minimize their dimensions to facilitate easy, minimally invasive implantation. As with all detection techniques, it is important to optimize the signal-to-noise ratio and sensitivity of the dual mode neural information sensors.

Based on the above considerations, it is easy to understand that micro machined MEAs are well-suited for multichannel detection of electrophysiology and neurotransmitters in vivo. The spacing of the microelectrodes is controllable to tens of micrometers to capture the dual mode activities from the same neuron. Through electrochemical methods and surface modifications with bio-nano functional materials, various neurotransmitters can be monitored concurrently with electrophysiological recordings with millisecond time resolution.

The remainder of this chapter will introduce our recent work on a series of silicon based micro-nano-electrode array (MNEA) sensors. They were developed for multichannel, *in vivo*, real-time, parallel detection of the dual modes of neural communication in the context of neurological diseases. The dual modes refer to the electrochemical signals of the neurotransmitters and the electrophysiological signals from the neurons in deep brain tissue. Specifically, the electrophysiology signals recorded by the micro-nano-electrodes in our work include the microscopic extracellular action potentials fired by neuron and the mesoscopic local field potentials integrated from the local neural network. They are distinguished by different frequencies and can be separated by signal filtering from the raw electrophysiology data (Seymour *et al.*, 2017). In subsequent sections, the sensor fabrication, methods of detection, and applications in the brain of rodents will be described in detail. Our objective is to provide effective tools for advancing knowledge of fundamental neuroscience and neurological disease.

7.2 Dual mode MNEA sensors for *in vivo* detection

7.2.1 Design and fabrication of MNEA sensors

Different types of MNEA sensors have been designed according to the detection goals, including the location and structure of the target brain area, the types of neurotransmitters to be monitored and the number of channels desired. Figure 7.1 shows some commonly used MNEA prototypes for experiments in rodents. The 16-channel MNEAs with four silicon shanks in Figures 7.1a and 7.1b) were designed for recordings in brain structures at the same depth. The shanks are 100 μm in width, 5~8 mm in length, and 200 μm in spacing.

The MNEA with a single shank in Figure 7.1c is for monitoring neural activity in different layers beneath the brain surface. The silicon shanks are 15~30 μm

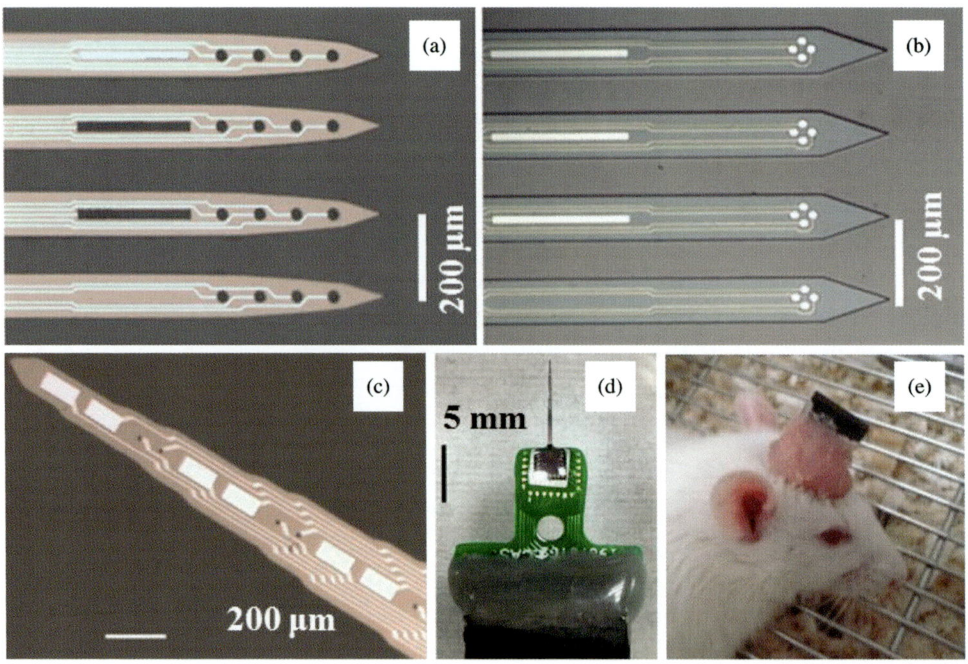

Figure 7.1 Prototypes of the silicon based MNEA sensors: (a) A 4-shank, 16-channel MNEA with four microelectrodes arranged in linear fashion on each shank. The microelectrodes are 20 μm in diameter and spaced 100 μm apart. The three rectangular electrodes (25 μm × 300 μm) were integrated as reference and counter electrodes for electrochemical recording. (b) A MNEA sensor with a tetrode on each of four shanks. The spacing within the quartet of electrodes is 6 μm edge-to-edge and 26 μm center-to-center (c) A 14 channel MNEA sensor with round sensors (diameter = 15 μm) for electrophysiological recording and rectangular sensors (60 μm × 125 μm) for electrochemical recording arranged in a linearly alternating pattern on a single shank. (d) A MNEA sensor assembled onto a PCB connecter. (e) An assembled MNEA sensor implanted into a freely moving rat (Wei *et al.*, 2015).

thick and have pointed tips for easy implantation. The diameter, spacing, and array patterns of the thin film platinum microelectrodes on the shanks are designed according to the size and distribution of the neurons in the targeted brain area. It must be recognized that the soma of neurons vary in size from 5~100 μm and action potentials are not effectively captured by microelectrodes more than 50 μm away from the neuron (Nisch *et al.*, 1994). Thus, the linearly arrayed circular microelectrodes with diameters of around 20 μm and spacing of 100 μm on four shanks (Figure 7.1a) are used for recording neurons distributed over a region 500 μm × 800 μm. The tetrodes on each shank in Figure 7.1b are arranged with a spacing of 26 μm center-to-center to discriminate neural action potentials from single neurons or several closely neighboring neurons. Each of the microelectrodes can also be used for electrochemical detection through suitable surface modifications and the larger rectangular microelectrodes (Figure 7.1c) increase the current response for electrochemical recording. However, it must be noted that, as demonstrated by previous studies (Yu and Wilson, 2000), when two adjacent microelectrodes were used for enzyme-based electrochemical detection simultaneously, their spacing must be at least 100 um. Otherwise, overlap of the diffusion layers may cause crosstalk between the electroanalytical electrodes, and there can be crosstalk if the sensors are also consuming oxygen (oxidases). So, for the tetrode type MNEA sensor in Figure 7.1 b, the electrode closest to the tip of the shank was an electroanalytical microelectrode, and the other three electrodes on the same silicon shank were electrophysiological microelectrodes. Different kinds of neurochemicals will be sensed by electroanalytical microelectrodes on different shanks with the spacing of 200 μm to avoid crosstalk. On the other hand, since the adjacent electrophysiological microelectrodes are used for sensing neural spikes from single neurons, their spacing can be less than 100 μm. As the electrophysiological signal frequency is relatively low (<5kHz), crosstalk is negligible. Besides, several rectangular microelectrodes (25 μm×300 μm) shown in Figures 7.1a and 7.1b) are for optional use as reference or counter electrodes. The microelectrodes are individually connected via leads (4 μm in width) on the backside of the shanks to bonding pads (200 μm × 200 μm). The leads are covered with thin insulating film of SiO_2/Si_3N_4.

The MNEA sensors are fabricated on silicon-on-insulator (SOI) wafers by procedures explained thoroughly in previously published papers (Wei *et al.*, 2015; Fan *et al.*, 2017; Xiao *et al.*, 2017; Li *et al.*, 2018). Briefly, the procedure involves six steps and three lithographic masks: (1) After a standard cleaning procedure, SiO_2 (500 nm) insulator is grown on the SOI wafer by thermal oxidation; (2) Ti/Pt (30 nm/250 nm) is then sputtered and patterned with a photo-resist lift-off process to form the conducting layer of the microelectrodes and their leads on the top silicon surface; (3) SiO_2/Si_3N_4 (300 nm/500 nm) insulation is then deposited by plasma enhanced chemical vapor deposition: the insulation is selectively removed by SF_6 reactive ion etching through a resist mask to uncover the microelectrode sites and the bonding pads; (4) The top silicon layer (15~30 μm) of the SOI wafer is then selectively etched down to the buried oxide using inductively coupled plasma deep reactive ion etching with a resist mask to define the geometry of the shanks; (5) The structures on the top silicon surface are protected with picein wax, the wafer is attached to a stainless steel plate, the bottom side of the wafer (400~550 μm) is wet etched in KOH solution (50%, 80 °C), and the buried oxide (2 μm) is removed with HF buffer; (6) The sensors are released by dissolving the picein wax in negative photoresist developer and are soaked in fuming nitric acid to remove any organic residue. After ultrasonic cleaning in deionized water, the sensors are assembled onto printed circuit boards by wire bonding. The bonding pads, bonding wires, and metal lines on the board are embedded in silicone rubber for the purpose of mechanically protecting and electrically isolating the connections (Figure 7.1d).

7.2.2 Modification of the MNEA sensors

To facilitate parallel detection of dual mode neural signals on one MNEA sensor, modification of the microelectrodes is indispensable. First, the bare thin film platinum microelectrodes exhibit high impedance due to their small surface area, which would likely introduce high frequency background noise for electrophysiological recording and produce inherently low current responses for electrochemical recording. Modifications with nanomaterials or nanostructures, such as platinum nano particles (PtNPs), graphene, and carbon nanotubes, greatly increases the effective surface area of the microelectrodes, thus decreasing their impedance. Second, there are many different kinds of neurochemicals in brain tissue. Electrochemical detection of neurotransmitters is easily interfered with by the similarly electroactive species found in brain tissues, such as uric acid (UA), ascorbic acid (AA), etc. Modification with Nafion or 1,3-phenylenediamine (mPD) attenuates these possible interferents. In addition, enzymes must be coated on to the microelectrodes to convert non-electroactive neurotransmitters, such as Glu, to an electroactive reporter molecule, typically H_2O_2. As Figure 7.2a shows, with electrochemical deposition and micro-coating under a microscope, it is possible to modify individual microelectrodes on individual shanks with different bio-nano materials. The detailed modification processes are explained next.

7.2.2.1 Electrode modification for electrophysiological recordings

Microelectrodes for electrophysiological recording are modified with PtNPs (see Figure 7.2c) to improve their signal-to-noise ratios. The electrodes are first cleaned in an oxygen plasma for 2~3 min to remove any organic contaminants and to enhance adhesion of the PtNPs. The electrodes are next immersed in 10 mL of

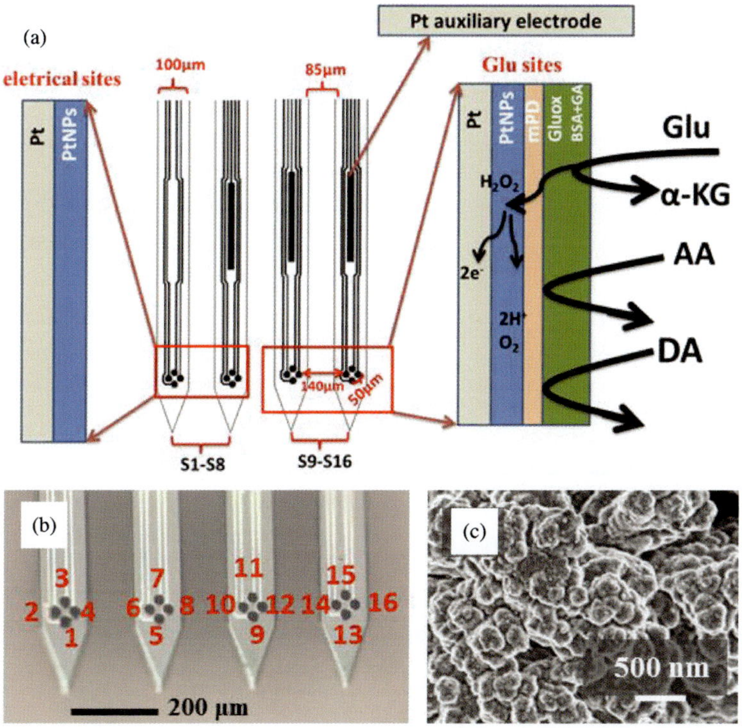

Figure 7.2 Modification of the 16 channel MNEA sensor with four shanks: (a) The electrical sites (S1–S8) on the left two shanks were modified with PtNPs and used for electrophysiological recording. The electrical sites (S9–S16) on the right two shanks could be used for electrochemical detection of glutamate. They were modified with three different layers (PtNPs/Gluox/mPD) of bio-nano materials. (b) Micrograph of the 16 sites modified with Pt nanoparticles. (c) The scanning electron microscope (SEM) photograph of the modified Pt nanoparticles (Xiao *et al.*, 2017).

the bath solution (24 mM H_2PtCl_6 with 2.1 mM $Pb(CH_3COO)_2$) and connected in parallel to an electrochemical workstation (Gamry Reference 600, USA) as working electrodes. Chronoamperometry is applied for 2 min with a constant working potential of −1.0V *versus* Ag|AgCl. This potential is appropriate for the microelectrodes with the diameter of 15~30 μm but needs to be adjusted for larger electrodes. During electrochemical deposition an ultrasonic field is applied, as it appears to improve the compactness of the deposited PtNP layer. Figures 7.2 (b) and 7.2 (c) show the morphology of the PtNP layer.

7.2.2.2 Electrode modification for glutamate detection

For glutamate detection, the microelectrodes are modified with three different bio-nano material layers, including PtNPs, glutamate oxidase (Gluox), and mPD (Xiao *et al.*, 2017; Li *et al.*, 2018). The PtNPs are deposited as explained, above. Then, covalent cross-linking with glutaraldehyde is employed to immobilize the Gluox onto the microelectrode, as follows. A glass capillary connected to a micro-syringe is used to microeject a solution of Gluox (1%), bovine serum albumin (1%), and glutaraldehyde (0.125%) onto the target microelectrodes under a microscope. The bio-nano layer is dried for 1 min, the coating process is repeated three times, and the layer is cured at room temperature for at least 72 h. Finally, the microelectrodes are coated with electrochemically polymerized mPD. To do this, the microelectrodes are immersed in a mPD solution (5 mM) and the polymerization is carried out by cyclic voltammetry (0~1 V *versus* Ag|AgCl; 100 mV/s; 10 cycles). The polymerized mPD membrane prevents larger molecules such as DA and AA from reaching the microelectrode surface. However, the small H_2O_2 molecules produced during the enzymatic oxidation of Glu penetrate the mPD layer, thereby producing a Glu-selective response.

7.2.2.3 Electrode modification for dopamine detection

For dopamine detection, the microelectrodes are modified with PtNPs (as explained, above) and Nafion. Nafion is negatively charged, so it concentrates positively charged ions, including DA because it is a cation at physiological pH, and attenuates negatively charged interferents, such as uric acid and ascorbic acid. Nafion (5%, Sigma-Aldrich, USA) is first diluted to 1% in ethanol and aspirated into a glass capillary connected to a micro-syringe. Under a microscope, the capillary tip is position over the MNEA shank with a micromanipulator and 0.5 μL of the solution is ejected to cover the target microelectrodes. To assure complete coverage and adequate film thickness, the coating is repeated three times at 1-min intervals to allow the film to dry. Finally, the sensor is placed in oven at 100 °C for 30 min to bake the Nafion membrane.

7.2.3 Electrode characterization and calibration

Before *in vivo* experiments, the impedance of the microelectrodes is measured to predict its capability for electrophysiological recording. Meanwhile, calibrations must be carried out to evaluate the sensitivity and selectivity of the microelectrodes for their target neurotransmitters. For in vitro tests, the microelectrode is connected to the electrochemical workstation (Gamry Reference 600, USA) as the working electrode and a Ag|AgCl wire is used as the reference electrode.

The electrophysiological microelectrodes are assessed with electrochemical impedance spectroscopy over a frequency range of 1 Hz to 1 MHz with a

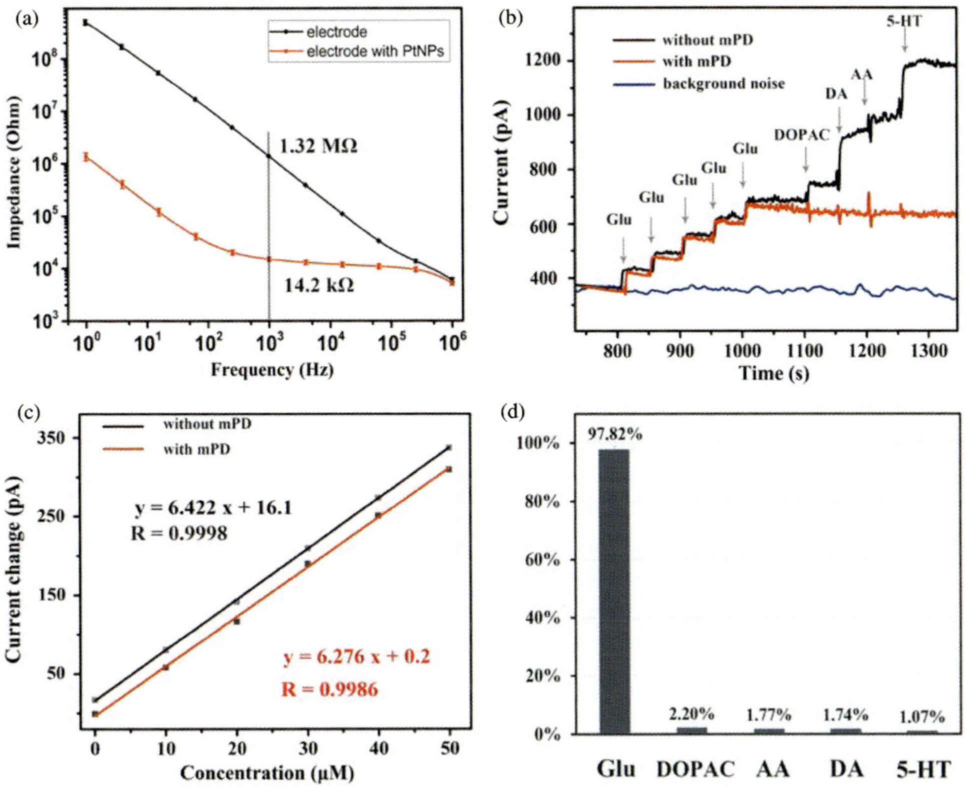

Figure 7.3 Electrophysiological and electrochemical properties of the MNEA sensor. (a) The electrochemical impedance spectroscopy of the microelectrodes before and after modification with PtNPs. The modification decreased the mean impedance of the sensors from 1.32 ± 0.03 MΩ to 14.2 ± 1.3 kΩ (n = 12) at 1 kHz. The impedance dropped by roughly 90 times on average. (b) A typical pre-calibration of one electrochemical site with and without a layer of mPD. Each addition of Glu or interferent (AA, DA, 5-HT, or DOPAC) increased the concentration by 10 μM. (c) The Glu Concentration-Current response curve of the microelectrodes with and without the mPD layer. (d) The selectivity of sensors with the mPD membrane for Glu compared to potential interferences (Li *et al.*, 2018).

sinusoidal potential of 20 mV peak-peak amplitude in 0.9% saline. Figure 7.3a shows how the mean impedance of a microelectrode (20 μm in diameter) decreased from 1.32 MΩ to 14.2 kΩ at 1 kHz after modification with PtNPs. This 90-fold drop in impedance is due to the greatly increased surface area of the contact between the solution and the porous nanostructure of the PtNPs. This low impedance guarantees a lowered background noise of about ±10μV for neural action potential recordings.

Calibrations of the microelectrodes designed to detect neurotransmitters are usually conducted with electrochemical chronoamperometry. The shanks of the

MEA sensor are immersed into 0.1M PBS and a magnetic stirrer is used to keep the solution homogeneous.

Glutamate detection (using H_2O_2 as the electrochemical messenger) is carried out at a working potential of +0.6 V *versus* Ag|AgCl. As Figure 7.3b shows, once the baseline current had stabilized, Glu was added to the solution every 50 s, to produce Glu concentrations of 10 μM, 20 μM, 30 μM, 40 μM, and 50 μM. After that, interferences (AA, DA, 5-HT, and DOPAC) were added in turn, each with a final concentration of 10 μM. After subtracting the baseline current, the current at each concentration is used for linear fitting to produce the Concentration-Current response curve (Figure 7.3c). As Figure 7.3 shows, microelectrodes with the PtNPs/Gluox/mPD layer exhibit a sensitivity for Glu detection of 6.27 pA/μM, a linearity (r^2) of 0.9986, and essentially no response to AA, DA, 5-HT, and DOPAC.

Similar procedures are used to detect DA except that the working electrode potential is held at 0.5 V *versus* Ag|AgCl. Microelectrodes (20 μm in diameter) modified with PtNPs/Nafion for DA detection usually have a sensitivity in the range of 17~19 pA/μM.

7.3 Dual mode detection in rats

Since dual mode detection is valuable for investigating neurological diseases, the dual mode MNEA sensors are implanted into disease-related areas of the rat brain. For example, the striatum is a brain area strongly implicated in Parkinson's disease, whereas the hippocampus is a frequent source of epileptic seizures. The MNEA sensors allow real time, parallel detection of electrophysiological activity and Glu or DA in these brain areas. The goal of these measurements is to probe the dynamic interactions between the electrophysiological and chemical modes of neural communication.

7.3.1 Surgical and dual mode detection methods

All animal experiments were performed in a professional biological laboratory and all procedures were conducted according to the National Institutes of Health Guide for The Care and Use of Laboratory Animals (NIH Publications No.8023, revised 1978). Sprague-Dawley (SD) rats weighing 250–300 g were selected for *in vivo* experiments. Individual rats were anesthetized by isoflurane (1%–2%) inhalation and securely fixed in a stereotaxic frame. After scalp incision, craniotomies were conducted according to the coordinates of the target brain area (Paxinos and Watson, 2005). To facilitate MNEA sensor implantation, dura mater was resected. A Ag|AgCl wire (0.3 mm in diameter) as a reference electrode was placed between the dura mater and the skull. Stainless steel bone screws were tapped into the skull for ground connection and further long-term mounting of the sensor. Then the rat, still anesthetized and still in the stereotaxic frame, was placed inside a grounded Faraday cage.

The MNEA sensors are inserted slowly into the brain tissue at 1~2 μm/s with a direct drive micropositioner (model 2662, David KOPF, USA).

The microelectrodes for electrophysiological recordings were connected to 128-channel data acquisition system (Cerebus, Blackrock Microsystems, USA). A band-pass filter (500–5k Hz) was applied to the raw data to extract extracellular action potentials (neural spikes). A low-pass filter (0–250 Hz) was used to extract local field potentials (LFPs).

The microelectrodes for neurotransmitter detection were connected to a single-channel electrochemical workstation (Gamry Reference 600, USA) and chronoamperometry was used to record Glu or DA. The rectangular microelectrodes were used as the counter electrode and the Ag|AgCl wire fixed on the dura mater was used as the reference electrode. The electrochemical current recorded on top of the dura covered by saline (0.9%) was considered to be the baseline current and was subtracted from the currents recorded in the brain tissue for conversion to concentration according to the prior calibrations.

7.3.2 Detection of glutamate and electrophysiology in the rat striatum

To explore the possibility of detecting dynamic changes in glutamate and electrophysiological signals in parallel, a single shank dual-mode MNEA sensor (Figures 7.1c and 7.4a) was implanted into the striatum of a rat. A single channel of Glu electrochemical current and six channels (Ch1~Ch6) of electrophysiology from different depths in the striatum were recorded simultaneously. After a period of baseline recording, KCl (100 mM, 1 mL/kg) was injected into the abdominal cavity (Wei *et al.*, 2015).

As Figure 7.4b shows, all six electrophysiology channels captured neural spikes with an amplitude of about 100 μV. The spike waveforms derived from Ch3 and Ch6 were different from those recorded by Ch1, Ch2, Ch4 and Ch5, indicating that they might come from two types of neurons. After KCl injection (T_1), the spike frequencies tended to increase (T_0). The average firing rate captured by Ch3 and Ch6 (Figure 7.4f) increased from 7.24/s (T_0) to 19.60/s (T_1). The average firing rate at the other channels increased from 2.72/s (T_0) to 9.01/s (T_1). Fluctuations of the local field potentials recorded by the six channels at different depths were nearly the same (Figure 7.4g). Correlated with the spike frequencies, the most obvious increase in the LFP power at Ch1 appeared in the frequency band from 12 to 18 Hz at the time before T_1 (Figure 7.4e).

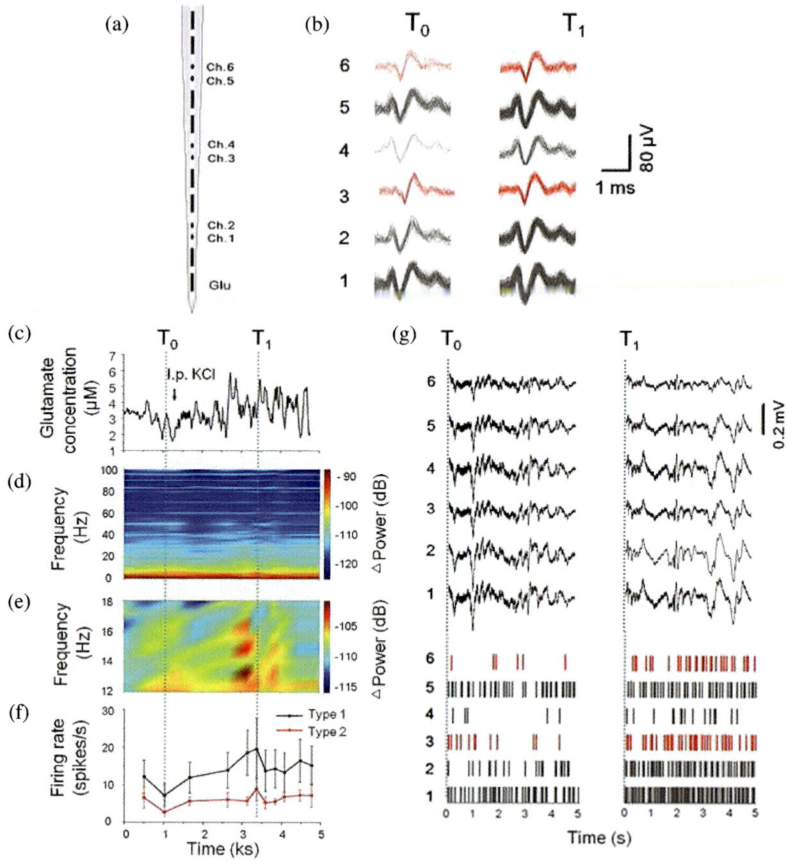

Figure 7.4 The electrophysiological signals and Glu concentration concurrently recorded in rat striatum before and after i.p. injection of KCl (100 mM, 1 mL/kg). (a) The distribution of the microelectrodes on the shank for electrophysiology (Ch1~Ch6) and Glu detection. (b) The neural spikes recorded by different microelectrode channels before (T_0) and after (T_1) KCl injection. Two different types of spike waveforms were derived from Ch 3,6 (in red) and Ch 1,2,4,5 (in black). (c) Variation of glutamate concentration elicited by i.p. injection of KCl (indicated by an arrow). (d) The power spectrogram of LFP recorded by Ch1 from 0 to 100 Hz and (e) from 12 to 18 Hz. (f) Variation of the mean firing rates of the two types of spikes. The same timeline was used in (c~f). (g) LFPs (upper) and spike trains (lower) recorded by the six microelectrode channels before (T_0) and after (T_1) i.p. KCl injection (Wei *et al.*, 2015).

On the other hand, Figure 7.4c shows that the average concentration of Glu increased from 3.01 μM to 4.67 μM after KCl injection (Figure 7.4c). Since Figures 7.4c–7.4f share the same timeline, it is easy to observe the dynamic relationships between the dual mode signals. It reflects the mechanism that by which the injected high concentration of K^+ diffuses into the brain to cause massive depolarization of neurons which consequently leads to the release of glutamate from vesicular cytosolic pools and astrocytes.

These results show that, through the dual mode MNEA sensor, multi-channel and multi-scale neural information can be obtained simultaneously, including the spatial distribution of the neurons and their types, the exciting or inhibiting states of the neurons, and the dynamical changes of the ambient neurotransmitters. It provides a high-efficiency tool for the research of neural mechanisms.

7.3.3 Detection of glutamate and electrophysiology in the rat hippocampus

The hippocampus plays important roles in the consolidation of information related to learning and memory. Injuries of the hippocampus may cause many neurological diseases such as Alzheimer's disease (AD) and epilepsy. The hippocampus has layered inner structures exhibiting a high density of neurons. To investigate the electrical and chemical properties of these neurons, a 16-channel four-shank MNEA sensor (Figure 7.2) was applied for dual mode detection in the hippocampal sub-regions (Xiao *et al.*, 2017).

As Figure 7.5a shows, when the sensor was implanted with a constant speed of 2 μm/s passing through the hippocampal sub-regions, the microelectrodes encountered four neuron layers, including the pyramidal cell layer of CA1 (DV: –2250 μm), the molecular layer of the dentate gyrus (MoDG, DV: –2840 μm), the granule cell layer of the dentate gyrus (GrDG, DV: –3020 μm), and the pyramidal cell layer of CA3 (DV: –3400 μm). The minimum vertical distance between the layers was about 200 μm. With the high spatiotemporal resolution MNEA sensor, the Glu concentration also varied dynamically with the implant depth. Four peaks of Glu concentration appeared concomitantly with the piercing of high-density neuron layers (CA1: 4.39 μM; MoDG: 5.36 μM; GrDG: 10.34 μM; CA3: 10.16 μM). Combined with the spike firing patterns concurrently recorded at different depths (marked by the points a~h), it was found that the high concentration of Glu accompanied high spike firing rates. The different waveforms recorded by the 8-channel microelectrodes in the hippocampal sub-regions are visible in Figure 7.5b. These findings verify that, based on the small diameter of the microelectrodes, it is possible to characterize the fine neural structures of the hippocampus both electrically and electrochemically with high spatiotemporal resolution.

7.3.4 Detection of glutamate and electrophysiology during epilepsy in rats

Temporal Lobe Epilepsy (TLE) is a chronic neurological disorder, characterized by sudden, repeated, and transient central nervous system dysfunction. To better

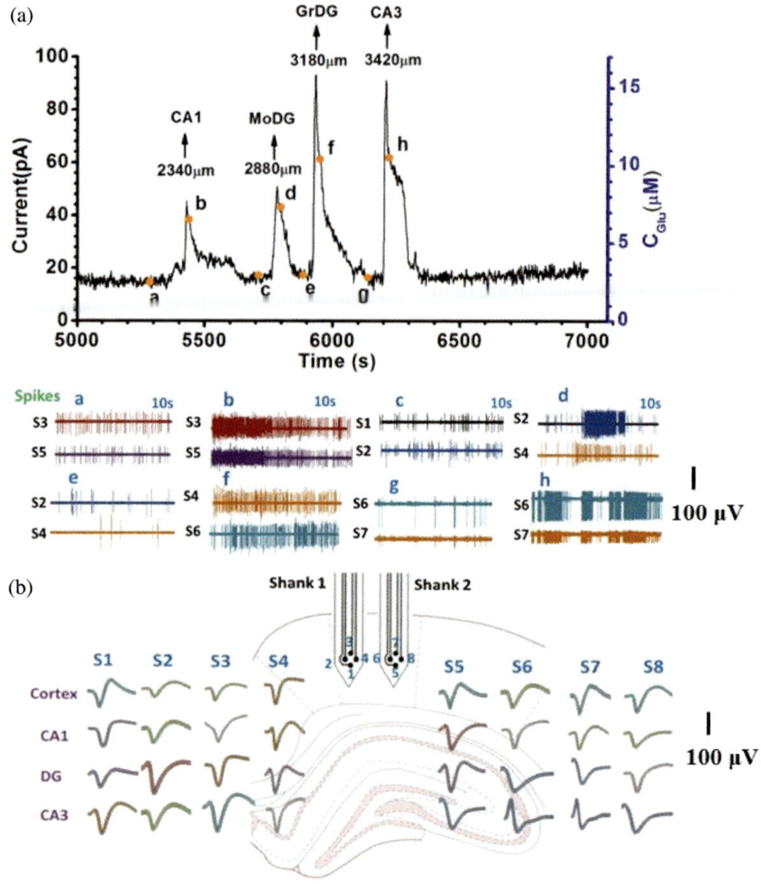

Figure 7.5 The Glu concentration and neural action potentials concurrently recorded in rat hippocampus sub-regions. (a) The electrochemical current continuously recorded by the microelectrode for Glu detection during the process of implanting the MEA sensor into the hippocampus with the speed of 2 μm/s. Four peaks were detected in sub-regions of the hippocampus including CA1, MoDG, GrDG and CA3. The corresponding spike firing patterns concurrently recorded at each time point (a~h) are shown below. (b) The typical spike waveforms recorded by the eight electrophysiological microelectrodes in cortex and hippocampus sub-regions (Xiao *et al.*, 2017).

understand TLE, we implanted 16-channel, 4-shank MNEA sensors (Figure 7.2) for simultaneous detection of Glu and multichannel electrophysiology in the hippocampus of normal and TLE rats (Li *et al.*, 2018).

The MNEA sensor was implanted into the hippocampus of anesthetized rats. As Figure 7.6a shows, after a period of baseline recording (phase A), pilocarpine was injected (at T_0) into the abdominal cavity to initiate TLE. About 600s later, when the drug had diffused into the brain and taken effect (phase B), the recorded

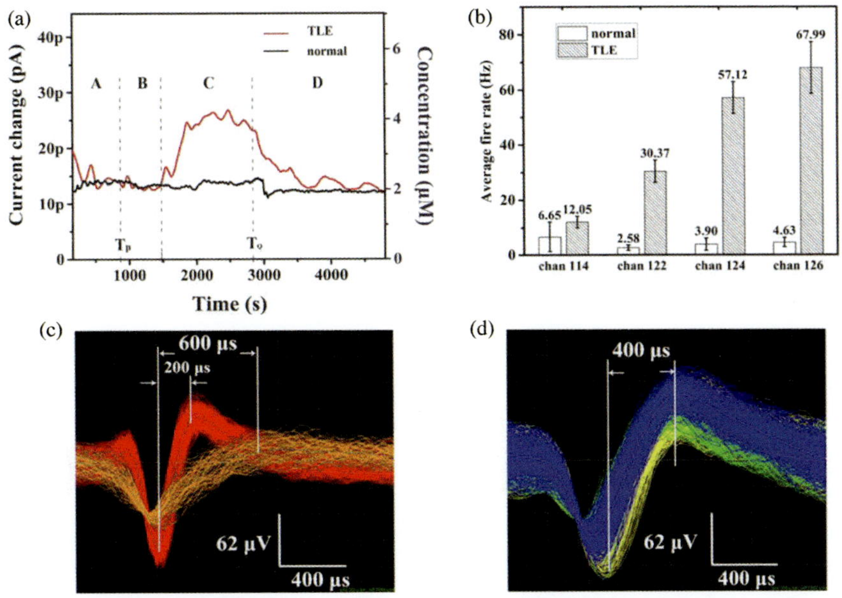

Figure 7.6 The glutamate and neural action potentials recorded in hippocampus CA1 of normal and Temporal Lobe Epilepsy (TLE) rats. (a) The glutamate signals in normal and TLE rats. After a period of stabilization (phase A), pilocarpine was injected (i.p.) in the rat at T_P to stimulate the onset of TLE. After a period of 600s for drug diffusion (phase B), the Glu concentration rapidly rose up to 26.7 ± 6.9 pA (4.22 ± 1.07 μM, n = 4) and remained elevated for a long period (phase C). Then diazepam was injected (i.p.) at T_Q to terminate the TLE (phase D). Normal control rats (black curve) received injections of 0.9% saline. (b) The average firing rates of action potentials recorded by four channels in normal control and TLE rats. (c) Normal rats exhibited two different discharge waveforms and peak latencies. (d) In the TLE rats all recording sites exhibited the same discharge waveform and peak latency: the waveforms are uniform with an increased amplitude compared to controls. The neurons were in a synchronous discharge state during TLE (Li *et al.*, 2018).

Glu concentration rapidly increased to 26.7 ± 6.9 pA (4.22 ± 1.07 μM, n = 4) and remained elevated (phase C). Later at T_Q, diazepam was injected to terminate TLE (phase D) and avoid deadly damage caused by severe super-synchronous discharge. The Glu concentration returned to its pre-TLE baseline during phase D. Normal rats in the control group received injections of 0.9% saline, which had essentially no effect on Glu concentrations.

Four channels of electrophysiological signals were compared between the normal and TLE rats. As Figure 7.6b shows, the average firing rates of the spikes during TLE seizure were increased to varying degrees compared to control rats. In the normal control rats, two different spike waveforms were found with different amplitude and the peak latency (200 μs and 600 μs) (Figure 7.6c). All recording sites in the TLE rats exhibited the same spike waveforms with the same latency (400 μs) and amplitude. This finding indicates that neurons tend to discharge synchronously in the state of TLE seizures. This study demonstrates that MNEA sensors are an efficient, reliable, and novel tool for the characterization of neurological disease states.

7.3.5 Detection of dopamine and electrophysiology in the Parkinsonian rat

Parkinson's disease (PD) is the most common serious movement disorder in the world. It produces motor symptoms such as rest tremor, bradykinesia, and rigidity. It was recognized that the degenerative death of dopamine-generating neurons is the key cause of PD. To investigate the changes of neural activities caused by the disease, the dual mode MNEA sensor was implanted into striatum of normal and PD rats.

The 8-channel electrophysiological signals and 1-channel dopamine signal were continuously recorded in normal and PD rats. As Figure 7.7a shows, compared to the normal rats, the LFPs of the PD rats fluctuate rapidly with the amplitude of 2.2 mV peak to peak. The average dopamine concentration in normal rats was 44.3±13.9 μM and greatly decreased to about 9.7±6.1 μM in PD rats (n = 5) (Figure 7.7b). These results are consistent with the mechanism of PD and verify the reliability of the MNEA sensor for recording the pathological features of signals in neurological diseases.

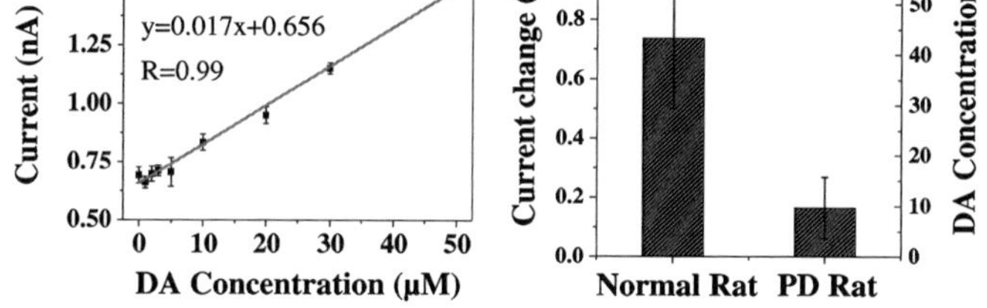

Figure 7.7 Comparison of electrophysiology and dopamine in normal and Parkinsonian rats. (a) Typical local field potentials recorded in the striatum of normal and PD rats; (b) The Current-Concentration calibration curve of the microelectrode for DA; (c) The average DA concentration recorded in the striatum of normal and PD rats (n = 5 respectively).

7.4 Summary and future prospects

A series of silicon based MNEA sensors were developed for real-time, parallel detection of neurotransmitter and electrophysiology in vivo. By modification of the MNEA sensor elements with specialized and specific bio-nano materials, the background noise of the microelectrodes for the detection of neural action potential was decreased to below 20 μV. The sensitivity of the microelectrodes (20 μm in diameter) were 6~7 pA/μM for Glu and 17~19 pA/μM for DA. The MNEA sensors were implanted into rat striatum and hippocampus. Multichannel recordings of neural action potentials, local field potentials, and electrochemical currents corresponding to Glu or DA concentration were obtained concurrently. The MNEA sensors are valuable for analyzing the dynamic relationships between the dual mode neural signals and represent new avenues for investigating neurological disorders including Parkinson's disease and epilepsy.

In the future, we hope to explore several additional aspects of this work. First, additional techniques for the selective modification of microelectrodes with enzyme-bearing layers will be of interest. These will lead to microelectrodes able to detect a wider range of different neurotransmitters simultaneously at neighboring microelectrode sites on one MNEA sensor. Second, faster electrochemical detection methods are needed to capture the events of single-cell quantum release of neurotransmitters using MNEA sensors. Third, the real-time detection of dual mode neural information under neural modulation needs to be explored in order to delve deeper into the mechanisms of neurological diseases and to eventually identify novel therapies to relieve patients of their symptoms.

Acknowledgments

This work was sponsored by the National Key R&D Program of Nano Science and Technology of China (2017YFA0205902), the NSFC (No. 61527815, No. 61673024, No. 61775216, No. 61771452), and the Key Research Programs (QYZDJ-SSW-SYS015) of Frontier Sciences, CAS.

References

Albuquerque EX, Pereira EF, Alkondon M (2009). Mammalian nicotinic acetylcholine receptors: from structure to function. *Physiol Rev*, 89: 73–120.

Bartheld CS, Bahney J, Herculanohouzel S (2016) The search for true numbers of neurons and glial cells in the human brain: a review of 150 years of cell counting. *J Comp Neurol*, 524: 3865–3895.

Barnett MW, Larkman PM (2007) The action potential. *Pract Neurol*, 7: 192–197.

Brazell MP, Kasser RJ, Renner KJ, *et al.* (1987) Electrocoating carbon fiber microelectrodes with Nafion improves selectivity for electroactive neurotransmitters. *J Neurosci Methods*, 22: 167–172.

Budai D (2010) Carbon fiber–based microelectrodes and microbiosensors, intelligent and biosensors, Vernon S. Somerset (Ed.), InTech Open Publishers, Croatia, pp 269–274.

Burmeister JJ, Moxon K, Gerhardt GA (2000) Ceramic–based multisite microelectrodes for electrochemical recordings. *Anal Chem*, 72: 187–192.

Burmeister JJ, Gerhardt GA (2001) Self–referencing ceramic–based multisite microelectrodes for the detection and elimination of interferences from the measurement of L–glutamate and other analytes. *Anal Chem*, 73: 1037–1042.

Burmeister JJ, Pomerleau F, Huettl P, *et al.* (2008) Ceramic–based multisite microelectrode arrays for simultaneous measures of choline and acetylcholine in CNS. *Biosens Bioelectron*, 23: 1382–1389.

Cepeda C, André VM, Yamazaki I (2010) Differential electrophysiological properties of dopamine D1 and D2 receptor–containing striatal medium–sized spiny neurons. *Eur J Neurosci*, 27: 671–682.

Cogan SF (2008) Neural stimulation and recording electrodes. *Ann Rev Biomed Engin*, 10: 275–309.

Chaurasia CS, Müller M, Bashaw ED, *et al.* (2007) AAPS–FDA Workshop White Paper: microdialysis principles, application, and regulatory perspectives. *AAPS Journal*, 9: E48–E59.

Defelipe J (2011) The evolution of the brain, the human nature of cortical circuits, and intellectual creativity. *Front Neuroanat*, 5: 29.

Delgado JM, DeFeudis FV, Roth RH, *et al.* (1971) Dialytrode for long term intracerebral perfusion in awake monkeys, *Archives Internationales de Pharmacodynamie et de Therapie*, 198: 9–21.

Desai SA, Rolston JD, Guo L, *et al.* (2010) Improving impedance of implantable microwire multi-electrode arrays by ultrasonic electroplating of durable platinum black. *Front Neuroeng*, 3: 5.

Ernesto F, Paolo M, Alessandro S, *et al.* (2001) Microdialysis in parkinsonian patient basal ganglia: acute apomorphine–induced clinical and electrophysiological effects not paralleled by changes in the release of neuroactive amino acids. *Exp Neurol*, 167: 356–365.

Fagg GE, Foster AC (1983) Amino acid neurotransmitters and their pathways in the mammalian central nervous system. *Neurosci*, 9: 701–719.

Fan X, Song Y, Ma Y, *et al.* (2017) In situ real–time monitoring of glutamate and electrophysiology from cortex to hippocampus in mice based on a microelectrode array. *Sensors*, 17: 61.

Fekete Z. (2015) Recent advances in silicon-based neural microelectrodes and microsystems: a review. *Sensor Actuat* B, 215: 300–315.

Follett KA, Weaver FM, Stern M, *et al.* (2010) Pallidal *versus* subthalamic deep–brain stimulation for Parkinson's disease. *N Engl J Med*, 362: 2077–2091

Gonon F, Cespuglio R, Ponchon JL, *et al.* (1978) *In vivo* continuous electrochemical determination of dopamine release in rat neostriatum, *C R Acad Sci Hebd Seances Acad Sci D*, 286: 1203–1206.

Gonon F, Buda M, Cespuglio R, *et al.* (1980) *In vivo* electrochemical detection of catechols in the neostriatum of anaesthetized rats: dopamine or DOPAC? *Nature*, 286: 902–904.

Gonon F, Buda M, Cespuglio R, *et al.* (1981) Voltammetry in the striatum of chronic freely moving rats: detection of catechols and ascorbic acid. *Brain Res*, 223: 69–80.

Hornykiewicz O (2002) Dopamine miracle: from brain homogenate to dopamine replacement. *Movement Disorders: Official Journal of the Movement Disorder Society*, 17: 501–508.

Hoogerwerf AC, Wise KD (2002) A three-dimensional microelectrode array for chronic neural recording. *IEEE Trans Biomed Engin*, 41: 1136–1146.

Hoyer D, Hannon JP, Martin GR (2002) Molecular, pharmacological and functional diversity of 5-HT receptors. *Pharmacol Biochem Behav*, 71: 533–554.

Huber G, Huber R, Jones BE, *et al.* (1992) Hyperthermophilic archae- and eubacteria occurring within Indonesian hydrothermal areas. *Syst Appl Microbiol*, 14: 397–404.

Huffman ML, Venton BJ (2008) Carbon-fiber microelectrodes for *in vivo* applications. *Analyst*, 134: 18–24.

Idil Cavus, Ba WS, Cassaday MP, *et al.* (2010) Extracellular metabolites in the cortex and hippocampus of epileptic patients. *Ann Neurol*, 57: 226–235.

Johnson MD, Franklin RK, Scott KA, *et al.* (2005) Neural probes for concurrent detection of neurochemical and electrophysiological signals *in vivo*. International Conference of the Engineering in Medicine & Biology Society, 7: 7325–7328.

Johnson MD, Franklin RK, Gibson MD, *et al.* (2008) Implantable microelectrode arrays for simultaneous electrophysiological and neurochemical recordings. *J Neurosci Methods*, 174: 62–70.

Jun JJ, Steinmetz NA, Siegle JH, *et al.* (2017) Fully integrated silicon probes for high-density recording of neural activity. *Nature*, 551: 232–236.

Kandel ER, Schwartz JH, Jessell TM (2000) *Principles of Neural Science*, 4th ed. McGraw–Hill, pp. 823–839.

Kenneth TK, Stefan GS, Terry L, *et al.* (2011) Sub-second dopamine detection in human striatum. *PLoS ONE*, 6: 1–5

Kewley DT, Hills MD, Borkholder DA, *et al.* (1997) Plasma-etched neural probes. *Sensor Actuat A*, 58: 27–35.

Krüger J, Bach M (1981) Simultaneous recording with 30 microelectrodes in monkey visual cortex, *Exp Brain Res*, 41: 191–194.

Li CL, Herbert J (1953) Microelectrode studies of the electrical activity of the cerebral cortex in the cat. *J Physiol*, 121: 117–140.

Li Z, Song Y, Xiao G, *et al.* (2018) Bio-electrochemical microelectrode arrays for glutamate and electrophysiology detection in hippocampus of temporal lobe epileptic rats. *Anal Biochem*, 550: 123–131

Lugomorales LZ, Loziuk PL, Corder AK, *et al.* (2013) Enzyme-modified carbon–fiber microelectrode for the quantification of dynamic fluctuations of nonelectroactive analytes using fast-scan cyclic voltammetry. *Anal Chem*, 85: 8780–8786.

Marg E, Adams JE (1967) Indwelling multiple micro-electrodes in the brain. *Electroencephal Clin Neurophysiol*, 23: 277–280.

Medina CL, Morales VA, Tapia R (2000) Action of 4-aminopyridine on extracellular amino acids in hippocampus and entorhinal cortex: a dual microdialysis and electroencehalographic study in awake rats. *Brain Res Bull*, 53: 255–262.

Millar J (1991) Simultaneous *in vivo*, voltammetric and electrophysiological recording with carbon fiber microelectrodes. *Methods Neurosci*, 4: 143–154.

Moralesvillagrán A, Medinaceja L, Lópezpérez SJ (2008) Simultaneous glutamate and EEG activity measurements during seizures in rat hippocampal region with the use of an electrochemical biosensor. *J Neurosci Methods*, 168: 48–53.

Mundroff ML, Wightman RM (2002) Amperometry and cyclic voltammetry with carbon fiber microelectrodes at single cells. In *Current Protocols in Neuroscience*. John Wiley & Sons, Inc. 6.14.1–6.14.22.

Nandi P, Lunte SM (2009) Recent trends in microdialysis sampling integrated with conventional and microanalytical systems for monitoring biological events: a review. *Anal Chim Acta*, 651: 1–14.

Nicolelis MA, Ghazanfar AA, Faggin BM, *et al.* (1997) Reconstructing the engram: simultaneous, multisite, many single neuron recordings. *Neuron*, 18: 529–537.

Nisch W, Böck J, Egert U, *et al.* (1994) A thin film microelectrode array for monitoring extracellular neuronal activity *in vitro*. *Biosensor Bioelectron*, 9: 737–741.

Phillips PE, Wightman RM (2003) Critical guidelines for validation of the selectivity of *in-vivo* chemical microsensors, *Trends Anal Chem*, 22: 509–514.

Pike CM, Grabner CP, Harkins AB (2009) Fabrication of amperometric electrodes. *J Vis Exp* 27: e1040–e1040.

Renshaw B, Forbes A, Morison BR (1940) Activity of isocortex and hippocampus: electrical studies with micro-electrodes. *J Neurophys*, 3: 74–105.

Robinson DL, Venton BJ, Heien ML, *et al.* (2003) Detecting subsecond dopamine release with fast-scan cyclic voltammetry *in vivo*, *Clin Chem*, 49: 1763–1773.

Rousche PJ, Normann RA (1998) Chronic recording capability of the Utah Intracortical Electrode Array in cat sensory cortex. *J Neurosci Methods*, 82: 1–15.

Seymour JP, Wu F, Wise KD, *et al.* (2017) State-of-the-art MEMS and microsystem tools for brain research. *Microsyst Nanoeng*, 3: 16066.

Song Y, Lin N, Hong J, *et al.* (2011) Application of microelectrode arrays with nano platinum black deposition on neural electrophysiological recording. *Journal of Southeast University*, 30(1): 29–32.

Stephens ML, Pomerleau F, Huettl P, *et al.* (2010) Real-time glutamate measurements in the putamen of awake rhesus monkeys using an enzyme–based human microelectrode array prototype. *J Neurosci Methods*, 185: 264–272.

Ulbert I, Maglóczky Z, Erőss L, *et al.* (2004) *In vivo* laminar electrophysiology co-registered with histology in the hippocampus of patients with temporal lobe epilepsy. *Exp Neurol*, 187: 310–318.

Urrestarazu E, Jirsch JD, Levan P (2010). High-frequency intracerebral EEG activity (100–500 Hz) following interictal spikes. *Epilepsia*, 47: 1465–1476.

Urrestarazu E, Chander R, Dubeau F, *et al.* (2007) Interictal high-frequency oscillations (100–500 Hz) in the intracerebral EEG of epileptic patients. *Brain*, 130: 2354–2366.

Woldring S, Dirken MN (1950) Spontaneous unit activity in the superficial cortical layers. *Acta Physiologica Et Pharmacologica Neerlandica*, 1: 369–379.

Wolbarsht ML, Macnichol EF, Wagner HG (1960) Glass insulated platinum microelectrode. *Science*, 132: 1309–1310.

Wei W, Song Y, Wang L, *et al.* (2015) An implantable microelectrode array for simultaneous L-glutamate and electrophysiological recordings *in vivo*, *Microsyt Nanoeng*, 1: 15002.

Xiao G, Song Y, Zhang S, *et al.* (2017) A high-sensitive nano-modified biosensor for dynamic monitoring of glutamate and neural spike covariation from rat cortex to hippocampal sub-regions. *J Neurosci Methods*, 291: 122–130.

Yang J, Martin DC (2004) Microporous conducting polymers on neural microelectrode arrays: II. physical characterization. *Sensor Actuat A* 101: 133–142.

Yu P, Wilson GS (2000) An independently addressable microbiosensor array: What are the limits of sensing element density? *Faraday Discuss*, 116: 305–317.

CHAPTER 8

ADVANCES IN TAILORING FUNCTIONAL SURFACES FOR *IN VIVO* MONITORING OF REACTIVE OXYGEN SPECIES AND RELATED MOLECULES INVOLVED IN OXIDATIVE STRESS IN THE BRAIN

Limin Zhang and Yang Tian*

Shanghai Key Laboratory of Green Chemistry and Chemical Processes, School of Chemistry and Molecular Engineering, East China Normal University, Dongchuan Road 500, Shanghai 200241, P. R. China

*ytian@chem.ecnu.edu.cn

8.1 Introduction

Reactive oxygen species (ROS), such as $O_2^{\bullet-}$ and H_2O_2, are side products of normal oxygen metabolism in all aerobic cells (Rhee 2006; Frinkel 2011; Fijikawa *et al.*, 2013). Imbalances between the overproduction of ROS and antioxidant defense mechanisms give rise to oxidative stress, which is harmful to tissues. Metal ions, such as Cu^{2+}, are efficient catalysts for the production of ROS (Hewitt and Raul, 2009; Giacovazzi *et al.*, 2014). Cu^{2+} bound to peptides catalyzes the production of H_2O_2 in the presence of oxygen and biological reducing agents, such as ascorbic acid (AA) (Huang *et al.*, 1999; Hewitt and Rauk, 2009). Production of ROS also decreases pH values, which leads to brain acidosis and augments cell death under pathophysiological conditions. Therefore, ROS and related molecules involved in oxidative stress attract much attention in a wide range of brain research fields, because they mediate the biochemistry of cellular pathology and may be involved in the etiology of aging and progressive neurodegenerative diseases, such as Alzheimer's and Parkinson's disease (Huang *et al.*, 1999; Wilson and Johnson, 2008; Hewitt and Rauk, 2009).

The *in vivo* analysis of chemical signals in brain extracellular fluid (ECF) with implanted electrochemical biosensors is a vital way to study brain function and to map brain activity (Khan and Michael, 2005; Wilson and Gifford, 2005; Robinson *et al.*, 2008; Wilson and Johnson, 2008; Zhang *et al.*, 2012; Li *et al.*, 2014). Electrochemical biosensors offer excellent spatial (10–200 μm) and temporal (s) resolution. By implanting a microelectrode in a specific brain region, changes in the concentration of a variety of chemical species in the ECF can be monitored by applying a suitable electrical signal and, typically, recording the resulting Faradaic current. However, it is challenging to develop directly implantable biosensors with high analytical performance for *in vivo* determinations in the brain. The first challenge is to improve the selectivity of the biosensors to fulfill the requirements for real-time monitoring in the brain, because a large variety of biomolecules coexist and interact with each other in the ECF. To this end, fast-scan cyclic voltammetry (FSCV) at high scan rates (> 100 V s^{-1}) and carbon fiber microelectrodes (CFME) were pioneered by Wightman and co-workers (Robinson *et al.*, 2008) to achieve selective detection of several neurotransmitters, including dopamine (DA), and 5-hydroxytryptamine (Venton and Wightman, 2003; Zhang *et al.*, 2017). The modification of electrode surfaces with electrocatalytic nanomaterials, which lower the overpotentials of certain species, improves biosensor selectivity. For example, the selective determination of AA has been achieved with carbon nanotube-based electrodes (Zhang *et al.*, 2017). Moreover, several enzyme-based biosensors have been developed for the selective determination of small molecules, such as the superoxide anion ($O_2^{\bullet-}$), H_2O_2, glucose, and lactate (Wilson and Hu, 2000; McMahon *et al.*, 2006; Zhang *et al.*, 2012). Nevertheless, biosensors continue to suffer from insufficient selectivity for *in vivo* measurements.

Further developments of functional electrode surfaces are urgently needed to enable *in vivo* monitoring of important species in the living brain with high analytical performance (McMahon *et al.*, 2006; Zhang *et al.*, 2012). Through careful tailoring of functionalized surfaces that facilitate direct electron transfer between electrodes and enzymes or proteins, such as superoxide dismutase (SOD) and cytochrome *c* (Cyt *c*), sensors have been created for the *in vivo* monitoring of $O_2^{\bullet-}$ and H_2O_2 (Liu *et al.*, 2008; Li *et al.*, 2010; Wang *et al.*, 2013; Chai *et al.*, 2013, 2014; Shao *et al.*, 2013; Zhou *et al.*, 2014; Luo *et al.*, 2015; Yu *et al.*, 2015; Zhang *et al.*, 2015; Milton *et al.*, 2016; Zhao *et al.*, 2016; Dong *et al.*, 2017). Unfortunately, only a limited number of naturally occurring biocatalysts and recognition molecules is available for such purposes. To overcome this limitation, the design and synthesis of new recognition molecules (or biomolecules) is necessary. In recent years, Cu-free derivatives of bovine erythrocyte SOD (E_2Zn_2SOD) and protoporphyrin IX have been rationally prepared for specific recognition of Cu^{2+} and Cd^{2+}, respectively (Chai *et al.*, 2013, 2014). Meanwhile, a series of organic molecules including N-(2-aminoethyl)-N,N′,N′-tris(pyridine-2-yl-methyl)ethane-1,2-diamine (AE-TPEA), 2,2′,2″-(2,2′,2″-nitrilotris(ethane-2,1-diyl)-tris((pyridine-2-ylmethyl)azanediyl)-triethanethiol (TPAASH), and N,N-di-(2-picoly) ethylenediamine (DPEA) have been synthesized for the selective recognition of Cu^{2+} (Shao *et al.*, 2013; Luo *et al.*, 2015; Zhang *et al.*, 2015), and N-(6-aminopyridin-2-yl)ferrocene (Fc-Py) has been designed for the selective determination of pH (Zhao *et al.*, 2016). Integration of these specific molecules with nanostructured surfaces has produced a variety of biosensors for highly selective *in vivo* determinations of metal ions and pH in the brain ECF (Chai *et al.*, 2013, 2014; Shao *et al.*, 2013; Luo *et al.*, 2015; Zhang *et al.*, 2015; Zhao *et al.*, 2016).

Another challenge for electrochemical devices is the difficulty of accurately quantifying target molecules, which arises from the differences between the environments in which *in vivo* experiments and *in vitro* calibrations are performed. In contrast to the low solution resistance in a typical 0.1 M electrolyte solution (~100 Ω), the solution resistance in cerebrospinal fluid is typically 3000–5000 Ω due to the presence of proteins and other biological species. Consequently, IR drop for a microelectrode with a current of ~1 μA is ~3–5 mV in the brain that, combined with potential drift of reference electrodes, leads to inaccuracies of electrical signals in the brain. Thus, the calibration curves determined *in vitro* may not always be applicable to measurements *in vivo*. To solve this problem, mathematical analysis using information from the background charging current has been used to exactly calibrate the data obtained by FSCV for *in vivo* measurements (Roberts *et al.*, 2013). More recently, principal component analysis in tandem with inverse least-squares regression has been introduced for resolving and quantifying overlapping compounds in FSCV data (Rodeberg *et al.*, 2015). Also, a two-channel ratiometric electrochemical biosensor was developed for determination of Cu^{2+} in rat brains with a stable built-in correction to avoid the error between *in vitro* and *in vivo* environments. However, this biosensor still suffers some inaccuracies because it employs two different working electrodes. To improve the situation further, a single ratiometric biosensor was designed and constructed to measure Cu^{2+} and L-cysteine (CySH) with high selectivity and accuracy: it has been applied successfully for real-time evaluation of Cu^{2+} and CySH in live rat brains with Alzheimer's disease (AD). More recently, this ratiometric strategy has been extended to monitor pH and chloride ion (Cl^-) in the brain using inner reference molecules with stable potential outputs (Dong *et al.*, 2017). In this case, the half-wave potential difference ($\Delta E_{1/2}$) between recognition and reference molecules was used as the measurable signal, remarkably enhancing the accuracy of potential-dependent biosensors.

Usually, it will not be sufficient to monitor a single biomolecule to comprehensively investigate physiological and pathological phenomena. For instance, previous studies have suggested that an increase in Cu^{2+} concentration causes an overproduction of ROS, resulting in a decrease in pH (Huang *et al.*, 1999). Meanwhile, the acidification of brain tissues leads to more serious oxidative stress (Atwood *et al.*, 1998). Consequently, the development of approaches for the simultaneous detection of multiple species with high selectivity, sensitivity, and accuracy is urgently needed. Recently, ratiometric biosensors with both current and potential signal outputs have been designed to simultaneously detect two species, such as Cu^+ and pH, pH and O_2, or glucose and pH (Li *et al.*, 2017; Liu *et al.*, 2017, 2017a). This concept of dual signal outputs has established a flexible and promising approach for further analysis of multiple species.

One key disadvantage of electrochemical techniques is the possibility that either the applied potentials or the electrochemically generated currents of implanted microelectrodes may have unexpected influence on the functions of adjacent neurons, resulting in false electrophysiological responses (Ewing *et al.*, 1983; Crespi *et al.*, 1995; Crespi, 2002, 2009). Although several early studies did not mention any effects on unit-spike activity, the effects of constant potential amperometry on neuronal activity *in vivo* remain unclear. Consequently, electrochemical strategies without the possibility of electrical stimulation are in urgent demand. Relying on measurements of open-circuit potentials has been realized for monitoring the levels of multiple metal ions in the brain.

This chapter focuses on new strategies for improving the selectivity and accuracy of sensors for the simultaneous detection of two species. We give a brief overview of the surface engineering strategies and molecular design principles that guide the development of biosensors with high analytical performance for *in vivo* monitoring of ROS and related molecules involved in oxidative stress in the living brain. The last section outlines potential future directions for synthesizing recognition molecules and tailoring functional surfaces from the development of single-component biosensors towards the design of advanced devices with multiple targets and the multi–functional nanodevices that combine the benefits of multiple sensing capacities and the incorporation of therapeutic modules.

8.2 *In vivo* monitoring of ROS and dopamine (DA) with high selectivity in the brain

8.2.1 Selective detection of $O_2^{\bullet-}$ based on direct electron transfer to specific enzymes

Superoxide anion ($O_2^{\bullet-}$), the primary ROS, is generated in substantial quantities as a result of normal intracellular metabolism. Reactions of $O_2^{\bullet-}$ are of great importance because they form other ROS, including ·OH and $ONOO^-$. Under normal metabolic conditions, $O_2^{\bullet-}$ is produced at a rate matched by the capacity of tissue to catabolize it. When $O_2^{\bullet-}$ production exceeds the body's natural ability to deal with this potentially cytotoxic species, a variety of pathological conditions may arise, including neurodegeneration (Auchere and Rusnak, 2002). The difficulties in elucidating the mechanism of $O_2^{\bullet-}$-induced oxidative stress arise from the lack of selective and reliable analytical methods for real-time determination of $O_2^{\bullet-}$ because of the unique properties of $O_2^{\bullet-}$, including its relatively high reactivity, short life-time, and broad concentration range. The typical electrochemical method for $O_2^{\bullet-}$ determination is based on direct electron transfer with an enzyme or protein. The two main biocatalysts are Cyt *c* and SOD. Taniguchi *et al.* reported the first direct electron transfer reaction of Cyt *c* at an electrode, promoted by monolayers of electrode modifiers (Taniguchi *et al.*, 1982). Since then, the direct electron transfer of Cyt *c* has been extensively studied and much effort has been put into the design of a third-generation biosensor for $O_2^{\bullet-}$. Recently, a biosensor was developed via covalently co-immobilizing Cyt *c* with an amino lipid onto a conducting polymer (poly-5,2-terthiophene-3-carboxylic acid) for monitoring the change in extracellular levels of $O_2^{\bullet-}$ in the live rat brain upon repeated administrations of an abused drug, cocaine (Rahman *et al.*, 2012). However, Cyt *c* is not a specific biocatalyst for $O_2^{\bullet-}$, as many reductants present in the brain, such as ascorbate, glutathione, and H_2O_2, can also reduce Cyt *c*. Therefore, the inherent properties of Cyt *c* cause the biosensor to suffer from interferences, which greatly limits the application of Cyt *c*-based $O_2^{\bullet-}$ biosensors in living systems.

Superoxide dismutase is well known to specifically catalyze the dismutation of $O_2^{\bullet-}$ to O_2 and H_2O_2, with a rate constant approaching 10^9 M^{-1} s^{-1}. The unique properties of SOD motivated researchers to investigate direct electron transfer reactions of SOD molecules at electrode surfaces. A variety of SOD-based third-generation biosensors have been developed for selective determination of $O_2^{\bullet-}$. The direct electrochemistry of bovine erythrocyte Cu,Zn-SOD was first achieved using a cysteine-modified electrode and the promoted direct electron transfer has been considered to be associated with the redox reaction of the copper site, which is also responsible for the dismutase activity (Tian *et al.*, 2002, 2004; Deng *et al.*,

2008; Luo *et al.*, 2009). The electron transfer of SODs was also well-promoted on gold electrodes modified with mercaptobenzoic acid (MPA). These SOD-based biosensors showed bifunctional electrocatalysis toward $O_2/O_2^{\bullet-}$ with high selectivity. With the further development of nanomaterials, the electron transfer of SOD was facilitated on gold nanostructures with different shapes, as well as on TiO_2 nanoneedles (Tian *et al.*, 2002, 2004; Deng *et al.*, 2008; Luo *et al.*, 2009). In particular, rapid electron transfer with Cu,Zn–SOD has been realized at highly conductive ZnO nanodisks with a formal redox potential ($E^{0'}$) located between $E^{0'}$ values of the redox couples $O_2/O_2^{\bullet-}$ and $O_2^{\bullet-}/H_2O_2$ (Deng *et al.*, 2008). The bidirectional electron transfer mediation observed on ZnO nanodisks is essentially based on the inherent specificity of SOD for dismutation of $O_2^{\bullet-}$. This ZnO/SOD electrode is capable of sensing $O_2^{\bullet-}$ cathodically at 0 mV (vs. Ag/AgCl), without interferences in living systems. This biosensor has demonstrated higher selectivity, higher sensitivity, and a wider dynamic range for detection of $O_2^{\bullet-}$ than those using TiO_2 and gold nanomaterials. Finally, the ZnO/SOD microelectrode was successfully applied to the determination of $O_2^{\bullet-}$ in bean sprouts grown under O_2 hyperoxia.

However, this type of ZnO nanodisk electrode is still difficult to miniaturize for *in vivo* analysis of $O_2^{\bullet-}$ in the brain. Meanwhile, it is also challenging to stably immobilize SOD on the electrode. Recently, we developed a new $O_2^{\bullet-}$ biosensor using a nitrilotriacetic acid (NTA)/histidine tag (HT) to anchor the histidine residues in SOD onto a CFME with a diameter of 10 μm through metal–chelate affinity (Figures 8.1A and 8.1B) (Wang *et al.*, 2013). This scheme has provided a more stable assembly of SOD on the CFME and the direct electron transfer to SOD has been greatly enhanced with a high rate constant (k_s) of $24 \pm 1.1\ s^{-1}$. The optimized $O_2^{\bullet-}$ biosensor exhibits a broad dynamic range from 10^{-7} to 10^{-4} M, a low detection limit of 21 nM, high selectivity, and long-term stability. This durable and reliable biosensor has been successfully applied to monitor $O_2^{\bullet-}$ in the rat brain during ischemia and reperfusion (Figure 8.1C). The obvious increased current suggests a burst of $O_2^{\bullet-}$ during both ischemia and reperfusion.

8.2.2 Selective and sensitive colorimetric visualization of cerebral dopamine based on double molecular recognition

Dopamine (DA) is one of the most important neurotransmitters in neural systems, because it plays a central role in several brain functions such as reward, movement, and mood (Bassareo and Di Chiara, 1999; Castro and Zigmond, 2001). Alterations in neuronal release and uptake of DA are implicated in psychiatric disorders, addiction, and Parkinson's disease. Over the past 30 years, a larger number of

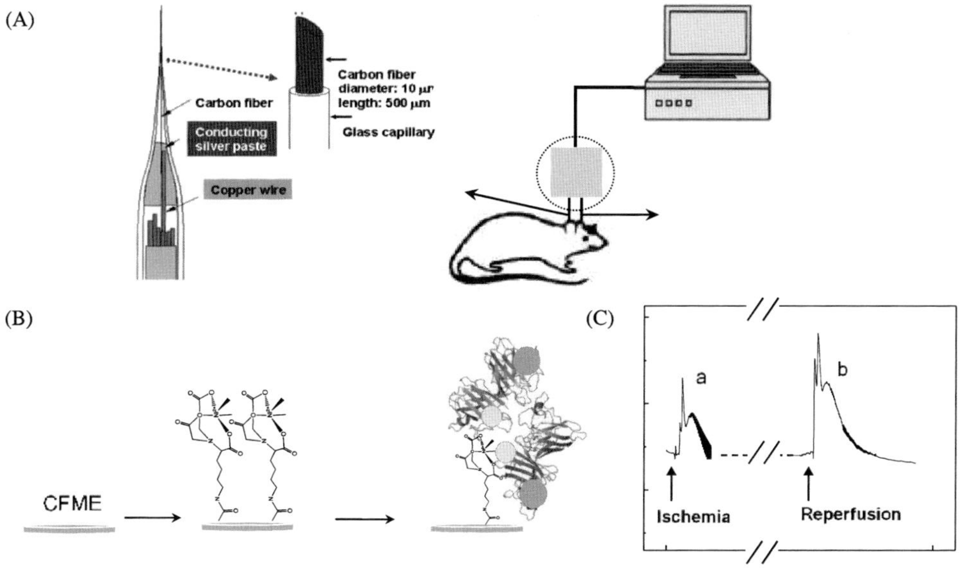

Figure 8.1 (A) Schematic diagram of an electroanalytical system for *in vivo* monitoring of $O_2^{\bullet-}$ in a rat brain with CFME. (B) Schematic illustration of the modification processes of NTA and SOD. (C) Amperometric responses obtained at a SOD-immobilized CFME electrode implanted in the rat brain upon (a) ischemia and (b) reperfusion. Reproduced with permission from Wang *et al.*, 2013.

elegant methods have been employed for DA detection, even though the complexity of cerebral systems presents a great challenge to provide theoretically and technically simple DA detection (Hyman and Malenka, 2001; Phillips *et al.*, 2003; Venton and Wightman, 2003). Boronic acids from stable boronate complexes with diols, and the resulting complexes, were employed to develop optical and electrochemical or electronic sensors for DA. Although the benefits of these biosensors are already apparent, the selectivity of DA biosensors over biomolecules such as glucose and lactate based on the same diols are still limited.

We first reported a direct, selective, and sensitive strategy for the colorimetric visualization of cerebral DA through the design and functionalization of gold nanoparticles (AuNPs) with 4-mercaptophenylboronic acid (MBA) and dithiobis (succinimidylpropionate) (DSP) (Figure 8.2A) (Kong *et al.*, 2011). These two molecules act as stabilizers for AuNPs and also interact with the diol and amine functional groups in DA to doubly recognize DA with high specificity. The possible interfering species that commonly exist in cerebral systems, such as AA, epinephrine, norepinephrine, glucose, lactate, and 3,4-dihydroxyphenylacetic acid cause negligible interference with detection of DA. Furthermore, to improve the

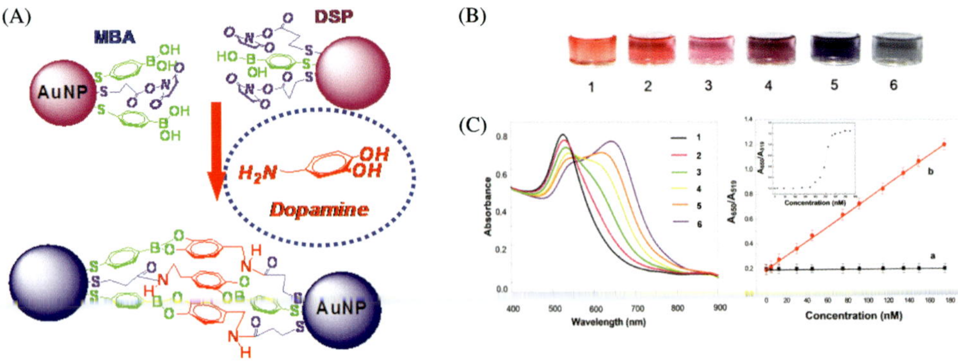

Figure 8.2 (A) Colorimetric detection using functionalized AuNPs (the MBA-DSP-AuNPs probe). (B) Colorimetric visualization and (C) UV/Vis spectra of MBA-DSP-AuNPs probe with the addition of DA at different concentrations, and corresponding linear plots of A_{650}/A_{519} *versus* DA concentration. Reproduced with permission from Kong *et al.*, 2011.

sensitivity of this probe, 175 nM DA was added to the MBA-DSP-AuNPs solution in advance. The optimized sensitivity is approximately 35 times larger in the presence compared to the absence of the additional DA (Figures 8.2B and 8.2C). The detection limit for DA was approximately 0.5 nM with a relative standard deviation of 1.0% (n = 5). With this remarkable performance, MBA-DSP-AuNPs provide a simple and direct platform for assaying DA in the rat brain based on *in vivo* microdialysis. The basal level of DA in the brain dialysate concentration has been estimated to be approximately 10 nM.

8.3 A ratiometric strategy based on specific recognition for selective and accurate determination

8.3.1 Accurate determination of Cu^{2+} based on molecular design

As one of the essential components utilized in all domains of life, the copper ion (Cu^{2+}) can induce the overproduction of ROS and drive oxidative damage to lipids, proteins, and nucleic acids (Hewitt and Rauk, 2009; Giacovazzi *et al.*, 2014). High amounts of Cu^{2+} are found within the senile plaques of AD patients, where it is directly bound to the amyloid–β peptide, and are linked to neurotoxicity and oxidative stress (Hewitt and Rauk, 2009). Relying on the inherent redox signal of Cu^{2+}, our group has established a series of electrochemical methods for measurement of Cu^{2+} by integrating specific recognition with efficient nanomaterials to form organic-inorganic hybrid systems. We first designed a two-channel ratiometric electrochemical strategy for the accurate detection of Cu^{2+} in rat brains (Chai *et al.*, 2013). The native bovine erythrocyte SOD contains two metal centers, Zn^{2+} and Cu^{2+}. When

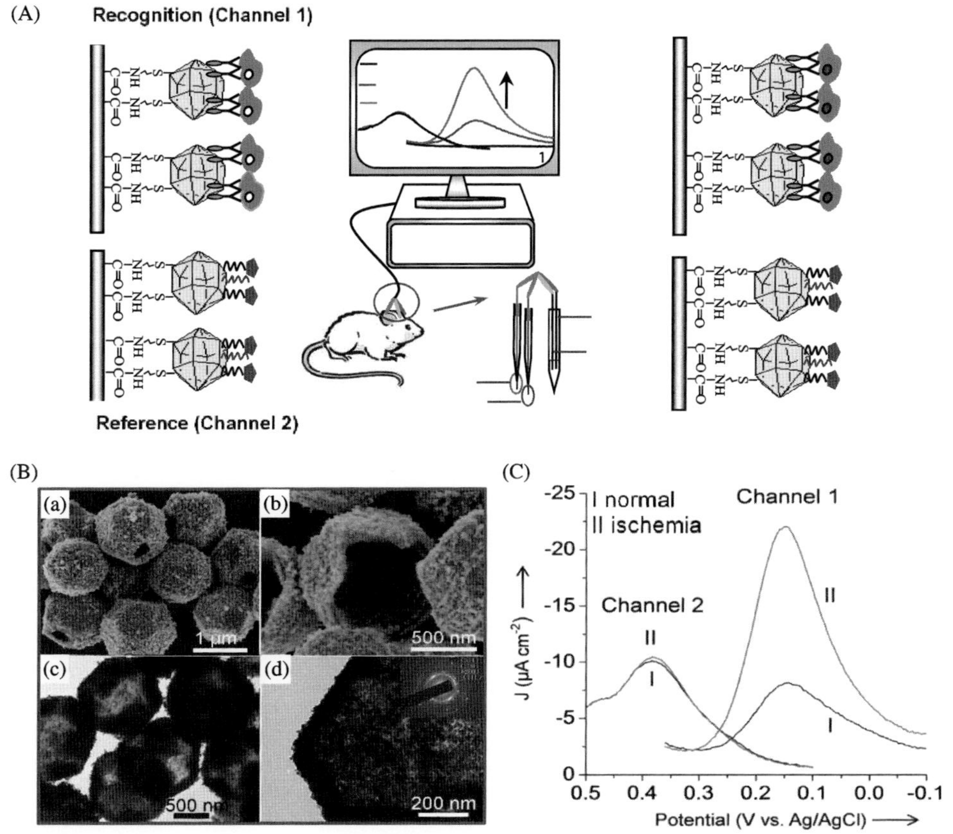

Figure 8.3 (A) The two-channel biosensor for *in vivo* monitoring of Cu^{2+} in a rat brain. (B) SEM (a, b) and TEM (c, d) images of the Au microcages. (C) Differential pulse voltammetry (DPV) responses obtained using the two-channel ratiometric biosensor for determination of Cu^{2+} in (I) a normal rat brain, and (II) a rat brain after cerebral ischemia. Reproduced with permission from Chai *et al.*, 2013.

Cu^{2+} is removed from SOD, E_2Zn_2SOD (which represents a Cu-free derivative of bovine erythrocyte copper-zinc superoxide dismutase), retains a specific recognition ability for Cu^{2+}, resulting in the selective determination of Cu^{2+}. Meanwhile, 6-(ferrocenyl) hexane thiol (FcHT) was used as reference element and modified at another electrode to provide a stable built–in correction (Figure 8.3A). In addition, gold truncated octahedral microcages full of nanograins were synthesized to enhance the sensitivity ~7-fold, due to their large surface area and high electrocatalytic activity (Figure 8.3B). This biosensor shows a high sensitivity of 30.2 mA M^{-1} cm^{-2} and a detection limit of 3 nM. Combined with the microelectrode technique, this biosensor has been directly implanted into a normal rat brain and a rat brain followed by cerebral ischemia (Figure 8.3C). The basal level of Cu^{2+} was estimated to be 1.67 ± 0.56 μM, while the Cu^{2+} increased 5-fold after global cerebral ischemia.

However, the two independent working electrodes give rise to a more complicated operation, a need for more instrumentation, and cause more injury to the animal's brain tissue. More importantly, there is still some error in the quantitative determination between the two electrodes. To this end, an accurate and selective ratiometric electrochemical sensor has been developed. The specific molecule, TPAASH (defined above), was designed and synthesized for specific recognition of Cu^{2+}. Meanwhile, the electroactive molecule, FcHT, was co-immoblized with TPAASH onto one electrode as an inner reference molecule. This ratiometric electrochemical biosensor was successfully used to detect Cu^{2+} in microdialysates from rat brain before and after global cerebral ischemia (Zhang *et al.*, 2015). We further synthesized a new molecule, DPEA, and developed a single biosensor for the determination of Cu^{2+} followed by CySH (Figure 8.4A) (Luo *et al.*, 2015). The electrochemical reduction peak of Cu^{2+} increased with increasing concentrations of Cu^{2+} (Figure 8.4B). More interestingly, this peak current decreased after the addition of CySH, which could be attributed to the release of Cu^{2+} from the DEPA-Cu^{2+} complex when CySH bonds to the Cu^{2+} center (Figure 8.4C). Meanwhile, 5′-MB-GGCGCGA$(T)_{13}$-SH-3′ (HS-DNA-MB, MB = methylene blue) with a separate peak potential was designed as an internal standard to avoid complicated interferences in the brain. This ratiometric biosensor with a single electrode has been successfully applied to evaluate the levels of Cu^{2+} and CySH in a live rat brain with AD. We found that the concentration of Cu^{2+} increased ~5-fold, and that of CySH decreased ~3-fold, in the brain of rats with AD compared to the brain of normal rats.

It is noteworthy that most endogenous antioxidants are able to reduce Cu^{2+} to Cu^{+}, which triggers the reduction of molecular oxygen to $O_2^{\bullet-}$ or H_2O_2 (Cooper, 2011). The overproduction of ROS causes a decrease in pH, which exacerbates oxidative stress (Webb *et al.*, 2011). Therefore, the development of strategies for the simultaneous monitoring of Cu^{+} and pH *in vivo* is essential for progress in our understanding of the roles that Cu ions and pH play in pathological and physiological events in the brain. Recently, we created an efficient biosensor for the ratiometric monitoring of Cu^{+} and pH in the brain using both current and potential outputs (Figure 8.5) (Liu *et al.*, 2017a). A series of N,N-bis(2-(2- (ethylthio)ethyl) (NS4) derivatives were designed for specific recognition of Cu^{+} ions (Figure 8.5C). After systematically evaluating the electrochemical parameters of Cu^{+} oxidation by tuning alkyl chain length, polyaromatic structure, and substitute group site of NS4, N, N-bis(2-2((ethylthio)ethyl)-2-naphthamide (NS4-C1) was optimized to detect Cu^{+} ions with the most negative potential and the largest current density. Meanwhile, 9, 10-anthraquinone was designed for selective pH recognition, while 2, 2′-azino-bis(3-ethylbenzthiazoline-6-sulfonic acid was employed as an inner reference (Figure 8.5D). This single biosensor, with both current and potential

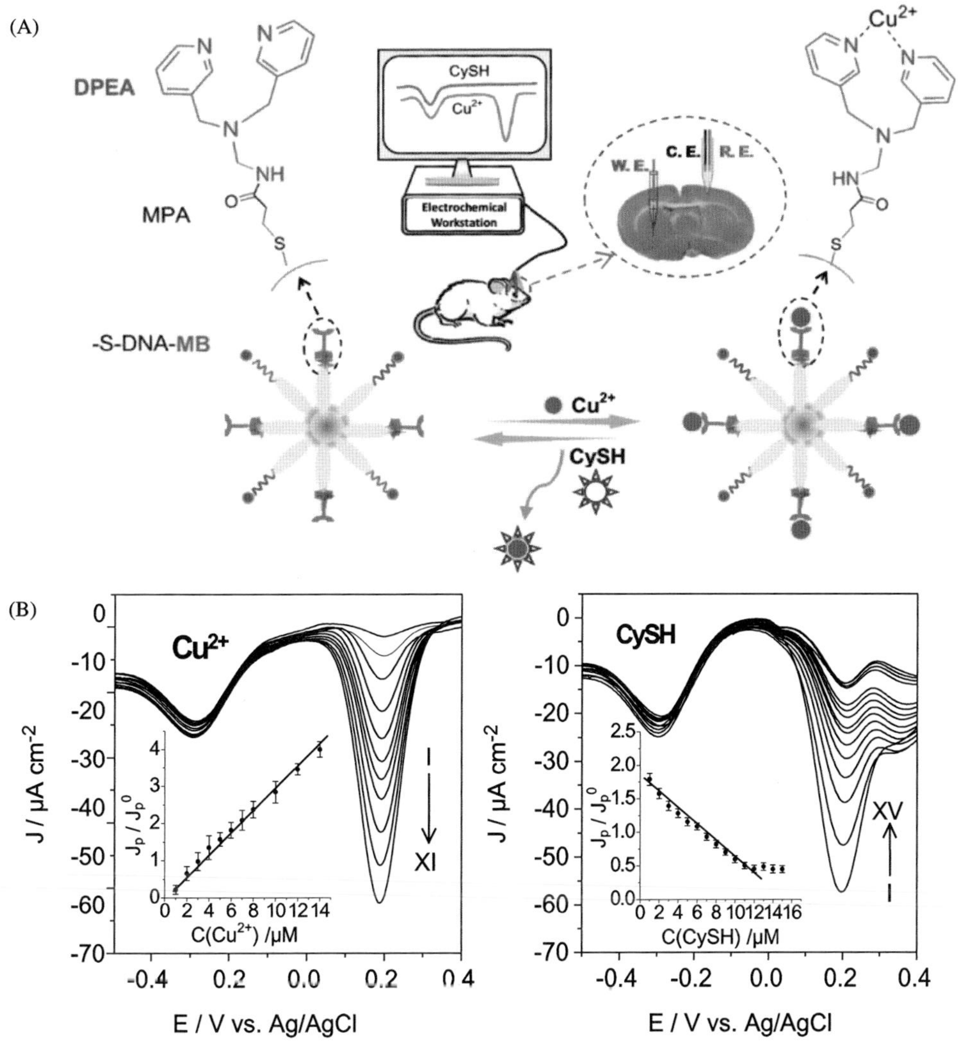

Figure 8.4 (A) Illustration of the single electrochemical biosensor for the *in vivo* ratiometric monitoring of Cu^{2+} and CySH in live rat brains with AD. (B) Right: DPVs obtained at a microelectrode immobilized by DPEA and MB (DPEA+MB/Au/CFME) in aCSF (pH 7.4) with different concentrations of Cu^{2+}. Inset: The linear relationship between the peak current density ratio obtained at 195 mV and −290 mV (J_p/J_p^0) and the Cu^{2+} concentration. Left: DPVs obtained at DPEA+MB/Au/CFME after reaction with Cu^{2+} in aCSF. Inset: The linear relationship between J_p/J_p^0 and the CySH concentration. Reproduced with permission from Luo *et al.*, 2015.

signal outputs simultaneously, determines Cu^+ concentrations from 0.5 to 9.5 μM and pH values ranging from 6.0 to 8.0. Eventually, the efficient biosensor was used to simultaneously detect Cu^+ and pH in the live brain. Concentrations of Cu^+ in the mouse model of AD increased by 1.5-fold in the cortex and 2.0-fold in the

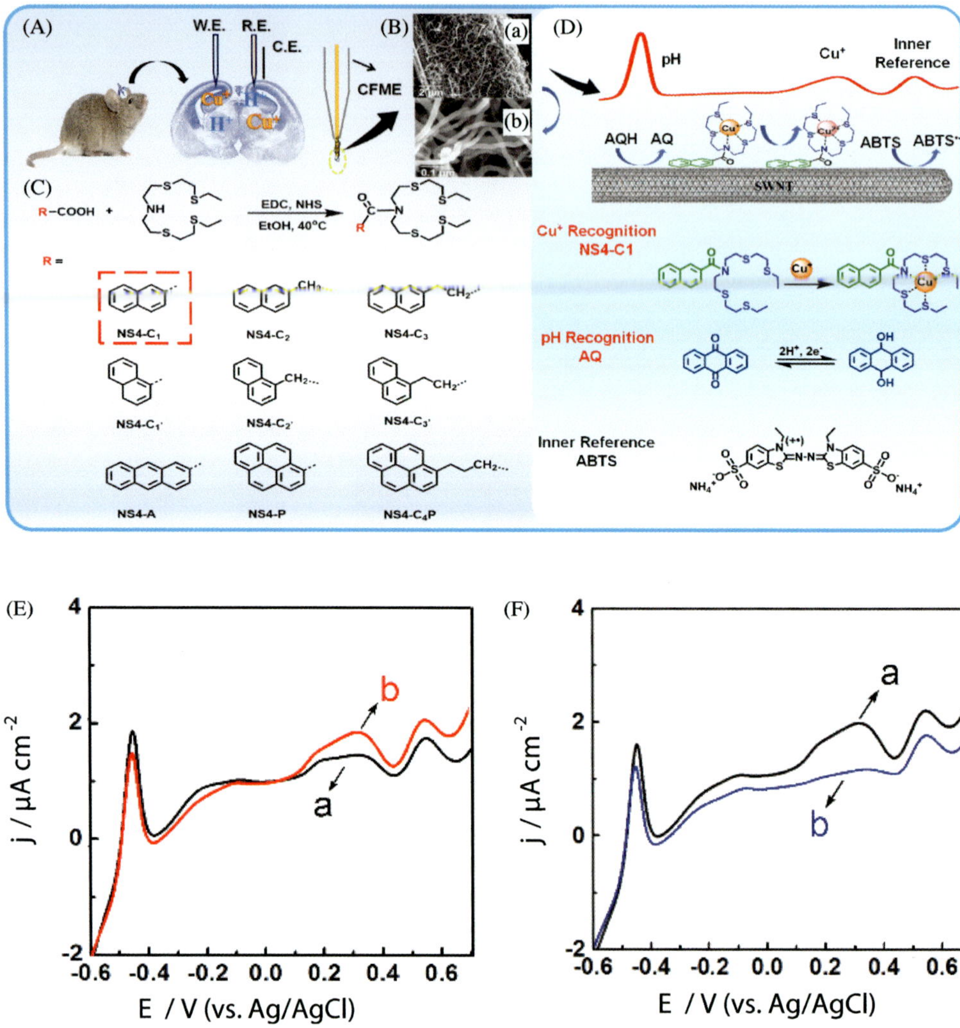

Figure 8.5 (A) The developed ratiometric single biosensor for simultaneously evaluating the levels of Cu^+ and pH in a live mouse brain. (B) SEM images of (a) CFME/SWNT and (b) the enlarged image of SWNTs. (C) The general synthesis procedure of NS4 and the corresponding molecular structures. (D) Preparation procedures for the functionalized electrodes. (E) Linear sweep voltammograms (LSVs) obtained at CFME/SWNT/AQ+NS4-C1 +ABTS electrode in (a) normal mouse brain and (b) mouse brain with AD. (F) LSVs of CFME/SWNT/AQ+NS4-C1+ABTS electrode in (a) mouse brain with AD, and (b) followed by microinjection of 0.1 mM EDTA (pH 9.0). Scan rate: 20 mV s^{-1}. Reproduced with permission from Liu *et al.*, 2017a.

striatum and hippocampus compared to normal mice (Figure 8.5E). Surprisingly, negligible pH changes were observed. The increased Cu^+ concentration observed in the mouse model of AD might be generated from the reduction of Cu^+ by overexpressed amyloid precursor protein (APP). APP is involved in Cu homeostasis and possesses a copper-binding site in its NH_2-terminal cysteine-rich domain, which can reduce Cu^{2+} to Cu^{+}.

8.3.2 Selective and reliable detection of cerebral ascorbic acid

Ascorbic acid (AA) is a vital antioxidant molecule in the brain, playing a major role as a ROS scavenger and defending the brain against high levels of oxidative stress (May *et al.*, 1998; Harrison and May, 2010). Thus, alterations to the AA concentration in the brain might have diagnostic value for neurodegenerative diseases. AA also has potentially therapeutic functions in ischemia, stroke, Alzheimer's disease, and Parkinson's disease (Rice, 2000; Kontos, 2001; Rebec *et al.*, 2003; Dorner *et al.*, 2009; Acuna *et al.*, 2013;). However, the exact mechanisms and functions of AA are still poorly understood due to the lack of accurate and selective methods for *in vivo* AA detection. AA is a water-soluble, hexonic sugar acid and has two dissociable protons with pK_a values of 4.2 and 11.8. At physiological pH, AA exists as monovalent anion, ascorbate.

In electrochemistry, AA is oxidized through a two-electron, one-proton pathway followed by a rapid, irreversible hydrolysis to produce an electroinactive 2,3-diketogulonic acid: the rate constant of the hydrolysis reaction is 1.2×10^3 s^{-1}. This reaction is responsible for the irreversible electrochemical redox process of AA oxidation with a high overpotential at most types of electrode. The high overpotential essentially invalidates selective electrochemical detection of AA because of the great interference from other electroactive species in the brain. Since AA oxidation proceeds via an inner-sphere electron transfer, it is sensitive to the nature of the electrode surface (McCreery, 2008).

Carbon nanotubes (CNTs) have electronic and structural properties distinct from other kinds of carbon materials. Mao and coworkers (Zhang *et al.*, 2005) found that AA oxidation was accelerated on heat-treated single-wall nanotubes (SWNTs). They further developed *in vivo* voltammetry with the CNT-modified CFMEs as the *in vivo* probe for monitoring AA in real time in the rat brain (Zhang *et al.*, 2007). *In vivo* voltammetry was used to follow the *in vivo* homeostatic regulation of ascorbate in the rat striatum upon infusions of exogenous AA. AA detection was further enhanced by leveraging the ability of vertically-aligned carbon nanotubes (VACNTs) to catalyze AA oxidation. A pristine carbon fiber microelectrode was coated with VACNTs (VACNT-CF) and employed to monitor AA in the brain (Xiang *et al.*, 2014). The VACNT-CFs effectively avoided the person-to-person and electrode-to-electrode variations associated with manual electrode modification procedures, leading to rapid and real-time *in vivo* monitoring of AA with good reproducibility and high stability.

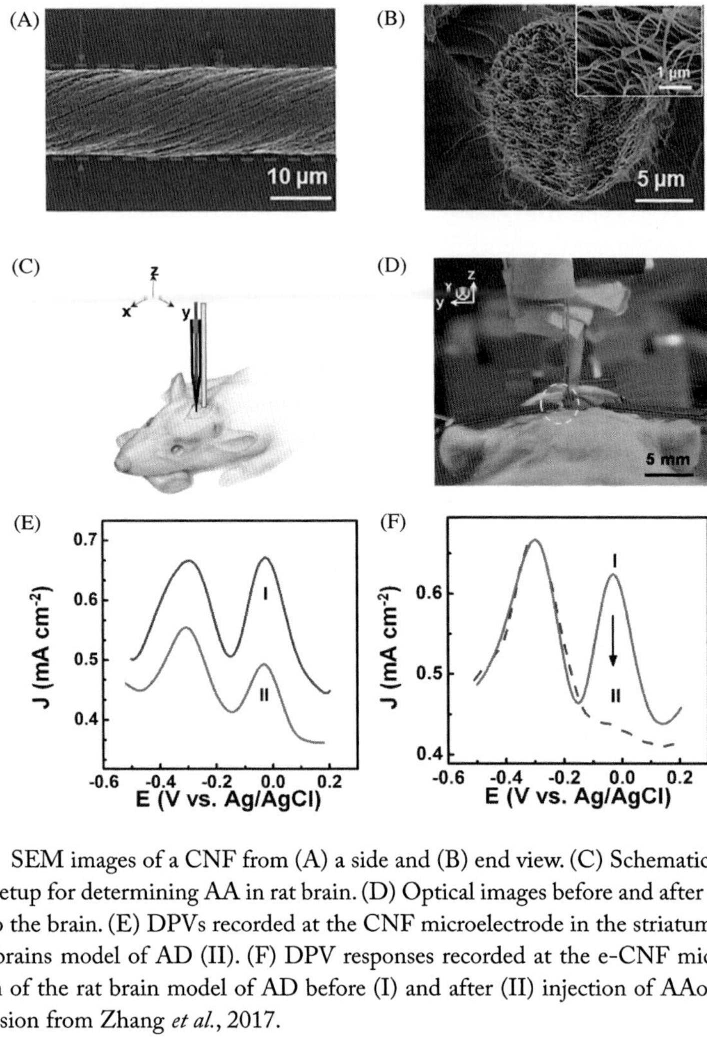

Figure 8.6 SEM images of a CNF from (A) a side and (B) end view. (C) Schematic illustration of the *in vivo* setup for determining AA in rat brain. (D) Optical images before and after the stereotaxic implant into the brain. (E) DPVs recorded at the CNF microelectrode in the striatum of normal rat (I) and rat brains model of AD (II). (F) DPV responses recorded at the e-CNF microelectrode in the striatum of the rat brain model of AD before (I) and after (II) injection of AAox. Reproduced with permission from Zhang *et al.*, 2017.

Recently, great effort has been devoted to the development of ratiometric electrochemical sensors. We have found that a novel CNT fiber (CNF) greatly facilitates AA oxidation at –50 mV *versus* Ag/AgCl with enhanced current intensity by engineering tunable defects and oxygen-containing species in CNTs (Figures 8.6A and 8.6B) (Zhang *et al.*, 2017). The low potential enables determination of AA levels with high selectivity. More interestingly, a separate peak was observed unexpectedly at –0.27 V after the CNF-based microelectrode was treated via electrochemical activation, providing a built-in correction to avoid complex environmental effects in the brain (Figures 8.6C and 8.6D). The ratiometric microelectrode was successfully applied to detect AA in normal rat brain and in a rat brain model of AD. AA levels were estimated to be 259 ± 6 μM in the cortex,

264 ± 20 μM in the striatum, and 261 ± 21 μM in the hippocampus of normal rats. However, AA levels decreased to 210 ± 30 μM in the cortex, 182 ± 5 μM the striatum, and 136 ± 20 μM in the hippocampus of AD rats (Figures 8.6E and 8.6F). This work is the first to accurately detect AA concentrations in the brains of a live animal model of AD. Similarly, Cheng *et al.* (2015) demonstrated a rationally designed ratiometric electrochemical biosensor for *in vivo* selective and reliable detection of cerebral AA taking carbon black as the electrode substrate due to its good conductivity, high surface-to-area ratio, and low cost. Carbon black was able to readily facilitate the oxidation of AA at a relatively negative potential (*ca.* –0.14 V) without particular physical or chemical pretreatment, forming the basis of selective measurement of AA. The designed sensor has been successfully applied to *in vivo* monitor the dynamic change of cerebral AA associated with pathological processes.

8.4 A potentiometric biosensor with an internal reference for determination of pH

The local pH of the brain microenvironment plays a significant role in signal transduction, enzymatic activity, and ion transport (Cardone *et al.*, 2005; Webb *et al.*, 2011). The production of ROS leads to a decrease in pH, which is implicated in many chronic degenerative diseases, such as AD and ischemia (Mutch and Hanson, 1984). The most popular pH determination strategy is based on the Nernst equation, which yields a well-defined linear correlation between the potential and the pH using ion-sensitive microelectrodes (ISMEs) (Zigmond and Harvey, 1970; Ammann *et al.*, 1981; Makos *et al.*, 2010). Unfortunately, ISMEs made from glass are difficult to miniaturize for measurements in the brain. More importantly, as explained above, the solution resistance in cerebrospinal fluid can be as high as 3000–5000 Ω (May *et al.*, 1998), which creates the possibility of problems with IR drop. An IR drop of 3–5 mV is comparable to the change in potential induced by a pH change of 0.1 pH unit for an electrochemical process with a linear variation of 59 mV/pH unit. However, a pH variation as small as 0.2–0.5 pH units in living organisms has been attributed to a variety of severe cellular problems and diseases. Thus, pH measurements in the brain that rely on potential variations without built-in corrections lead to determination errors of ~20-50%. Hence, real-time determinations of brain pH with high selectivity and accuracy remain challenging.

To address this challenge, we recently developed a two-channel electrochemical pH biosensor (Figure 8.7) with an internal reference for the real-time monitoring of pH in different regions of rat brains upon ischemia (Zhao *et al.*, 2016). In one channel, Fc-Py was an electrochemical probe for the selective recognition of

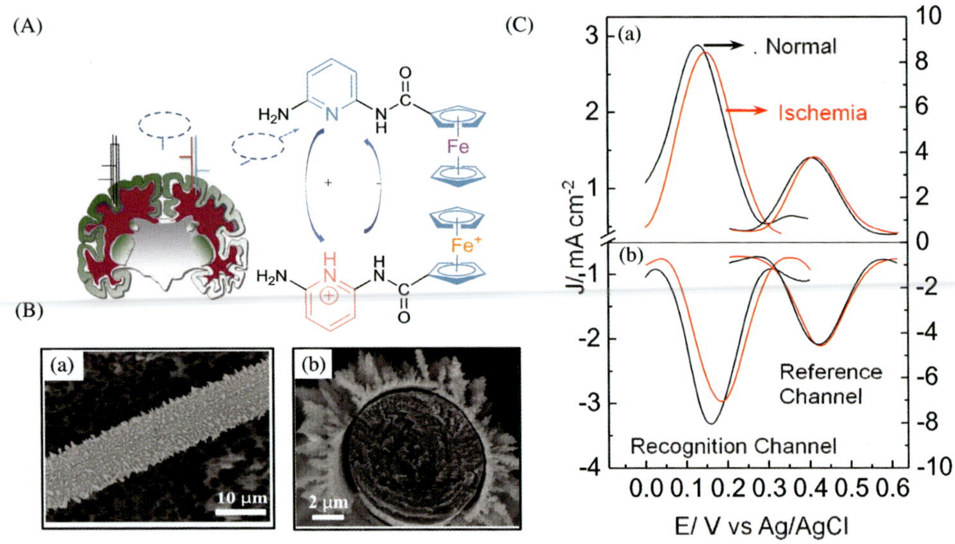

Figure 8.7 (A) Illustration of the two–channel electrochemical biosensor for *in vivo* determination of pH in a live rat brain. (B) SEM images of CFME/Au: (a) top view and (b) cross section. (C) DPV responses for (a) oxidation and (b) reduction processes obtained at the two-channel ratiometric biosensor (Recognition channel: N-(6-aminopyridin-2-yl)ferrocene (Fc-Py)-modified CFME/Au; Response channel: 6–(ferrocenyl) hexanethiol (FcHT)-modified CFME/Au. Reproduced with permission from Zhao *et al.*, 2016.

pH, in which the ferrocene group is electrochemically redox active and the pyridine group responds to protons. On the other channel, FcHT with a well-separated redox potential, inert to pH, was optimized as an internal reference to improve accuracy. The pH calibration curve was obtained from the half-wave potential ($E_{1/2}$) difference between the response and the reference signal ($\Delta E_{1/2}$). The $\Delta E_{1/2}$ value is less dependent on the potential of the outer reference electrode and the potential shift caused by the IR drop, thus greatly improving the accuracy of the pH determination *in vivo*.

This two-channel biosensor was directly implanted into brain regions including the striatum, the hippocampus, and the cortex and was used successfully for the real-time monitoring of pH values after global cerebral ischemia. However, the sensitivity of this two-channel pH sensor is 0.13 pH units. A more sensitive pH sensor is necessary because the pH fluctuations in living systems are generally smaller. Thus, we developed a novel pH biosensor (Figure 8.8) by co-immobilizing the specific recognition agent, 1,2-naphthoquinone (1,2-NQ), which has good electrochemical redox behavior, and FcHT at one electrode (CFME/Au/1,2-NQ+FcHT) (Zhou *et al.*, 2016). This design established a $2H^+/2e^-$ approach over a wide pH range

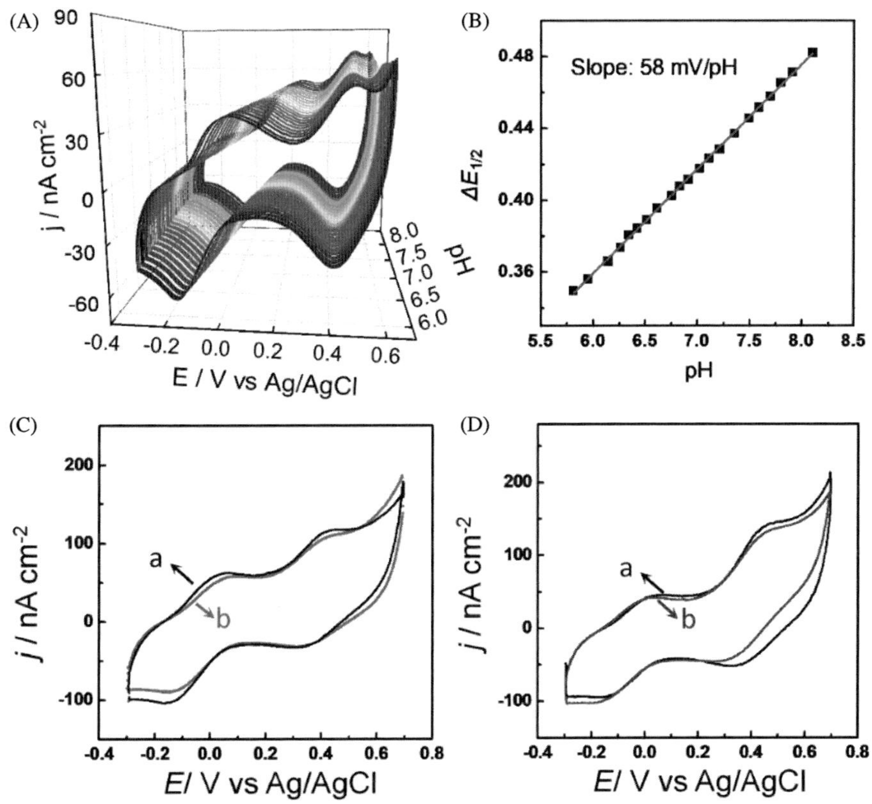

Figure 8.8 (A) CVs obtained at CFME/Au/1,2-NQ+FcHT in 10 mM of PBS with pH changes from 5.8 to 8.0. (B) The calibration plot of the half-wave potential difference between the 1,2-NQ and FcTH redox peaks ($\Delta E_{1/2}$) versus pH. (C) CVs obtained at a ratiometric CFME/Au/1,2-NQ+FcHT electrode for the determination of pH in the striatum of (a) a normal rat brain, and (b) a rat brain followed by cerebral ischemia for 5 min, and (D) (a) a rat brain upon ischemia for 5 min, and (b) further micro-injection of a 100 μM Na_2CO_3 solution. Reproduced with permission from Zhou *et al.*, 2016.

from 5.8 to 8.0 while decreasing the limit of detection to 0.07 pH units (Figures 8.8A and 8.8B). The *in vivo* results demonstrate that pH values are 7.21 ± 0.05 in the striatum, 7.13 ± 0.09 in the hippocampus, and 7.27 ± 0.06 in the cortex of the normal rat brains. However, pH decreased to 6.75 ± 0.07 and 6.52 ± 0.03 in the striatum and the hippocampus, respectively, upon global cerebral ischemia, with a negligible pH change in the cortex (Figures 8.8C and 8.8D). These results indicate that ischemia leads to variations in the acidity of the brain microenvironment. To the best of our knowledge, this is the first report of accurate pH values in the different regions of normal rat brains and those after global cerebral ischemia.

8.5 Dual signal outputs for *in vivo* analysis of two species involved in oxidative stress in the brain

Both O_2 and pH are critically important species correlated with brain ischemia. During ischemia, when O_2 supply is limited, the electron transport chain of the inner mitochondrial membrane is extensively reduced. This reduced state may result in the production of ROS. Then, when respiration is inhibited but glycolysis persists, protons and lactate may be generated, leading to a decrease in pH (Cheer *et al.*, 2006; Ariansen *et al.*, 2012). On the other hand, both O_2 and pH are vital for cancer growth and the lack of O_2 can lead to hypoxia-induced cell death (Venton *et al.*, 2003). However, an analytical method is still the bottleneck for understanding the mechanism of ischemic brain damage. Over the past few decades, many elegant approaches have been developed for monitoring either O_2 or pH (Xiang *et al.*, 2014; Roussakis *et al.*, 2015; Zhao *et al.*, 2016; Zhou *et al.*, 2016). However, the simultaneous recognition and accurate determination of O_2 and pH in the live brain remains challenging.

To accomplish the simultaneous determination of O_2 and pH, an organic molecule, Hemin–aminoferrocene (Hemin–Fc) was designed and synthesized (Figure 8.9A) (Liu *et al.*, 2017). Hemin–Fc has three functions. The cathodic current of the Hemin group was employed as the specific response to determine

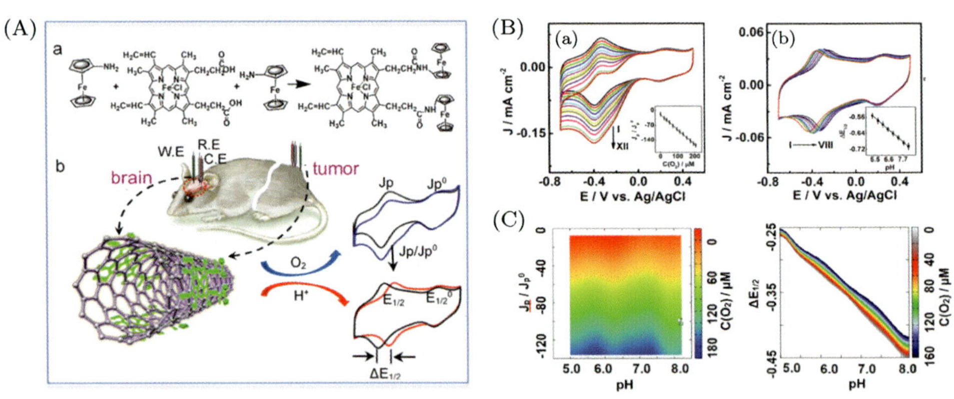

Figure 8.9 (A) (a) Scheme for the formation of the conjugate Hemin–Fc from Hemin and Fc. (b) Illustration of the working mechanism of the present biosensor for the simultaneous determination of O_2 and pH. (B) CVs obtained with a Hemin-Fc-modified carbon nanotube fiber (Hemin–Fc/CNF) microelectrode (a) in 0.1 M PBS bubbled with different concentrations of pure O_2 and (b) different pH values. (C) (a) Relationship among the reduction peak current density ratio of Hemin and Fc ($J_p/J_p{}^0$), the concentration of O_2, and the pH; (b) Relationship among the half-wave potential difference between the Hemin and Fc redox peaks ($\Delta E_{1/2}$), the concentration of O_2, and the pH. Reproduced with permission from Liu *et al.*, 2017.

the O_2 concentration, taking advantage of the catalytic activity of Hemin for the four-electron reduction of O_2. Meanwhile, the $E_{1/2}$ of the redox peaks was used to determine the pH, since the redox process of the porphyrin ring is pH-dependent. In addition, the Fc group is inert to both O_2 and pH and so provides a built-in correction for the quantification of O_2 and pH. To improve the stability and sensitivity of the electrode, a carbon nanotube fiber was fabricated as a stable and biocompatible substrate to immobilize Hemin–Fc via π–π stacking. Notably, cubic spline interpolation was used to smooth the experimental measurements. Relative images were employed to quantify the levels of O_2 and pH in live systems (Figures 8.9B and 8.9C). This straightforward bifunctional biosensor was first used for the simultaneous determination of O_2 and pH in a live brain with two ischemic animal models: one with middle cerebral artery occlusion (MCAO) and the other with carotid artery ligation (CAL). The pH value was estimated to decrease by 0.34 units and 0.47 units in the cortex and the striatum of mouse brain, respectively, after 1.5 h of MCAO. The initial concentration of O_2 was estimated to be 61.0 ± 6.8 and 51.0 ± 5.9 μM in the cortex and the striatum, respectively, and gradually decreased to 17.0 ± 2.8 and 11.0 ± 2.5 μM, respectively, 1.5 h after MCAO. After MCAO for 2 h, the longer time ischemia induced neuronal death. The present study provided direct evidence that the severity of neuronal death strongly depends on the increasing duration of brain acidity. From the results of the CAL ischemic model, we can see that the pH values in the striatum and the hippocampus remarkably decreased to 6.72 ± 0.03 and 6.58 ± 0.04, respectively, after a short ischemia time of 21 min.

This methodology, employing dual signal outputs, has provided design criteria for artificial organic molecules and has opened a route to developing biosensors for determination of two targets in the live brain. Then, further employed glucose oxidase as a specific recognition element for the simultaneous quantification of glucose and pH in a diabetic rat brain, because the active center (FAD) undergoes a $2H^+/2e^-$ process (Li *et al.*, 2017). The experimental data demonstrate that the basal pH level decreased to 6.9 ± 0.1 and 7.1 ± 0.1 in the striatum and the cortex of rat brains of a diabetic model, respectively. In contrast, glucose concentrations were 2.22 ± 0.18 mM and 1.44 ± 0.12 mM in the striatum and the cortex of normal rat brains and increased to 4.53 ± 0.40 mM and 6.65 ± 0.31 mM, respectively, in the rat brains of the diabetic model. Detailed information about the levels of pH and glucose in these rat brains may help us understand damage to brain function in diabetics. The simplicity of operation and instrumentation of the biosensor should promote its use in a broad range of biochemical applications.

Hydrogen peroxide (H_2O_2) is an important, membrane permeable ROS in the brain that is most often regarded as potentially toxic because it forms highly reactive hydroxyl radicals that can irreversibly alter DNA, lipids, and proteins (Halliwell, 2006). Thus, an imbalance between H_2O_2 generation and metabolism could result in pathological oxidative stress and neuronal degeneration in disease and aging. The production of H_2O_2 and other reactive oxygen species continues to be implicated in the etiology of neurodegenerative disorders, such as Parkinson's disease (Smythies, 2000; Kulagina and Michael, 2003; Rhee, 2006; Kishida and Klann, 2007). Moreover, emerging evidence suggests that H_2O_2, much like nitric oxide, plays an important role as a diffusible neuromodulator in various aspects of brain function. The electrochemical measurement of H_2O_2 is of widespread interest to bioanalytical scientists because it is often enzymatically generated at biosensors to serve as an electroactive reporter molecule in the indirect detection of non-electroactive biomolecules such as glucose, glutamate, and acetylcholine. At most frequently used electrodes, the electron-transfer kinetics of H_2O_2 are rather sluggish. To this end, enzymes are often attached to the carbon fiber surface to facilitate the detection of compounds with poor electroactivity. Use of horseradish peroxidase (HRP) as a biocatalyst for H_2O_2 serves as effective protocols for H_2O_2 biosensing through either direct or mediated electron transfer of HRP (Cooper, 2011; Webb *et al.*, 2011). HRP enables biosensors to operate at a relatively low potential (–0.3 V *versus* Ag/AgCl) so they normally suffer from interference from O_2. In contrast, Cyt *c* is an excellent choice for H_2O_2 sensing since Cyt *c* does not react with O_2. On the basis of the direct electron transfer of Cyt *c*, the developed biosensor for H_2O_2 has demonstrated high selectivity, good sensitivity, and wide detection linear range. Generally, the immobilization of Cyt *c* in designed biocompatible structures can significantly facilitate its electron transfer and improve the performance of biocatalytic processes. By confining Cyt *c* in hydrogels, we achieved direct electron transfer of Cyt *c* with a formal potential of 100.7 ± 3.2 mV *versus* Ag/AgCl. More interestingly, the 3D scaffold of hydrogels provides the immobilization of Cyt *c* with long-term stability of up to two weeks and good reproducibility. However, these protocols are as-yet difficult to apply to *in vivo* measurement. Based on the strategy with dual-signal outputs, we recently developed a ratiometric biosensor with high sensitivity and selectivity for detecting of H_2O_2 and pH in microdialysates from a rat brain after global cerebral ischemia (Li and Tian, 2018). Catalase is capable of converting H_2O_2 into H_2O through an electrocatalytic reaction with one electron and one proton transfer. The basal level of H_2O_2 was estimated to be 1.71 ± 0.22 μM in striatum, 2.12 ± 0.14 μM in hippocampus, 1.84 ± 0.13 μM in cortex. H_2O_2 concentrations changed to 10.03 ± 1.40 μM, 13.53 ± 1.65 μM, and 9.65 ± 0.98 μM in the rat brain after global cerebral ischemia.

8.6 Simultaneous detection of multiple metal ions in the brain based on open-circuit potential

Electrophysiological techniques are a powerful way to study various aspects of brain function, such as the electrical activity of neurons, synaptic plasticity, cranial nerve circuit loops, and so on (Hefti and Felix, 1983; Liu *et al.*, 2013; Jackowska and Krysinski, 2013; Wei *et al.*, 2015). Electrophysiologists have won the Nobel Prize in Physiology and Medicine many times. However, electrophysiological technology records the electrical signals of neuron-to-neuron communication, whereas the generation of those electrical signals depends on changes of neurotransmitters and ions. Therefore, to fully understand brain physiological and pathological processes, it is essential to monitor the variations of chemical substances in the brain as well. However, a key disadvantage of electrochemical techniques is the possibility that either the applied potentials or the electrochemically generated currents at implanted microelectrodes may unexpectedly influence the function of adjacent neurons, resulting in false electrophysiological responses. Consequently, approaches without the flow of electrochemical currents that could influence nearby neurons are highly desirable.

To this end, we recently created an open-circuit potential method to measure the equilibrium concentrations of four ions including K^+, Ca^{2+}, Na^+ and pH. The recognition molecules for ions have been designed and coated with PVC (polyvinyl chloride) to form ion-selective membranes. The generated membrane potential relies on the separation of charge. In this case, there is no current flow across brain tissue, so the possibility of electrical stimulation of that tissue is minimized. In addition, the membranes containing the recognition molecules have good selectivity for their target ions and the offer antifouling properties that act against protein adsorption. Since there is almost no current during the potentiometric analysis, electrochemical measurements were combined with electrophysiology to obtain both chemical and electrical signals of neuron-to-neuron communications. An inner-reference electrode has been also constructed to provide a built-in response correction method. In order to achieve electrochemical brain imaging and simultaneous analysis of $K^+/Ca^{2+}/Na^+$/pH, an ion-selective microelectrode array (ISMEA) was fabricated. A 5-channel ISMEA was developed to monitor $K^+/Ca^{2+}/Na^+$/pH at the same time and electrical signals simultaneously in normal rat brain and rat brain upon ischemia. Due to the high antifouling property and long-term stability, this biosensor was successfully employed for determination of multiple ions in free-moving rats under seizure status. The experimental results have provided reliable and direct observations leading to deeper understanding of the mechanism of seizure, which is very beneficial for further therapy.

8.7 Conclusion and future prospects

The development of *in vivo* monitoring strategies for reactive oxygen species and related molecules involved in oxidative stress in brain is challenging work but required for progress in understanding the roles that biomolecules play in pathological and physiological processes. Rationally tailoring functional surfaces, as well as designing specific molecules, has greatly improved the sensitivity and selectivity of biosensors for distinguishing between biomolecules and fulfilling the demanding requirements for *in vivo* measurements. The electrochemical ratiometric strategy has established fascinating approaches for *in vivo* quantification of Cu^{2+}, pH, O_2, and glucose with high accuracy. The development of biosensors with remarkable analytical performance has produced accurate and reliable information regarding multiple species in live brains, individually and in combination with each other; this information should be very beneficial for evaluating brain processes at the molecular level. The design principles guiding the development of biosensors for biomolecules in the brain presented in this study could potentially to be extended to other targets of interest. Moreover, recent advances in molecular science, nanoscience, nanoengineering, and biological chemistry could potentially drive the development of additional multi-target biosensors. The possibility of designing and tailoring surfaces to construct implantable multi-functional nanodevices that promise to combine the benefits of multiple sensing capabilities and incorporate therapeutic modules might eventually be realized.

Acknowledgments

Support from the NSFC (21635003, and 21675054), and Supported by innovation program of Shanghai Municipal Education Commission (201701070005E00020).

References

Acuna A, Esparza M, Kramm C, Beltran F, Parra A, Cepeda C, Toro C, Vidal R, Hetz C, Concha I, Brauchi S, Levine M, Castro M (2013) A failure in energy metabolism and antioxidant uptake precede symptoms of Huntington's disease in mice. *Nat Commun*, 4: 1.

Ammann D, Lanter F, Steiner RA, Schulthess P, Shijo Y, Simon W (1981) Neutral carrier based hydrogen ion selective microelectrode for extra- and intracellular studies. *Anal Chem*, 53: 2267–2269.

Ariansen JL, Heien ML, Hermans A, Philips PE, Hernadi I, Bermudez M, Schultz W, Wightman RM (2012) Monitoring extracellular pH, oxygen, and dopamine during reward delivery in the striatum of primates. *Front Behav Neurosci*, 6: 36.

Atwood CS, Moir RD, Huang XD, Scarpa RC, Bacarra N, Romano DM, Hartshorn MK, Tanzi RE, Bush AI (1998) Dramatic aggregation of Alzheimer abeta by Cu(II) is induced by conditions representing physiological acidosis. *J Biol Chem*, 273: 12817–12826.

Auchere F, Rusnak F (2002) What is the ultimate fate of superoxide anion *in vivo*. *J Biol Inorg Chem*, 7: 664–667.

Bassareo V, Di Chiara G (1999) Differential responsiveness of dopamine transmission to food-stimuli in nucleus accumbens shell/core compartments. *Neuroscience*, 89: 637–641.

Cardone R, Casavola V, Reshkin S (2005) The role of disturbed pH, dynamics and the Na^+/H^+ exchanger in metastasis. *Nat Rev Cancer*, 5: 786–759.

Castro SL, Zigmond MJ (2001) Stress-induced increase in extracellular dopamine in striatum: Role of glutamatergic action via N-methyl-D-aspartate receptors in substantia nigra. *Brain Res*, 901: 47–54.

Chai X, Zhou X, Zhu A, Zhang L, Qin Y, Shi G, Tian, Y (2013) A two-channel ratiometric electrochemical biosensor for *in vivo* monitoring of copper ions in a rat brain using gold truncated octahedral microcages. *Angew Chem Int Ed*, 52: 8129–8133.

Chai X, Zhang L, Tian Y (2014) Ratiometric electrochemical sensor for selective monitoring of cadmium ions using biomolecular recognition. *Anal Chem*, 86: 10668–10673.

Cheer JF, Wassum KM, Wightman RM (2006) Cannabinoid modulation of electrically evoked pH and oxygen transients in the nucleus accumbens of awake rats. *J Neurochem*, 97: 1145–1154.

Cheng H, Wang X, Wei H (2015) A ratiometric electrochemical sensor for effective and reliable detection of ascorbic acid in living brains. *Anal Chem*, 87: 8889–8893.

Cooper GJS (2011) Therapeutic potential of copper chelation with triethylenetetramine in managing diabetes mellitus and Alzheimer's disease. *Drugs*, 71: 1281–1320.

Crespi F (2002) *In vivo* voltammetry and concomitant electrophysiology at a single micro-biosensor to analyse ischemia, depression and drug dependence. *J Neurosci Methods*, 119: 173–184.

Crespi F (2009) Anxiolytics antagonize yohimbine-induced central noradrenergic activity: A concomitant *in vivo* voltammetry-electrophysiology model of anxiety. *J Neurosci Methods*, 180: 97–105.

Crespi F, England T, Ratti E, Trist DG (1995) Carbon-fiber microelectrodes for concomitant *in-vivo* electrophysiological and voltammetric measurements-no reciprocal influences. *Neurosci Lett*, 188: 33–36.

Deng Z, Rui Q, Xia Y, Liu H, Tian Y (2008) *In vivo* detection of superoxide anion in bean sprout based on ZnO nanodisks with facilitated activity for direct electron transfer of superoxide dismutase. *Anal Chem*, 80: 5839–5846.

Dong H, Zhang L, Liu W, Tian Y (2017) Label-free electrochemical biosensors for monitoring of chloride ion in an animal model of Alzhemier's disease. *ACS Chem Neurosci*, 8: 339–346.

Dorner J, Miller BR, Klein EL, Murphy-Nakhnikian A, Andrew RL, Barton SJ, Rebec GV (2009) Corticostriatal dysfunction underlies diminished striatal ascorbate release in the R6/2 mouse model of Huntington's disease. *Brain Res*, 1290: 111.

Ewing AG, Alloway KD, Curtis MA, Dayton RM, Wightman RM, Rebec GV (1983) Simultaneous electrochemical and unit recording measurements: Characterization of the effects of d-amphetamine and ascorbic acid on neostriatal neurons. *Brain Res*, 261: 101–108.

Fijikawa Y, Morgan B, Dick TP. *Oxidative Stress and Redox Regulation*, Springer, 2013.

Frinkel T (2011) Signal transduction by reactive oxygen species. *J Cell Biol*, 194: 7–15

Giacovazzi R, Ciofini I, Rao L, Amatore C, Adamo C (2014) Copper-amyloid-β complex may catalyze peroxynitrite production in brain: Evidence from molecular modeling. *Phys Chem Chem Phys*, 16: 10169–10174.

Halliwell BJ (2006) Oxidative stress and neurodegeneration: Where are we now? *J Neurochem*, 97: 1634–1658.

Harrison F, May J (2010) Vitamin C function in the brain: Vital role of the ascorbate transporter (SVCT2). *Free Radical Bio Med*, 46: 719–730.

Hefti F, Felix D (1983) Chronoamperometry *in vivo*-does it interfere with spontaneous neuronal activity in the brain. *J Neurosci Methods*, 7: 151–156.

Hewitt N, Rauk A (2009) Mechanism of hydrogen peroxide production by copper-bound amyloid beta peptide: A theoretical study. *J Phys Chem B*, 113: 1202–1209.

Huang X, Atwood CS, Hartshorn MA, Multaup, G, Coldstein LE, Scarpa RC, Cuajungco MP, Gray DN, Lim J, Moir RD, Tanzi RE, Bush AI (1999) The A beta peptide of Alzheimer's disease directly produces hydrogen peroxide through metal ion reduction. *Biochemistry*, 38: 7609–7616.

Hyman SE, Malenka RC (2001) Addiction and the brain: The neurobiology of compulsion and its persistence. *Nat Rev Neurosci*, 2: 695–703.

Jackowska K, Krysinski P (2013) New trends in the electrochemical sensing of dopamine, *Anal Bioanal Chem*, 405: 3753–3771.

Khan AS, Michael AC (2003) Invasive consequences of using micro-electrodes and microdialysis probes in the brain. *Trends Anal Chem*, 22: 503–508.

Kishida KT, Klann E (2007) Sources and targets of reactive oxygen species in synaptic plasticity and memory. *Antioxid Redox Signaling*, 9: 233–244.

Kong B, Zhu A, Luo Y, Tian Y, Yu Y, Shi G (2011) Sensitive and selective colorimetric visualization of cerebral dopamine based on double molecular recognition. *Angew Chem Int Ed*, 50: 1837–1840.

Kontos H (2001) Oxygen radicals in cerebral ischemia. *Stroke*, 32: 2712–2716.

Kulagina NV, Michael AC (2003) Monitoring hydrogen peroxide in the extracellular space of the brain with amperometric microsensors. *Anal Chem*, 75: 4875– 4881.

Kuo T, McCreery RL (1999) Surface chemistry and electron-transfer kinetics of hydrogen-modified glassy carbon electrodes. *Anal Chem*, 71: 1553-1560.

Li S, Tian Y (2018) An electrochemical biosensor with dual signal outputs for ratiometric monitoring the levels of H_2O_2 and pH in the microdialysates from a rat brain. *Electroanalysis*, 30: 1–8.

Li S, Zhu A, Zhu T, Zhang J, Tian Y (2017) Single biosensor for simultaneous quantification of glucose and pH in a rat brain of diabetic model using both current and potential outputs. *Anal Chem*, 89: 6656–6662.

Li X, Liu Y, Zhu A, Luo Y, Deng Z, Tian Y (2010) Real-time electrochemical monitoring of cellular H_2O_2 integrated with *in situ* selective cultivation of living cells based on dual functional protein microarrays at Au–TiO_2 surfaces. *Anal Chem*, 82: 6512–6518.

Li Y, Zhang S, Wang L, Xiao R, Liu W, Zhang X, Zhou Z, Amatore A, Huang W (2014) Nanoelectrode for amperometric monitoring of individual vesicular exocytosis inside single synapses. *Angew Chem Int Ed*, 53: 12456–12460.

Liu C, Song N, Lin S, Zhou M, Wang X, Cai X (2013) Planar microelectrode chip for synchronous simulative neurochemical and neuroelectrial monitoring. *J Nanosci Nanotechnol*, 13: 736–740.

Liu H, Tian Y, Xia P (2008) Pyramidal, rodlike, spherical gold nanostructures for direct electron transfer of copper, zinc–superoxide dismutase: Application to superoxide anion biosensors. *Langmuir*, 24: 6359–6366.

Liu L, Zhao F, Liu W, Zhu T, Zhang J, Chen C, Dai Z, Peng H, Huang J, Hu Q, Bu W, Tian Y (2017) An electrochemical biosensor with dual signal outputs: Toward simultaneous quantification of pH and O_2 in brain upon ischemia and in tumor during cancer starvation therapy. *Angew Chem Int Ed*, 56: 10471–10475.

Liu W, Dong H, Zhang L, Tian Y (2017a) Development of an efficient biosensor for the *in vivo* monitoring of Cu and pH in the brain: Rational design and synthesis of recognition molecules. *Angew Chem Int Ed*, 56: 16328–16332.

Luo Y, Tian Y, Zhu A, Rui Q, Liu H (2009) Direct electron transfer of superoxide dismutase promoted by high conductive TiO_2 nanoneedles. *Electrochem Commun*, 11: 174–176.

Luo Y, Zhang L, Liu W, Yu Y, Tian Y (2015) A single biosensor for evaluating the levels of copper ion and L-cysteine in a live rat brain with Alzheimer's disease. *Angew Chem Int Ed*, 54: 14053–14056.

Makos MA, Omiatek DM, Ewing AG, Heien ML (2010) Development and characterization of a voltammetric carbon fiber microelectrode pH sensor. *Langmuir*, 26: 10386–10391.

May J, Qu Z, Mendiratta S (1998) Protection and recycling of alpha-tocopherol in human erythrocytes by intracellular ascorbic acid. *Arch Biochem Biophys*, 349: 281–289.

McCreery RL (2008) Advanced carbon electrode materials for molecular electrochemistry. *Chem Rev*, 108: 2646–2687

McMahon CP, Rocchitta G, Serra PA, Kirwan SM, Lowry JP, O'Neill RD (2006) Control of the oxygen dependence of an implantable polymer/enzyme composite biosensor for glutamate. *Anal Chem*, 78: 2352–2359.

Milton RD, Wang T, Knoche KL, Minteer SD (2016) Tailoring biointerfaces for electrocatalysis. *Langmuir*, 32: 2291–2301.

Mutch WAC, Hansen AJ (1984) Extracellular pH change during spreading depression and cerebral ischemia: Mechanisms of brain pH regulation. *J Cereb Blood Flow Metab*, 4: 17–27.

Phillips PE, Stuber GD, Heien ML, Wightman RM, Carelli RM (2003) Subsecond dopamine release promotes cocaine seeking. *Nature*, 422: 573–574.

Rahman MA, Kothalam A, Choe ES, Won M, Shim Y (2012) Stability and sensitivity enhanced electrochemical *in vivo* superoxide microbiosensor based on covalently co-immobilized lipid and cytochrome *c*. *Anal Chem*, 84: 6654–6660.

Rebec G, Barton S, Marseilles A, Collins K (2003) Ascorbate treatment attenuates the Huntington behavioral phenotype in mice. *Neuroreport*, 14: 1263–1265.

Rhee SG (2006) Cell signaling. H_2O_2, a necessary evil for cell signaling. *Science*, 312: 1882–1883.

Rice ME (2000) Ascorbate regulation and its neuroprotective role in the brain. *Trends Neurosci*, 23: 209–216.

Roberts JG, Toups JV, Eyualem E, McCarty GS, Sombers LA (2013) *In situ* electrode calibration strategy for voltammetric measurements *in vivo*. *Anal Chem* 85: 11568–11575.

Robinson DL, Hermans A, Seipel AT, Wightman RM (2008) Monitoring rapid chemical communication in the brain. *Chem Rev*, 108: 2554–2584.

Rodeberg NT, Johnson JA, Cameron CM, Saddoris MP, Carelli RM, Wightman RM (2015) Construction of training sets for valid calibration of *in vivo* cyclic voltammetric data by principal component analysis. *Anal Chem*, 87: 11484–11491.

Roussakis E, Li X, Nichols AJ, Evans CL (2015) Oxygen-sensing methods in biomedicine from the macroscale to the microscale. *Angew Chem Int Ed*, 86: 5017–5023.

Shao X, Gu H, Wang Z, Chai X, Tian Y, Shi G (2013) Highly selective electrochemical strategy for monitoring of cerebral Cu^{2+} based on a carbon dot–TPEA hybridized surface. *Anal Chem*, 85: 418–425.

Smythies J (2000) Redox aspects of signaling by catecholamines and their metabolites. *Antioxid Redox Signaling*, 2: 575–583.

Taniguchi I, Toyosawa K, Yamaguchi H, Yasukouchi K (1982) Voltammetric response for horse heart cytochrome *c* at a gold electrode in the presence of sulfur bridged bipyridines. *J Electroanal Chem Interfacial Electrochem*, 140: 187–193.

Tian Y, Mao L, Okajima T, Ohsaka T (2002) Superoxide dismutase-based third-generation biosensor for superoxide anion. *Anal Chem*, 74: 2428–2434.

Tian Y, Mao L, Okajima T, Ohsaka T (2004) Electrochemistry and electrocatalytic activities of superoxide dismutases at gold electrodes modified with a self-assembled monolayer. *Anal Chem*, 76: 4162–4168.

Venton BJ, Michael DJ, Wightman RM (2003) Correlation of local changes in extracellular oxygen and pH that accompany dopaminergic terminal activity in the rat caudate-putamen. *J Neurochem*, 84, 373–381.

Venton BJ, Wightman RM (2003) Psychoanalytical electrochemistry: Dopamine and behavior. *Anal Chem*, 75, 414A–421A.

Wang Z, Liu D, Gu H, Zhu A, Tian Y, Shi G (2013) NTA-modified carbon electrode as a general relaying substrate to facilitate electron transfer of SOD: Application to *in vivo* monitoring of $O_2^{\bullet-}$ in a rat brain. *Biosens Bioelectron*, 43: 101–107.

Webb B, Chimenti M, Jacobson M, Barber D (2011) Dysregulated pH: A perfect storm for cancer progression. *Nat Rev Cancer*, 11: 671–677.

Wei W, Song Y, Wang L, Zhang S, Luo J, Xu S, Cai X (2015) An implantable microelectrode array for simultaneous L-glutamate and electrophysiological recordings *in vivo*. *Microsyst Nanoeng*, 1: 15002.

Wilson GS, Gifford R (2005) Biosensors for real-time *in vivo* measurement. *Biosens Bioelectron*, 20: 2388–2403.

Wilson GS, Hu Y (2000) Enzyme-based biosensors for *in vivo* measurements. *Chem Rev*, 100: 2693–2704.

Wilson GS, Johnson MA (2008) *In vivo* electrochemistry: What can we learn about living systems? *Chem Rev*, 108: 2462–2481.

Xiang L, Yu P, Hao J, Zhang M, Zhu L, Dai L (2014) Vertically aligned carbon nanotube-sheathed carbon fibers as pristine microelectrodes for selective monitoring of ascorbate *in vivo*. *Anal Chem*, 86: 3909.

Xiang L, Yu P, Zhang M, Hao J, Wang Y, Zhu L, Dai L, Mao L (2014) Platinized aligned carbon nanotube-sheathed carbon fiber microelectrodes for *in vivo* amperometric monitoring of oxygen. *Anal Chem*, 86: 5017–5023.

Yu P, He X, Mao L (2015) Tuning interionic interaction for highly selective *in vivo* analysis. *Chem Soc Rev*, 44: 5959–5968.

Zhang L, Han Y, Zhao F, Shi G, Tian Y (2015) A selective and accurate ratiometric electrochemical biosensor for monitoring of Cu^{2+} ions in a rat brain. *Anal Chem*, 87: 2931–2936.

Zhang L, Liu F, Sun X, Wei G, Tian Y, Liu Z, Huang R, Yu Y, Peng H (2017) Engineering carbon nanotube fiber for real-time quantification of ascorbic acid levels in a live rat model of Alzheimer's disease. *Anal Chem*, 89: 1831–1837.

Zhang M, Liu K, Gong K, Su L, Chen Y, Mao L (2005) Ascorbate depletion in the rat striatum induced by global ischemia with carbon nanotube-modified glassy carbon electrode integrated into a thin-layer radial flow cell. *Anal Chem*, 77: 6234–6242.

Zhang M, Liu K, Xiang L, Lin Y, Su L, Mao L (2007) Carbon nanotube-modified carbon fiber microelectrodes for *in vivo* voltammetric measurement of ascorbic acid in rat brain. *Anal Chem*, 79: 6559–6565.

Zhang M, Yu P, Mao L (2012) Rational design of surface/interface chemistry for quantitative *in vivo* monitoring of brain chemistry. *Acc Chem Res*, 45: 533–543.

Zhao F, Zhang L, Zhu A, Shi G, Tian Y (2016) *In vivo* monitoring of local pH values in a live rat brain based on design and synthesis of specific electroactive molecule for H^+. *Chem Commun*, 52: 3717–3720.

Zhou J, Liao C, Zhang L, Wang Q, Tian Y (2014) Molecular hydrogel-stabilized enzyme with facilitate electron transfer for determination of H_2O_2 released from live cells. *Anal Chem*, 86: 4395–4401.

Zhou J, Zhang L, Tian Y (2016) Micro electrochemical pH sensor applicable for real-time ratiometric monitoring of pH values in rat brains. *Anal Chem*, 88: 2113–2118.

Zigmond MJ, Harvey JA (1970) Resistance to central norepinephrine depletion and decreased mortality in rats chronically exposed to electric foot shock. *J Neuro-Visceral Relations*, 31: 373.

CHAPTER 9

REAL TIME STUDIES OF ACETYLCHOLINE RELEASE FROM SINGLE SYNAPSES AND SINGLE CELLS WITH NANOMETER SPATIAL RESOLUTION

Mei Shen* and Ran Chen

Department of Chemistry,
University of Illinois at Urbana-Champaign, USA
*mshen233@illinois.edu

9.1 Introduction

Acetylcholine (ACh), the first neurotransmitter identified in 1914 (Dale, 1914; Loewi, 1921), is involved in a broad range of emotional responses and human behaviors, such as attention, learning, and memory. Defects in ACh transmission dynamics are implicated in a variety of diseases and disorders, such as neurodegeneration, diabetes, and drug abuse. Despite the biological and biomedical significance of cholinergic transmission, it remains challenging to measure ACh release dynamics in real time, quantitatively, at single synapses and single cells. Recent progress in our hands has addressed this challenge (Shen *et al.*, 2018; Well *et al.*, 2018). In this chapter, we present recent advances in the development of nanoITIES electrodes (where ITIES means **I**nterface between **T**wo **I**mmiscible **E**lectrolyte **S**olutions) for the quantification of ACh transmission in real time. We also present the experimental findings that elucidate ACh transmission dynamics with nanometer spatial resolution at single synapses and single cells.

9.2 Background in acetylcholine sensing from neuronal models

Prior to 2018, existing methods to detect ACh in neuronal systems included radiochemical assays (Johnson and Pilar, 1980), outside-out patch membrane probes (Sun and Poo, 1987), and micrometer sized enzyme-based sensors (Garguilo and Michael, 1995; Xin and Wightman, 1997; Mitchell, 2004; Bruno *et al.*, 2006; Burmeister *et al.*, 2008; Giuliano *et al.*, 2008; Wilson and Johnson, 2008; Mattinson *et al.*, 2011; Asri *et al.*, 2016). The enzymatic sensors were used for measuring ACh in the intact brain and brain slices of rats and mice but they have not yet been used to monitor ACh release from single cells. ACh release from the Leghorn chick ganglion has been studied with a radiochemical assay, where radioactive labeling and conversion of ACh to choline-$^{32}PO_4$ is the basis for the detection of ACh. This assay, which is not non-invasive, enabled the study of ACh release from ~3,000 cells. Outside-out patch probes of ~2 μm were employed to study ACh release from cultured xenopus neurons but without quantitative information. Additionally, Raman microscopy has been used to measure the concentration of cholinergic transmitter averaged from multiple frog neuromuscular junctions while no exocytosis dynamics were reported (Fu *et al.*, 2017). Two photon microscopy was also used to measure ACh through the insertion of green fluorescent protein (GFP) into muscarinic receptors, while the authors showed that the fluorescence signal is dependent upon the added ACh standard solution, concentration information of the ACh released during exocytosis / neurotransmission was not reported (Jing *et al.*, 2015).

None of the methods just mentioned have been demonstrated for interrogating ACh signaling at single synapses. Since synapses are less than 100 nm wide (Pickel *et al.*, 1981; Groves *et al.*, 1994; Moukhles *et al.*, 1997; McCann *et al.*, 2003; Brossard *et al.*, 2005; Dani *et al.*, 2010), it is desirable to create nanoprobes capable of probing chemical transmission with nanometer spatial resolution. The next section describes the recent developments of a different type of electrode, the nanoITIES electrode, for the detection and quantification of ACh signaling at single cells and single synapses. Critically, nanoITIES electrodes enable the measurements of both concentration and dynamics of cholinergic transmission in real time with high signal to noise ratio up to 130, nm spatial resolution and high temporal resolution.

9.3 NanoITIES electrodes: An emerging platform for acetylcholine detection and quantification at single cells and synapses

To address the pressing challenges in ACh detection, we introduced nanoITIES electrodes (Columbo *et al.*, 2015) to enable quantitative and dynamic measurements of ACh release from living neurons (Shen *et al.*, 2018; Well *et al.*, 2018) using *in vivo* amperometry. ACh sensing at nanoITIES electrodes is based on the ion transfer across the polarized / electrified nanointerface. This interface is formed between an organic electrolyte filling solution contained inside a nanopipet and an outside aqueous medium. The potential-driven transfer of ions across this interface generates a current. The theory for chemical sensing with ITIES electrodes is well established and the applied potential $\Delta_0^w\phi$, follows the Nernst equation:

$$\Delta_0^w\phi = \Delta_0^w\phi_i^0 + \frac{RT}{Z_iF}\ln\left(\frac{a_i^0}{a_i^w}\right)$$

where $\Delta_0^w\phi^0$ is the standard ion transfer potential, R is the gas constant, T is the temperature, z_i is the charge of the ion (i), F is the Faraday constant, a_i^0 and a_i^w are the activity of the ionic analyte in the organic and aqueous phase, respectively.

The standard ion transfer potential is related to the Gibbs free energy of transfer, which is a function of the structure and hydrophilicity of each analyte (Samec *et al.*, 1977; Taylor and Girault, 1986; Shao and Mirkin, 1998; Reymond *et al.*, 2000; Rossier *et al.*, 2000; Qian *et al.*, 2001; Qian *et al.*, 2004; Berdduque *et al.*, 2005; Zhan *et al.*, 2007; Vanysek and Ramirez, 2008; Liu *et al.*, 2011; Stockmann *et al.*, 2012; Vallejo *et al.*, 2012; Nestor *et al.*, 2014; Amemiya *et al.*, 2013; Colombo and Shen, 2015), thus analytes can be identified using ITIES electrodes. The mass transfer limiting current measured at nanoITIES electrodes is a function of the concentration of the analyte (Lefrou, 2006), which can thus be used for the quantification of the neurotransmitters. We have demonstrated both the detection and quantification of ACh, dopamine, GABA, tryptamine, and serotonin (Colombo *et al.*, 2015, 2016; Iwai, *et al.*, 2018), where the current increased linearly with the concentration of each neurotransmitter. We measured the current-time trace at the steady state detection potential selective for cholinergic transmitter detection against other substances that have been identified or suggested to be released from our neuronal model, the *Aplysia* neurons and their vesicles (Section 9.4.4), to learn about its release dynamics and concentration profiles in real time. This steady state detection potential we applied is E_{ACh} = −0.48 V with respect to the half wave transfer potential of a standard compound, tetrabutyl ammonium (TBA), $E_{1/2,\mathrm{TBA}}$.

We describe in the next section the fabrication, characterization of the nanoITIES electrodes for ACh detection.

9.4 Fabrication and characterization of the nanoITIES electrodes

The fabrication and characterization of the nanoelectrodes are much more challenging than macro- or microelectrodes. For instance, nanoelectrodes cannot be visualized using the optical microscope and need the high resolution of an electron microscope. We present below the experimental details in the fabrication, characterization of the nanoITIES electrodes, as well as strategies to ensure the formation of stable ITIES interfaces, and the selectivity of nanoITIES electrodes towards acetylcholine (ACh) detection. Additional details can be found from the publications in our group (Colombo *et al.*, 2015, 2016; Shen and Colombo, 2015; Iwai *et al.*, 2018; Shen *et al.*, 2018; Welle *et al.*, 2018).

9.4.1 Fabrication of the nanopipet tip

Nanopipets were fabricated by pulling a quartz capillary (O.D. 1 mm, I.D. 0.7 mm, Sutter Instrument, Novato, CA) using a P-2000 laser puller (Sutter Instrument, Novato, CA), where a combination of five pulling parameters is required to generate the right size. For instance, in interrogating synaptic chemical transmission dynamics, the nanoITIES pipet electrodes with radii around 15 nm were produced with the following pulling parameters: Heat = 730, Filament = 4, Velocity = 55, Delay = 120, Pull = 95. These pulling parameters may vary depending on the condition of the puller, capillary types, etc.

9.4.2 Surface treatment of the glass surface to ensure the formation of stable interface

It is important to make sure that a stable interface forms at the orifice of the nanopipet. To accomplish this, it is essential to convert the glass surface from hydrophilic into hydrophobic, thus the organic electrolyte filling solution is contained inside the nanopipet. This surface treatment process becomes very challenging at the nanoscale. The Amemiya group has developed pioneering procedures based on chemical vapor silanization for the surface treatment of the nanopipet, which was later adopted by other leading groups, i.e., the Mirkin group and our group. To modify the glass surface to be hydrophobic, the pulled nanopipets were placed inside a plastic desiccator, where a vacuum was created and the nanopipets were exposed to silane vapors. These chemical vapor silanization surface treatment process have been effective in producing nanoITIES electrodes with stable interfaces, where the oil phase is contained inside the pipet. The stable interfaces have been verified using electrochemical methods, including cyclic voltammetry and scanning electrochemical microscopy (SECM) approach curves (Shao and Mirkin, 1997; Ishimatsu *et al.*, 2010; Wang *et al.*, 2010; Kim *et al.*, 2012; Shen *et al.*, 2012; Chen *et al.*, 2017).

9.4.3 Characterization of the nanoITIES electrodes

We use the high-resolution electron microscope and electroanalytical methods to characterize the nanoITIES electrodes. We visualize the nanopipet using a

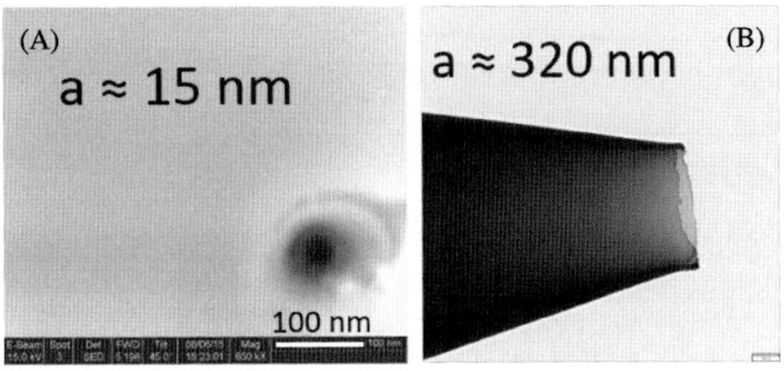

Figure 9.1 Visualization of the nanoITIES electrodes using SEM and TEM. (A) SEM picture of a nanopipet tip with a radius of around 15 nm used for the single synaptic quantification of cholinergic transmission. (Reprinted with permission from Shen *et al.*, 2018, *J Am Chem Soc*, 140: 7764–7768. Copyright (2018) American Chemical Society.) (B) TEM picture of a nanopipet tip with a radius around 320 nm for detecting gamma-aminobutyric acid in vitro. (Reprinted with permission from Iwai *et al.*, 2018, *Anal Chem*, 90: 3067–3072. Copyright (2018) American Chemical Society.)

scanning electron microscope (SEM). For SEM visualization, the nanopipets were coated with a thin Au/Pd film by a high-resolution sputter coater (Quorum Technologies LTD, Kent, UK) to avoid the surface charge, and the orifices were observed by high resolution field emission SEM (FEI dual-beam 235, FEI Co., Hillsboro OR, USA) under a 15/20 kV electron beam. We recently started to visualize the nanopipets using a transmission electron microscope (TEM), where the nanopipets were observed under a 120-kV electron beam (Philips CM200, FEI Co., Hillsboro OR) (Iowai *et al.*, 2018). SEM and TEM of the nanoITIES electrodes are shown in Figure 9.1. A nanoelectrode of ~15 nm in radius was used in the single synaptic cleft study.

9.4.4 Assembling of the nanoITIES electrodes and its selectivity

Nanopipets were backfilled with an organic electrolyte solution (volume: 2–5 μL) using a 10 μL Hamilton syringe. A typical organic phase is 1, 2-DCE (1,2 dichloroethane), nitrobenzene, etc. The filled organic electrolyte solution was brought to the tip end of the nanopipet by creating gentle vibrations. An etched Pt wire was inserted into the pipet and secured with wax. A Ag wire (250 μm diameter) coated with AgCl was placed in the aqueous phase as the outside reference electrode. A two-electrode configuration was used, where a potential was applied between the Pt wire inside the nanopipet and the outer Ag/AgCl reference electrode. Tetrabutyl ammonium chloride (TBACl) or tetraethylammonium chloride (TEACl) was added at the end of the experiments as the internal standard to verify the transfer potentials of the measured analyte. All electrochemical recordings were taken using CHI920D Scanning Electrochemical Microscope (SECM) (potentiostat rise time: < 1 μs, 0.8 μs typical) (CH Instruments, Austin, TX).

The selectivity of the analyte detected using the ITIES electrodes can be modulated by varying the composition of the organic electrolyte filling solution, such as the solvent type, added ionophore, etc. We performed *in vitro* control experiments to verify the selectivity of the nanoITIES pipet electrode for ACh detection against other substances that have been identified or suggested to be released from the neuronal model, pedal ganglion neurons of *Aplysia californica* and their vesicles. We carried out the selectivity study under the same conditions as *in situ* cholinergic transmission experiments, with the cell diagram shown in Cell 1. Under the conditions of Cell 1, gamma-amino butyric acid (GABA) (Figure 9.2B), glutamate (Figure 9.2C), dopamine (Figure 9.2D), serotonin (Figure 9.2E), pedal peptide, all amino acids, high concentration K^+ (Figure 9.2F) from the stimulating solution and H^+ change accompanying the exocytosis process (Figure 9.2G) do not generate an interfering current (Shen *et al.* 2018).

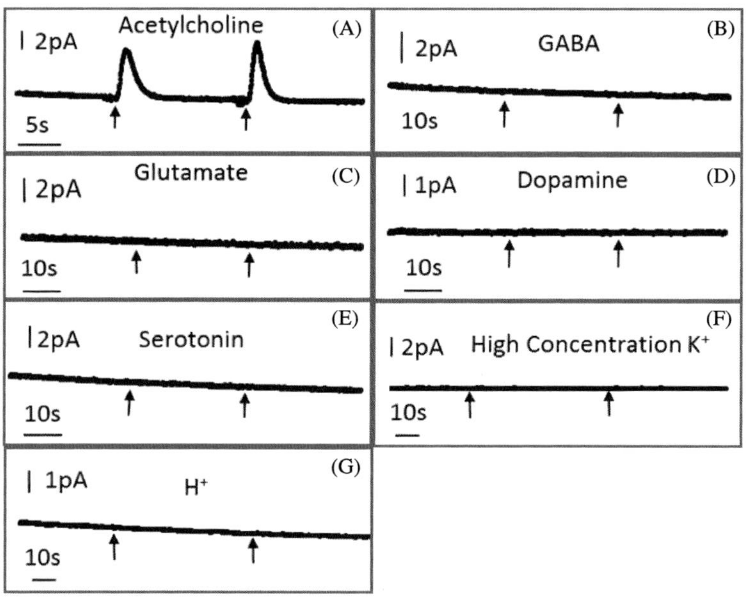

Figure 9.2 *In vitro* control experiments to demonstrate that the nanoITIES pipet electrode is selective towards acetylcholine detection. Possible transmitters known to be released from the pedal ganglion neuron of *Aplysia californica*, the high concentration of K^+ used in the stimulation solution, and pH variance accompanying exocytosis were tested for selectivity via flow injecting each of the species while monitoring the current-time trace. Appearance of current peaks (A) indicates the electrode's selectivity for acetylcholine detection. Lack of signal for gamma-amino butyric acid (GABA) (B), glutamate (C), dopamine (D), serotonin (E), high concentration K^+ (F) and H^+ (G) verifies that these species do not generate an interfering current. Applied potential is the diffusion limiting potential of ACh detection, $E_{ACh} = -0.48$ V *versus* $E_{1/2,\ TBA}$, which was used in the amperometric detection of ACh released from living neuronal structures in real time. (Reprinted with permission from Shen *et al.*, 2018, *J Am Chem Soc*, 140: 7764–7768. Copyright (2018) American Chemical Society.)

Cell 1: Pt | 5 mM TDDATFAB + 1, 2-DCE +1 mM DB18C6 || ASW + neurons | AgCl | Ag, where TDDATFAB was used as the electrolyte in the filling solution of the nanoITIES pipet electrode. 1,2 DCE corresponds to 1,2 dichloroethane and DB18C6 to dibenzo-18-crown-6. TDDATFAB is the TFAB (tetrakis(pentafluorophenyl)borate) salt of TDDA prepared by metathesis from tetradodecylammonium (TDDA) chloride and potassium tetrakis(pentafluorophenyl)borate (TFAB). Artificial seawater (ASW) is the cellular medium for *Aplysia californica* neurons, and contains the following: 460 mM NaCl, 10 mM KCl, 10 mM $CaCl_2$, 22 mM $MgCl_2$, 26 mM $MgSO_4$ and 10 mM HEPES (pH 7.8) in 18.3 MΩ cm deionized water.

9.5 Scanning Electrochemical Microscopy (SECM) to enable first-ever positioning of acetylcholine sensing probe with nanometer resolution

Positioning the ACh sensing probes close to the release sites of neurons was achieved at nanometer spatial resolution using scanning electrochemical microscopy (SECM). SECM was developed in Allen J. Bard's lab in 1989 (Bard *et al.*, 1989; Bard and Faulkner, 1994). Since then, it has emerged as a powerful platform for biological studies, energy related research, and mechanistic studies (Bard and Mirkin, 2012), such as unveiling the different electron transfer kinetics of the oxidation and reduction of $Ru(bpy)_3^{2+}$, where bpy = bipyridine (Shen and Bard, 2011). Briefly, a nanometer- or micrometer-sized electrode is mounted on a 3-dimensional positioning system that allows the electrode to be moved with nanometer spatial resolution while chemical information is simultaneously obtained in real time (Shen *et al.*, 2012; Chen *et al.*, 2014; Mirkin and Amemiya, 2015; Mirkin *et al.*, 2016; Chen *et al.*, 2017). We further lab-built and incorporated a side-view optical microscope into the SECM platform. The lab-built side view optical microscope provided rough positioning of the neurotransmitter sensing probe before the fine nanometer resolution positioning with SECM.

9.6 Interrogating ACh release at single synapses

Shen *et al.* (2018) demonstrated *in situ* simultaneous measurements of synaptic cholinergic transmitter concentration and release dynamics. We used nanoscale electroanalytical methods: nanoITIES electrode of 15 nm in radius and nanoresolved scanning electrochemical microscope (SECM). We positioned a 15-nm radius nanoITIES electrode around the synaptic cleft formed between the axon of one neuron and the neurite of another neuron (Figure 9.3) using the SECM nanopositioning platform described above (Shen *et al.*, 2018). The neuronal model we used is *Aplysia californica*, previously used by leading groups to unveil critical

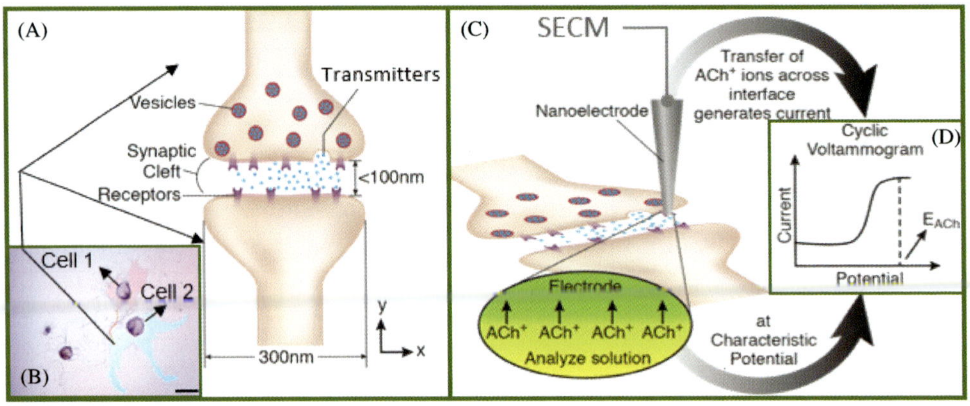

Figure 9.3 Study of cholinergic neurotransmission at single synaptic cleft with nanoelectrode and scanning electrochemical microscope. (A) Illustration of synaptic cleft dimensions. (B) Cultured living *Aplysia* pedal ganglion neurons used for the experiment, where the axon from cell 1 (pink) formed a synaptic connection with the body of cell 2. Scale bar: 200 μm. (C) A nanoITIES pipet electrode was positioned around the synaptic cleft to measure the concentration and release dynamics of acetylcholine (ACh^+) simultaneously using in vivo amperometry; the positioning of the nanoelectrode was achieved using the scanning electrochemical microscope with a spatial resolution of 5 nm. The zoom shows the nanoITIES formed at the tip of the nanoITIES pipet electrode, and ionic transmitter (ACh^+) transfers across the interface, generating a current and thus getting detected. (D) Cyclic voltammogram corresponding to ACh detection, where the detection potential follows the Nernst equation, and a steady state transfer potential, $E_{Acti} = -0.48$ V *versus* $E_{1/2}$, T_{BA}, selective for cholinergic neurotransmitter detection was used in amperometry to study its synaptic concentration dynamics. (Reprinted with permission from Shen *et al.*, 2018, *J Am Chem Soc*, 140: 7764–7768. Copyright (2018) American Chemical Society.)

information on neurobiology (Kandel, 2001; Le *et al.*, 2013). The steady state transfer potential, $E_{ACh} = -0.48$ V *versus* $E_{1/2,\mathrm{TBA}}$, selective for cholinergic neurotransmitter detection was used to study its synaptic concentration dynamics using *in vivo* amperometry. ACh release was evoked with high concentration K^+.

The evoked exocytotic events were observed in the form of current peaks at the detection potential for ACh. Diverse synaptic neurotransmitter concentration dynamics, consisting of singlets, doublets and multiplets, were observed with signal to noise ratios of 6 to 130. The occurrence frequency was approximately 50% for singlets, 25% for doublets, and 25% for multiplets. The measured ACh concentration near single synapses was in the mM range (with maximum peak concentrations of ~2.4 mM). The median number of ACh molecules was 1.0×10^6 for singlets, 2.0×10^6 for doublets, and 2.0×10^7 for multiplets. Quantitative analysis suggests that the singlets correspond to single vesicular exocytosis events, while the doublets and multiplets correspond to simultaneous multiple vesicular exocytosis events. In addition, the lower range of the number of molecules for doublets and

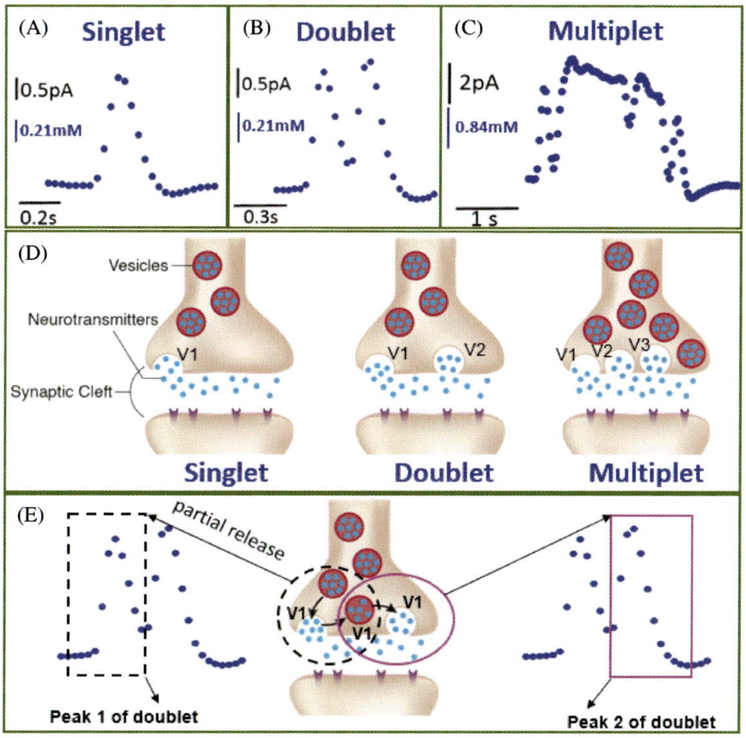

Figure 9.4 Single synaptic cholinergic neurotransmission measured *in situ.* (A–C) Current-time traces (amperometry) representing synaptic transmitter concentration and release dynamics simultaneously, where diverse cholinergic concentration dynamics were observed consisted of (A) singlets, (B) doublets, and (C) multiplets. (D) Proposed mechanism of the variation in synaptic transmitter release dynamics. (Left) Neurotransmitter is released into the synaptic cleft from a single vesicle. (Middle) Neurotransmitter is released into the synaptic cleft from two vesicles simultaneously. (Right) Neurotransmitter is released from multiple vesicles simultaneously, which are going through either different stages of exocytosis as shown here, or similar stages of exocytosis. (E) An alternative mechanism is possible for explaining doublets and multiplets based on the phenomenon of partial release. A vesicle goes through partial release twice, generating a doublet; the two individual peaks (Peak 1 and Peak 2) correspond to each partial release event. (Reprinted with permission from Shen *et al.*, 2018, *J Am Chem Soc*, 140: 7764–7768. Copyright (2018) American Chemical Society.)

multiplets were on the same order of magnitude as singlets, suggesting that some doublets and multiplets represent partial vesicular exocytosis events.

9.6.1 A new mechanism of exocytosis: Multi-vesicle-exocytosis

The mechanism of exocytosis is not well understood. Based on our measurements at the synapses, we proposed a multi-vesicle-exocytosis mechanism (Shen *et al.*, 2018), where multiple vesicles fuse with the presynaptic membrane concurrently and release the neurotransmitter packaged inside each vesicle. These multiple vesicles could go through either different stages of exocytosis or similar stages of exocytosis (Shen *et al.*, 2018). Prior to 2018, proposed single-vesicle-exocytosis mechanisms included full release, partial release, and kiss-and-run (Travis and Wightman 1998; Stevens and Willams, 2000; Aravanis *et al.*, 2003; Wightman and Haynes, 2004; Amatore *et al.*, 2005, 2006, 2007, 2009, 2010; Omiatek *et al.*, 2010; Keighron *et al.*, 2012; Mellander *et al.*, 2014; Li *et al.*, 2015, 2016; Ren *et al.*, 2016; Dunevall *et al.*, 2017; Phan *et al.*, 2017). Our proposed multi-vesicle-exocytosis mechanism complements these existing single-vesicle-exocytosis mechanisms.

9.7 ACh signaling at single cells

Neurotransmitter release is an essential part of brain communication (Michael and Borland, 2007; Phan *et al.* 2017). In addition to synaptic events, quantifying exocytotic transmitter release from neuronal soma is important to understanding neurotransmission and neurological disorders. We discuss below the first study on the quantification of ACh signaling at single cells, with nanometer spatial resolution.

9.7.1 Interrogating ACh release at single cells

We recently used SECM with 0.3 nm spatial resolution to accurately position nanoITIES electrodes ~140 nm away from the soma of Aplysia neurons and quantitative measured ACh release at single cells (Welle *et al.*, 2018). We used *in vivo* amperometry to study the concentration and release dynamics of ACh in response to stimulation with high concentrations of K^+ with a time resolution on the order of ms. We observed *in situ* release of ACh with signal to noise ratios of 6–19 on nanoITIES electrodes with radii of 750 nm to 860 nm.

These measurements elucidated key information related to cellular communication: the extracellular concentration of acetylcholine during somatic exocytosis (2.7 ± 1.0 μM), the number of molecules released ($4.8 \pm 1.3 \times 10^7$), the dynamics of exocytosis, vesicle density, Ca^{2+} dependence of somatic exocytosis, and the effect of the electrode-release site distances on the measured signals. Our study of the Ca^{2+} dependence of release contributes to the mechanistic understanding of somatic exocytosis, which currently is not fully understood. When Ca^{2+} was not present, no exocytosis events were observed in response to K^+-stimulation; in contrast, ACh exocytosis events were recovered once Ca^{2+} was restored. Ca^{2+} is known to be involved in the vesicular release of transmitters (Haynes *et al.*, 2006); these findings confirm that somatic ACh release involves vesicular exocytosis.

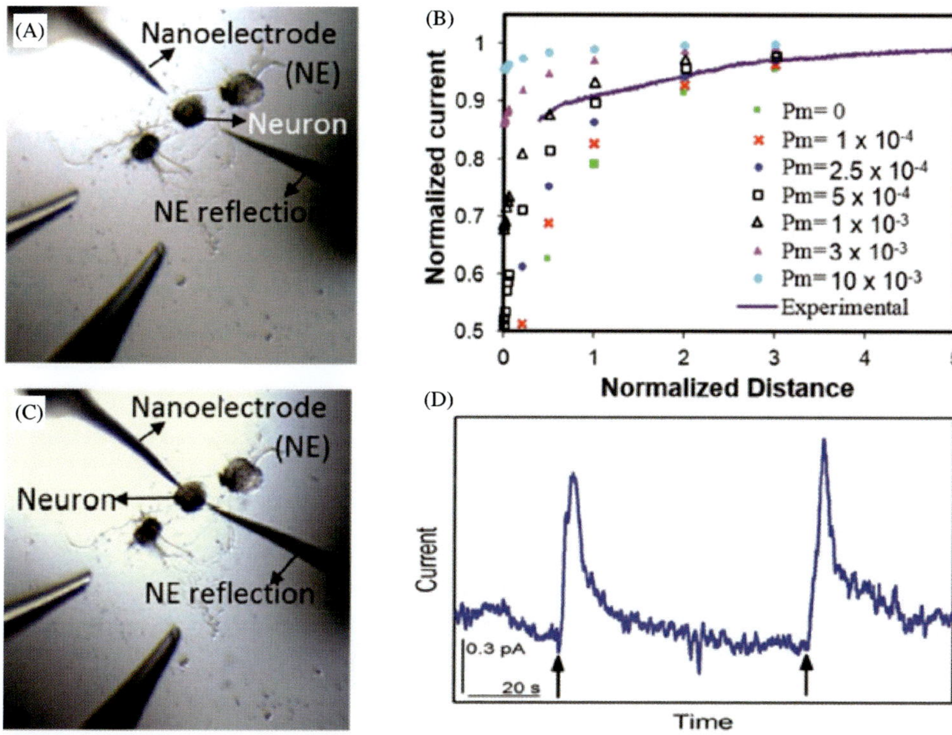

Figure 9.5 High resolution spatiotemporal observation of somatic ACh exocytosis from living neurons in real time. The nanoITIES electrode (radius ~860 nm) was positioned with scanning electrochemical microscope (SECM) to be ~140 nm from the soma. Optical microscope images (A) before and (C) after SECM positioning, where it can be seen that it is very hard to position the ACh-sensing nanoelectrode to be nanometer away from the release sites by visual and optical observation only. The use of a nanometer positioning platform, SECM, is needed. (B) Experimental SECM approach curve to the neuron soma (solid purple curve) overlaid with simulated approach curves with varying permeability coefficients (P_m) to accurately determine the distance between nanoelectrode and soma. Please see the original publication for a detailed explanation of the approach curve. (D) Amperometric recordings of ACh release from a single Aplysia neuron in response to high concentration K^+-stimulation, indicated by the arrows. We used the optical mirror images (reflection) of the nanoelectrode (NE) to roughly position the nanoelectrode before nanometer resolution positioning with SECM. (Reprinted with permission from Welle *et al.*, 2018, *Chem Sci*, 9: 4937–4941. Copyright (2018) The Royal Society of Chemistry.)

9.7.2 An amperometic strategy to measure vesicle density in living cells

We have created a powerful platform based on electroanalytical methods with nanometer spatial resolution to measure the vesicle density in living cells. This generates additional information complementary to that normally acquired by electron microscopy.

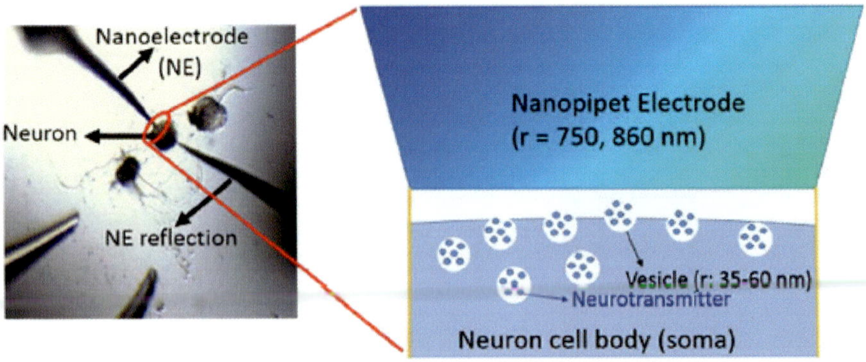

Figure 9.6 Measurement of vesicle density using *in vivo* amperometry and nanoITIES electrodes. The measured vesicle density is ~25 ± 2 vesicles per μm^2. The vesicle radius is reported in Gillette and Pomeranz, 1975. (Reprinted with permission from Welle *et al.*, 2018, *Chem Sci*, 9: 4937–4941. Copyright (2018) The Royal Society of Chemistry.)

Electron microscopy has been widely used to determine the vesicle density in cells (Parducz *et al.*, 1971; Yang *et al.*, 2005; Mantlla *et al.*, 2007; Nemani *et al.*, 2010; Wagner *et al.*, 2015; Khanmohammadi *et al.*, 2017; Dong *et al.*, 2018; Nguyen and Krainc, 2018). These techniques require the staining and fixation of the neurons, and the measurements are conducted under vacuum where there is no possibility of measuring exocytotic events. In our approach using *in situ* amperometry with nanoelectrodes (Welle *et al.*, 2018), multiple vesicular contents were measured from single neuronal soma via nanoITIES electrodes with radii of 750 nm to 860 nm during each amperometric peak event. Using the content per vesicle determined in our recent single synaptic study of acetylcholine release from the same animal model (Shen *et al.*, 2018), the releasable vesicle density was calculated to be 25 ± 2 vesicles/μm^2. This number is similar order of magnitude to the value of 90 vesicles/μm^2 measured based on 3-D reconstruction of synaptic boutons from serial electron microscopy sections (Yang, *et al.*, 2005). The platform we have created allows the measurements of both vesicle density information and exocytosis in real time.

9.8 Summary

In this chapter, we have presented recent advances in the development of nano-ITIES electrodes with size as small as 15 nm in radius for interrogating cholinergic exocytosis with unprecedented nanometer spatial resolution, and high temporal resolution (millisecond). We have shown the capability of nanoITIES electrodes to study neurotransmission in real time across a range of spatial dimensions, from single cells to single synapses. Besides sensing, the concentration and the number of the cholinergic transmitter molecules released during exocytosis have been measured in real time using nanoITIES electrodes. The development of NanoITIES electrodes expands the capability of voltammetry and amperometry to probe both redox active and non-redox active neurochemicals in real-time, in a non-invasive and label-free manner.

The combination of nanoITIES electrodes with the nanometer spatial resolution of SECM is emerging as a powerful platform to enable the interrogation of release dynamics in real-time from living neurons. Based on our findings, we propose the multi-vesicle-exocytosis mechanism, which suggests that multiple vesicles are involved in concurrent exocytosis events to generate doublets and multiplets in addition to singlets. *In situ* amperometry has been demonstrated as an alternative non-invasive method to determine the number of releasable vesicles per unit area of living cells while providing additional information about neurotransmission, e.g. the concentration and release dynamics of transmitters.

One can envision a broader application of the nanoITIES platform in future work, including the study of ACh release dynamics from different neuronal structures and models, comparing ACh release processes between healthy and diseased neurons, and studying exocytosis events in real time in response to other forms of stimulation, e.g. stimulate the neurons optogenetically and mechanically.

Acknowledgments

M. S. and R. C. acknowledge coworkers who contributed to the work presented in this chapter (Z. Qu, J. DesLaurier, T. Welle, K. Alanis, M. Colombo) and the collaboration from Professor J. Sweedler and his group. M. S. appreciates the support from the University of Illinois at Urbana-Champaign and the National Institute of Neurological Disorders and Stroke of the National Institute of Health (award number R21NS085665).

References

Amatore C, Arbault S, Bonifas I, Bouret Y, Erard M, Ewing AG, Sombers LA (2005) Correlation between vesicle quantal size and fusion pore release in chromaffin cell exocytosis. *Biophys J*, 88: 4411–4420.

Amatore C, Arbault S, Bouret Y, Guille M, Lemaitre F, Verchier Y (2006) Regulation of exocytosis in chromaffin cells by trans-insertion of lysophosphatidylcholine and arachidonic acid into the outer leaflet of the cell membrane. *ChemBioChem*, 7: 1998–2003.

Amatore C, Arbault S, Bonifas I, Guille M, Lemaitre F, Verchier Y (2007) Relationship between amperometric pre-spike feet and secretion granule composition in chromaffin cells: an overview. *Biophys Chem*, 129: 181–189.

Amatore C, Arbault S, Bonifas I, Guille M (2009) Quantitative investigations of amperometric spike feet suggest different controlling factors of the fusion pore in exocytosis at chromaffin cells. *Biophys Chem*, 143: 124–131.

Amatore C, Oleinick AI, Svir I (2010) Reconstruction of aperture functions during full fusion in vesicular exocytosis of neurotransmitters. *ChemPhysChem*, 11: 159–174.

Amemiya S, Kim J, Izadyar A, Kabagambe B, Shen M, Ishimatsu R (2013) Electrochemical sensing and imaging based on ion transfer at liquid/liquid interfaces. *Electrochim Acta*, 110: 836–845.

Aravanis AM, Pyle JL, Tsien RW (2003) Single synaptic vesicles fusing transiently and successively without loss of identity. *Nature*, 423: 643–647.

Asri R, O'Neill B, Patel JC, Siletti KA, Rice ME (2016) Detection of evoked acetycholine release in mouse brain slices. *Analyst*, 141: 6416–6421.

Bard AJ, Fan F-R F, Kwak J, Lev O (1989) Scanning electrochemical microscopy. Introduction and principles. *Anal Chem*, 61: 132–138.

Bard AJ, Faulkner RF (1994) *Electrochemical Methods. Fundamentals and Applications*. Hoboken, NJ: John Wiley & Sons.

Bard AJ, Mrikin MV (2012) *Scanning Electrochemical Microscopy*. Boca Raton, FL: CRC.

Berduque A, Sherburn A, Ghita M, Dryfe RAW, Arrigan DWM (2005) Electrochemically modulated liquid-liquid extraction of ions. *Anal Chem*, 77: 7310–7318.

Brossard C, Feuillet V, Schmitt A, Randriamampita C, Roman M, Raposo G, Trautmann A (2005) Multifocal structure of the T cell — dendritic cell synapse. *Eur J Immunol*, 35: 1741–1753.

Bruno JP, Gash C, Martin B, Zmarowski A, Pomerleau F, Burmeister J, Huettl P, Gerhardt GA (2006) Second-by-second measurement of acetylcholine release in prefrontal cortex. *Eur J Neurosci*, 24: 2749–2757.

Burmeister JJ, Pomerleau F, Huettl P, Gash CR, Werner CE, Bruno JP, Gerhardt GA (2008) Ceramic-based multisite microelectrode arrays for simultaneous measures of choline and acetylcholine in CNS. *Biosens Bioelectron*, 23: 1382–1389.

Chen R, Balla RJ, Lima A, Amemiya S (2017) Characterization of nanopipet-supported ITIES tips for scanning electrochemical microscopy of single solid-state nanopores. *Anal Chem*, 89: 7269–7272.

Chen Q, Luo L, Faraji H, Feldberg SW, White HS (2014) Electrochemical measurements of single H_2 nanobubble nucleation and stability at Pt nanoelectrodes. *J Phys Chem Lett*, 5: 3539–3544.

Colombo ML, Shen M (2015) Electrochemical nanoprobes for the chemical detection of neurotransmitters. *Anal Methods*, 7: 7095–7105.

Colombo ML, Sweedler JV, Shen M (2015) Nanopipet-based liquid-liquid interface probes for the electrochemical detection of acetylcholine, tryptamine, and serotonin via ionic transfer. *Anal Chem*, 87: 5095–5100.

Colombo ML, McNeil S, Iwai N, Chang A, Shen M (2016) Electrochemiical detection of dopamine via assisted ion transfer at nanopipet electrode using cyclic voltammetry. *J Electrochem Soc*, 163: H3072–H3076.

Dale HH (1914) The action of certain esters and ethers of choline and their relation to muscarine. *J Pharm Exp Ther*, 6: 147-190.
Dani A, Huang B, Bergan J, Dulac C, Zhuang X (2010) Superresolution imaging of chemical synapses in the brain. *Neuron*, 68: 843–856.
Dong Y, Guo CR, Chen D, Chen SM, Peng Y H, Shi JR (2018) Association between age-related hearing loss and cognitive decline in C57BL/6J mice. *Mol Med Rep*, 18: 1726–1732.
Dunevall J, Majdi S, Larsson A, Ewing A (2017) Vesicle impact electrochemical cytometry compared to amperometric exocytosis measurements. *Curr Op Electrochem*, 5: 85–91.
Fu D, Yang W, Xie XS (2017) Label-free imaging of neurotransmitter acetylcholine at neuromuscular junctions with stimulated Raman scattering. *J Am Chem Soc*, 139: 583–586.
Garguilo MG, Michael AC (1995) Enzyme-modified electrodes for peroxide, choline, and acetylcholine. *Trends Anal Chem*, 14: 164–169.
Gillette R, Pomeranz B (1975) Ultrastructural correlates of interneuronal function in the abdorminal ganglion of *Aplysia californica*. *J Neurobiol*, 6: 463–474.
Giuliano C, Parikh V, Ward JR, Chiamulera C, Sarter M (2008) Increases in cholinergic neurotransmission measured by using choline-sensitive microelectrodes: Enhanced detection by hydrolysis of acetylcholine on recording sites? *Neurochem Int*, 52: 1343–1350.
Groves PM, Linder JC, Young SJ (1994) 5-hydroxydopamine-labeled dopaminergic axns: Three-dimensional reconstruction of axons, synapses and postsynaptic targets in rat neostriatum. *Neuroscience*, 58: 593–604.
Haynes CL, Buhler LA, Wightman RM (2006) Vesicular Ca^{2+}-induced secretion promoted by intracellular pH-gradient disruption. *Biophys Chem*, 123: 20–24.
Ishimatsu R, Kim J, Jing P, Striemer CC, Fang DZ, Fauchet PM, McGrath JL, Amemiya S (2010) Ion-selective permeability of an ultrathin nanoporous silicon membrane as probed by scanning electrochemical microscopy using micropipet-supported ITIES tips. *Anal Chem*, 82: 7127–7134.
Iwai NT, Kramaric M, Crabbe D, Wei Y, Chen R, Shen M (2018) GABA detection with nano ITIES pipet electrode: A new mechanism, water/DCE-octanoic acid interface. *Anal Chem*, 90: 3067–3072.
Jing M, Zhang P, Wang G, Feng J, Mesik L, Zeng J, Jiang H, Wang S, Looby JC, Guagliardo NA, Langma LW, Lu J, Zuo Y, Talmage DA, Role LW, Barrett PQ, Zhang LI, Luo M, Song Y, Zhu JJ, Li Y (2018) A genetically encoded fluorescence acetylcholine indicator for *in vitro* and *in vivo* studies. *Nat Biotechnol*, 36: 726–737.
Johnson DA, Pilar G (1980) The release of acetylcholine from post-ganglionic cell bodies in response to depolarization. *J Physiol*, 299: 605–619.
Kandel ER (2001) The molecular biology of memory storage: A dialogue between genes and synapses. *Science*, 294: 1030–1038.
Keighron JD, Ewing AG, Cans AS (2012) Analytical tools to monitor exocytosis: A focus on new fluorescent probes and methods. *Analyst*, 137: 1755–1763.
Khanmohammadi M, Darkner S, Nava N, Nyengaard JR, Wegener G, Popoli M, Sporring J (2017) 3D analysis of synaptic vesicle density and distribution after acute foot-shock stree by using serial section transmission electron microscopy. *J Microsc*, 265: 101–110.

Kim J, Shen M, Nioradze N, Amemiya S (2012) Stabilizing nanometer scale tip-to-substrate gaps in scanning electrochemical microscopy using an isothermal chamber for thermal drift suppression. *Anal Chem*, 84: 3489–3492.

Le AP, Kang S, Thompson LB, Rubakhin SS, Sweedler JV, Rogers JA, Nuzzo RG (2013) Quantitative reflection imaging of fixed *Aplysia californica* pedal ganglion neurons on nanostructured plasmonic crystals. *J Phys Chem*, B 117: 13069–13081.

Lefrou C. (2006) A unified new analytical approximation for positive feedback currents with a microdisk SECM tip. *J Electroanal Chem*, 592: 103–112.

Li X, Majdi S, Dunevall J, Fathali H, Ewing AG (2015) Quantitative measurement of transmitters in indiudual vesicles in the cytoplasm of single cells with nanotip electrodes. *Angew Chem Int Ed*, 54: 11978–11982.

Li X, Dunvall J, Ewing AG (2016) Quantitative chemical measurements of vesicular transmitters with electrochemical cytometry. *Acc Chem Res*, 49: 2347–2354.

Liu S, Li Q, Shao Y (2011) Electrochemistry at micro- and nanoscopic liquid/liquid interfaces. *Chem Soc Rev*, 40: 2236–2253.

Loewi OP (1921) Über humorale Übertragbarkeit der Herznervenwirkung. *Pfluegers Arch Physiol*, 189: 239–242.

Mantlla CB, Rowley KL, Zhan WZ, Fahim MA, Sieck GC (2007) Synaptic vesicle pools at diaphragm neuromuscular junctions vary with motoneuron soma, not axon terminal, inactivity. *Neurosci*, 146: 178–189.

Mattinson CE, Burmeister JJ, Quintero JE, Pomerleau F, Huettl P, Gerhardt GA (2011) Tonic and phasic release of glutamate and acetylcholine neurotransmission in sub-regions of the rat prefrontal cortex using enzyme-based microelectrode arrays. *J Neurosci Methods*, 202: 199–208.

McCann ME, Vanherberqhen B, Eleme K, Carlin LM, Newsam RJ, Goulding D, Davis DM (2003) The size of the synaptic cleft and distinct distributions of filamentous actin, ezrin, CD43 and CD45 at activating and inhibitory human NK cell immune synapses. *J Immunol*, 170: 2862–2870.

Mellander LJ, Kurczy ME, Najafinobar N, Dunevall N, Ewing AG, Cans AS (2014) Two modes of exocytosis in an artificial cell. *Sci Rep*, 4: 3847.

Michael AC, Borland LM (2007) *Electrochemical Methods for Neuroscience*. Boca Raton, FL: CRC Press/Taylor & Francis.

Mirkin MV, Amemiya S (2015) *Nanoelectrochemistry*. Boca Raton, FL: CRC.

Mirkin MV, Sun T, Yu Y, Zhou M (2016) Electrochemistry at one nanoparticle. *Acc Chem Res*, 49: 2328–2335.

Mitchell KM (2004) Acetycholine and choline amperometric enzyme sensors characterized *in vitro* and *in vivo*. *Anal Chem*, 76: 1098–1106.

Moukhles H, Bosler O, Bolam JP, Vallee A, Umbriaco D, Geffard M, Doucet G (1997) Quantitative and morphometric data indicate precise cellular interactions between serotonin terminals and postsynaptic targets in rat substania nigra. *Neuroscience*, 76: 1159–1171.

Nemani VM, Lu W, Berge V, Nakamura K, Onoa B, Lee MK, Chaudhry FA, Nicoll RA, Edwards RH (2010) Increased expression of α-synuclein reduces neurotransmitter release by inhibiting synaptic vesicle reclustering after endocytosis. *Neuron*, 65: 66–79.

Nestor U, Wen H, Girma G, Mei Z, Fei W, Yang Y, Zhang C, Zhan D (2014) Facilitated Li+ ion transfer across the water/1,2-dichloroethane interface by the solvation effect. *Chem Commun*, 50: 1015–1017.

Nguyen M, Krainc D (2018) LRRK2 phosphorylation of auxilin mediates synaptic defects in dopaminergic neurons from patients with Parkinson's disease. *Proc Nat Acd Sci USA*, 115: 5576–5581.

Omiatek DM, Dong Y, Heien ML, Ewing AG (2010) Only a fraction of quantal content is released during exocytosis as revealed by electrochemical cytometry of secretory vesicles. *ACS Chem Neurosci*, 1: 234–245.

Parducz A, Feher O, Joo F (1971) Effects of stimulation and hemicholinium (HC-3) on the fine structure of nerve endings in the superior cervical ganglion of the cat. *Brain Res*, 34: 61–72.

Phan NTN, Li X, Ewing AG (2017) Measuring synaptic vesicles using cellular electrochemistry and nanoscale molecular imaging. *Nat Rev Chem*, 1: 0048.

Pickel VM, Beckley SC, Joh TH, Reis DJ (1981) Ultrastructural immunocytochemical localization of tyrosine hydroxylase in the neostriatum. *Brain Res*, 225: 373–385.

Qian Q, Wilson GS, Bowman-James K, Girault HH (2001) MicroITIES detection of nitrate by facilitated ion transfer. *Anal Chem*, 73: 497–503.

Qian Q, Wilson GS, Bowman-James K (2004) MicroITIES detection of adenosine phosphates. *Electroanalysis*, 16: 1343–1350.

Ren L, Mellander LJ, Keighron J, Cans AS, Kurczy ME, Svir I, Oleinick, A, Amatore C, Ewing AG (2016) The evidence for open and closed exocytosis as the primary release mechanism. *Q Rev Biophys*, 49: e12

Reymond F, Fermin D, Lee HJ, Girault HH (2000) Electrochemistry at liquid/liquid interfaces. *Electrochim Acta*, 45: 2647–2662.

Rossier JS, Gokulrangan C, Girault HH, Svojanovsky S, Wilson GS (2000) Characterization of protein adsorption and immunosorption kinetics in photoablated polymer microchannels. *Langmuir*, 16: 8489–8494.

Samec Z, Marecek V, Koryta J, Khalil MW (1977) Investigation of ion transfer across the interface between two immiscible electrolyte solutions by cyclic voltammetry. *J Electroanal Chem*, 83: 393–397.

Shao Y, Mirkin MV (1997) Scanning electrochemical microscopy (SECM) of facilitated ion transfer at the liquid/liquid interface. *J Electroanl Chem*, 439: 137–143.

Shao Y, Mirkin MV (1998) Probing ion transfer at the liquid/liquid interface by scanning electrochemical microscopy (SECM). *J Phys Chem B*, 102: 9915–9921.

Shen M, Bard AJ (2011) Localized electron transfer and the effect of tunneling on the rates of $Ru(bpy)_3^{2+}$ oxidation and reduction as measured by Scanning Electrochemical Microscopy. *J Am Chem Soc*, 133: 15737–15742.

Shen M, Ishimatsu R, Kim J, Amemiya S (2012) Quantitative imaging of ion transport through single nanopores by high-resolution scanning electrochemical microscopy. *J Am Chem Soc*, 134: 9856–9859.

Shen M, Qu Z, DesLaurier J, Welle, TM, Sweedler JV, Chen R (2018) Single synaptic observation of cholinergic neurotransmission on living neurons: Concentration and dynamics. *J Am Chem Soc*, 140: 7764–7768.

Stevens CF, Williams JH (2000) "Kiss and run" exocytosis at hippocampal synapses. *Proc Natl Acad Sci USA*, 97: 12828–12833.

Stockmann TJ, Montgomery A, Ding Z (2012) Determination of alkai metal ion transfers at liquid/liquid interfaces stabilized by a micropipette. *J Electroanal Chem*, 684: 6–12.

Sun Y, Poo M (1987) Evoked release of acetylchilone from the growing embryonic neuron. *Proc Natl Acad USA*, 84: 2540–2544.

Taylor G, Girault HHJ (1986) Ion transfer reactions across a liquid-liquid interface supported on a micropipette tip. *Electroanal Chem*, 208: 179–183.

Travis ER, Wightman RM (1998) Spatio-temporal resolution of exocytosis from individual cells. *Annu Rev Biophys Biomol Struct*, 27: 77–103.

Vallejo LJS, Ovejero JM, Fernandez RA, Dassie SA (2012) Simple ion transfer at liquid/liquid interfaces. *Int J Electrochem*, 2012: 462197.

Vanysek P, Ramirez L (2008) Interface between two immiscible liquid electrolytes. *J Chil Chem Soc*, 53: 1455–1463.

Wagner N, Laugks U, Heckmann M, Asan E, Neuser K (2015) Aging Drosophila melanogaster display altered pre- and postsynaptic ultrastructure at adult neuromuscular junctions. *J Comp Neurol*, 523: 2457–2475.

Wang Y, Velmurugan J, Mirkin MV (2010) Kinetics of charge-transfer reactions at nanoscopic electrochemical interfaces. *Isr J Chem*, 50: 291–305.

Wang Y, Kececi K, Velmurugan J, Mirkin MV (2013) Electron transfer/ion transfer of scanning electrochemical microscopy (SECM): A new tool for imaging and kinetic studies. *Chem Sci*, 4: 3606–3616.

Welle TM, Alanis K, Colombo ML, Sweedler JV, Shen M (2018) A high spatiotemporal study of somatic exocytosis with scanning electrochemical microscopy and nanoITIES electrodes. *Chem Sci*, 9: 4937–4941.

Wightman RM, Haynes CL (2004) Synaptic vesicles really do kiss and run. *Nat Neurosci*, 7: 321–322.

Wilson GS, Johnson MA (2008) *In-vivo* electrochemistry: What can we learn about living systems? *Chem Rev*, 108: 2462–2481.

Xin Q, Wightman RM (1997) Enzyme modified amperometric sensors for choline and acetylcholine with tetrathiafulvalene tetracyanoquinodimethane as the electron-transfer mediator. *Anal Chim Acta*, 341: 43–51.

Yang G, Gong YD, Gong K, Jiang WL, Kwon E, Wang P, Zheng H, Zhang XF, Gan WB, Zhao NM (2005) Reduced synaptic vesicle density and active zone size in mice lacking amyloid precursor protein (APP) and APP-like protein 2. *Neurosci Lett*, 384: 66–71.

Zhan D, Li X, Zhan W, Fan F, Bard AJ (2007) Scanning electrochemical microscopy 58. Application of a micropipet-supported ITIES tip to detect Ag^+ and study its effect on fibroblast cells. *Anal Chem*, 79: 5225–5231.

CHAPTER 10

LONGITUDINAL RECORDINGS OF DOPAMINE TRANSMISSION DURING BEHAVIOR WITH FAST-SCAN CYCLIC VOLTAMMETRY

Ryan D. Farero and Paul E. M. Phillips*

Center for Excellence in Neurobiology of Addiction, Pain and Emotion, Graduate Program in Neuroscience, Department of Psychiatry & Behavioral Sciences, and Department of Pharmacology, University of Washington, USA

***pemp@uw.edu**

10.1 Introduction

Fast-scan cyclic voltammetry (FSCV) is an electrochemical detection technique that applies rapid potential sweeps via an electrode to oxidize and reduce electroactive molecules, producing current which can be detected at the same electrode. The current-voltage relationship, the cyclic voltammogram, provides a chemical signature for the detected species, affording a degree of chemical resolution. This chemical resolution differentiates FSCV from other rapid electrochemical detection paradigms such as constant-potential amperometry and high-speed chronoamperometry (Michael and Borland, 2007). Consequently, FSCV has been instrumental in the detection of concentration fluctuations in complex chemical environments on sub-second time scales (Wightman, 2006). Specifically, this technique has been a valuable tool for studying neurotransmitter signaling in the brains of behaving animals (Phillips *et al.*, 2003; Roitman *et al.*, 2004; Stuber *et al.*, 2005; Cheer *et al.*, 2007; Day *et al.*, 2007; Gan *et al.*, 2010; Flagel *et al.*, 2011;

Oleson *et al.*, 2012; Howe *et al.*, 2013; Hamid *et al.*, 2016). Originally, these experiments were carried out with removable microelectrodes which were implanted transiently for each recording session. However, because of the desire to track changes in neurochemical signals that come about due to psychological or pathological processes, our lab developed an approach for making longitudinal measurements of dopamine with FSCV over the course of months (Clark *et al.*, 2010). In this chapter, we will describe chronic electrodes for FSCV and their utility. In particular, we will focus on knowledge gained in the realm of substance use disorders to provide concrete examples of the value of this approach.

10.2 Chronic electrodes

The primary objective underlying the development of electrodes to be chronically implanted for use with FSCV was to track long-term biological changes in neurotransmitter dynamics. More specifically, our goal was to enable the monitoring of neurotransmission throughout paradigms studying cognition and learning, and during the progression of clinical models of disease. Zygmunt Kruk, Julian Millar and colleagues carried out the first FSCV recordings with chronically implanted carbon-fiber microelectrodes (Kruk *et al.*, 1998). These microelectrodes were constructed from a single carbon fiber insulated in pulled borosilicate glass, similar to what is used for acute preparations. However, they concluded that, with their design, the window for longitudinal recordings was quite limited. The iteration of a chronic carbon-fiber microelectrode used in our lab is also fabricated from a single carbon fiber, but it is encased within a polyimide-coated fused-silica cylinder (Figure 10.1; Clark *et al.*, 2010). These devices are less fragile than their borosilicate-glass counterparts and are afforded better biocompatibility by the polymer coating (Hassler *et al.*, 2011).

Following the introduction of these electrodes, a number of anticipated benefits have come to fruition. Of course, the most overt advantage of chronic electrodes has been the ability to longitudinally record neurochemical transmission events. Specifically, chronic implantation avoids repeated electrode reinsertion and damage to brain tissue, permitting repeated recording at a single anatomical location. With the current design, electrodes are secured directly to the cranium, taking up less space than a microdrive, making them more amenable to multi-electrode recording in multiple brain regions of a single animal (Clark *et al.*, 2010; Wanat *et al.*, 2010; Willuhn *et al.*, 2012, 2014). Moreover, an ethical advantage of chronic electrodes is the reduction in animals needed for experiments. Chronic electrodes permit a greater number of recordings to be carried out in each subject, reducing the number of animals needed to obtain a dataset with adequate statistical power. The statistical power is further enhanced with the capacity for repeated-measure

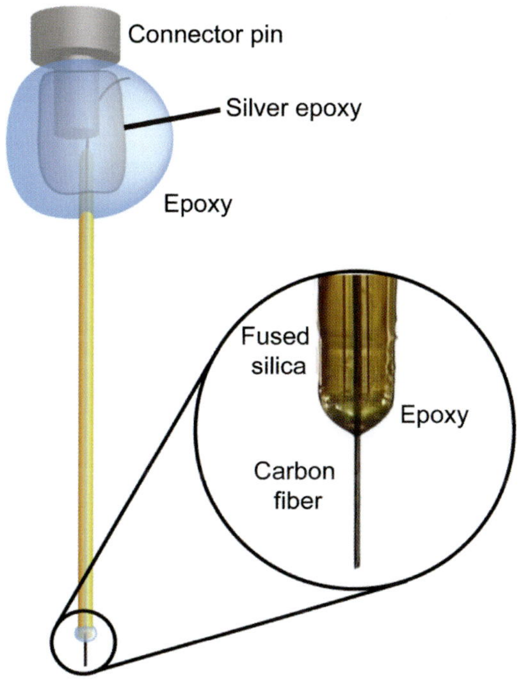

Figure 10.1 Schematic of the chronic carbon-fiber microelectrode. The electrode consists of a carbon fiber encased within a polyimide-coated fused-silica cylinder. The carbon fiber extends approximately 150 μm beyond the silica. The carbon fiber within the fused silica is sealed and electrically insulated using two-component epoxy. At the other end of the fused silica the carbon fiber is electrically connected to a female silver pin via silver epoxy. The silver epoxy and a portion of the silver pin are insulated with two-component epoxy. After Clark *et al.*, 2010.

experimental designs that allow within-subject comparison of experimental and control conditions. Another key advantage of chronically implanted electrodes is that the preparation for recording before a behavioral session is minimally invasive — essentially just plugging in a single (multi-conductor) connector. This procedure can be contrasted with the necessity to implant and position an acute electrode before each session. Doing so requires prolonged restraint of the subject which is potentially stressful and often produces an observable aversive state. In addition to being an important concern for animal welfare, this consequence can be a tangible confound when studying the neurobiology underlying stress-related and stress-sensitive behaviors.

10.2.1 Validation of chronic electrodes

A significant concern of the field for the utilization of chronically implanted electrodes is the precipitation of a neuroinflammatory response that results in glial encapsulation of the probe rendering it either unusable or only capable of detecting aberrant signals that are not representative of normal physiological conditions. To determine the extent of the glial reaction to implantation of fused-silica-insulated carbon-fiber microelectrodes, we carried out histological analysis one, two and four months after microelectrode implantation (Clark *et al.*, 2010). At none of these time points did we observe any evidence of encapsulation of the exposed carbon fiber (i.e., the detection site). On occasion, we did observe some gliotic reaction around the shaft of the electrode, primarily close to locations where the electrode had damaged vasculature. Furthermore, the number and morphology of glial cells observed around the microelectrode implantation sites appeared predominantly normal, indicating that significant recruitment of microglia has not occurred. Therefore, while there was a detectable glial reaction to the microelectrode, this response was nominal and did not produce encapsulation of the recording site. Moreover, there was no observable perturbation of the dopamine terminals at the optical microscope level around the recording site, as assessed by the density of antibody staining for the dopamine biosynthetic enzyme, tyrosine hydroxylase.

To validate neurochemical recordings with FSCV, we conducted multifaceted characterization of the electrochemical signal at chronically implanted carbon-fiber microelectrodes. This characterization used criteria from Phillips and Wightman (2003), which were updated from those originally laid out by pioneers of the field of neurochemical monitoring (Marsden *et al.*, 1988). In the first instance, we validated voltammetric signals evoked by the unsignaled delivery of food. We provided convergent evidence that dopamine can be resolved from this signal using electrochemical, anatomical/physiological, pharmacological and independent verification (Clark *et al.*, 2010). Subsequently, studies carried out in our lab also fulfilled these four criteria for voltammetric signals during cocaine self-administration (Willuhn *et al.*, 2012, 2014).

10.2.2 Utility of chronic electrodes

Fused-silica-based chronic electrodes have been utilized to record longitudinal changes in phasic dopamine transmission by several laboratories (e.g. Clark *et al.*, 2010; Oleson *et al.*, 2012, Wassum *et al.*, 2013; Gentry *et al.*, 2016; Collins *et al.*, 2016; Ko and Wanat, 2016; Syed *et al.*, 2016; Levy *et al.*, 2017; Schwerdt *et al.*, 2017). Many of the studies to date have focused on the processes underlying reinforcement learning, decision making, or substance use. Some of the highlights of those studies are summarized below.

At the outset of the use of these chronic electrodes was a demonstration of the dynamics of mesolimbic dopamine transmission over the course of a multi-day classical conditioning procedure (Clark *et al.*, 2010). As anticipated from earlier studies (Schultz *et al.*, 1997; Day *et al.*, 2007), phasic dopamine transmission conformed to the general characteristics of a reward-prediction-error signal where it was elicited by unexpected rewards or reward-predictive cues, but not expected rewards. The use of chronic electrodes to track these signals over many sessions has facilitated novel analyses of their characteristics. We found that phasic dopamine release initially reflects changes in stimulus-reward associations that can be inferred from the acquisition of conditioned approach behaviors, but they continue to be refined during post-asymptotic training as the underlying reward predictions incorporate additional task statistics such as the variance of reward timing (Clark *et al.*, 2013). The dopamine signal contains a sensory (possibly stimulus-salience) signal and a quantitative prediction-error signal (Hart *et al.*, 2014), and conveys a proxy of uncertainty that incorporates both learned uncertainty (i.e., risk) as well as uncertainties due to lack of experience (i.e., ambiguity) (Hart *et al.*, 2015). Moreover, it was determined that the prototypical pattern of dopamine transmission during classical learning is not ubiquitous. Instead, patterns of dopamine differ between individuals depending on the cognitive strategy they adopted to make the association between stimulus and reward (Flagel *et al.*, 2011). It is suggested that this difference reflects the different computation processes used in these alternative forms of learning (Clark *et al.*, 2012).

By studying reward-related dopamine signals using chronic electrodes in the context of operant behavior, and specifically economic decision making, the value functions underlying phasic dopamine signals have been investigated. A particular focus has been to deconstruct the phasic dopamine signal that is elicited by cues predicting the availability of rewards. This signal reflects the predicted reward magnitude (Gan *et al.*, 2010; Wanat *et al.*, 2013) and is modulated by the current motivational state of the subject (Aitken *et al.*, 2016; Papageorgiou *et al.*, 2016). However, it does not universally incorporate all economic elements that influence subjective value (Gan *et al.*, 2010; Hollon *et al.*, 2014). For example, raising the utility of a reward transaction by increasing the reward size elicits a more robust increase in dopamine release than an equal change in subjective value conferred by a reduction in the effort requirement to obtain the reward (Gan *et al.*, 2010). An additional nuance in the context of operant behavior is that the presentation of a cued option is not sufficient to elicit the dopamine signal; rather the initiation of a compatible action is also required (Syed *et al.*, 2016). Moreover, when these reward-seeking actions are prolonged to comply with spatial, temporal or sequential constraints of the task then elevated dopamine signaling is sustained across the appetitive phase (Howe *et al.*, 2013; Hollon *et al.*, 2014; Collins *et al.*, 2016). While most of the studies in this domain have examined positive reinforcement, dopamine signals have also been recorded with chronic electrodes under negative-reinforcement conditions (working to remove an aversive stimulus) where phasic dopamine was observed to cues that predict avoidable punishments (Oleson *et al.*, 2012).

All of these findings have capitalized upon the ability of chronic electrodes to track dopamine over multiple sessions in an individual and to use within-subject comparisons between different experimental conditions. This approach has also proved valuable for investigating how dopamine signals are perturbed by environmental manipulations, including the effects of stress (Wanat *et al.*, 2013) or abused substances. In one avenue of research, it was demonstrated that dopamine release in the nucleus accumbens elicited by cues predicting risky (i.e., probabilistic) rewards is enhanced in animals that voluntarily consumed alcohol during adolescence, and this neurochemical change is accompanied by a stronger economic risk preference (Nasrallah *et al.*, 2011). Other research has shown that increased dopamine transmission is concomitant with the elevated sensitivity to drug-associated stimuli that takes place following repeated cocaine administration (Ostlund *et al.*, 2014) and this signaling is under the control of the orexin/hypocretin system (Levy *et al.*, 2017). Over the past several years, our lab has undertaken the task of conducting a survey of striatal dopamine signaling across different phases of substance use in rat models. These experiments have especially benefited from the use of chronic electrodes by tracking changes in dopamine transmission across extended periods of drug use and correlating these neurochemical trajectories with changes in individual behavior. In the following section of this chapter, we will elaborate on this subset of studies to illustrate how the use of chronic electrodes provides tangible benefits for studying the neurobiological basis of behavior.

10.3 Investigation of neural mechanisms of addictive behaviors

A particular focus of our lab has been to use FSCV at chronic electrodes across the course of substance use to track changes in dopamine transmission in nucleus accumbens core (NAcc), a subregion of the ventral striatum, and dorsolateral striatum (DLS). This work is motivated by theories of drug addiction that implicate dopaminergic transmission in these striatal regions (Wise, 1987; Robinson and Berridge, 1993; White, 1996; Robbins and Everitt, 1999; Berke and Hyman, 2000; Dackis and O'Brien, 2001; Volkow *et al.*, 2012). The dopaminergic input from the ventral tegmental area to the NAcc is hypothesized to regulate drug intake as agents work to maintain elevated levels of tonic dopamine (Wise *et al.*, 1995; Suto and Wise, 2011). This pathway has also been implicated in the attribution of incentive salience to conditioned stimuli associated with drugs, and augmentation thereof ("incentive sensitization") following continued drug use (Robinson and Berridge, 1993). Other investigations have concluded that dopamine transmission in the DLS, a region linked to the formation of habitual behavior (Yin *et al.*, 2004;

Smith and Graybiel, 2016), subsumes control of drug seeking following chronic drug use (Everitt and Robbins, 2016). All of these accounts predict alterations in dopamine signaling during the development of addiction. However, the direction, timing and anatomical location of the changes seem somewhat disparate on the surface, and yet each of the theories are based on substantial empirical evidence. We will describe results from our laboratory that directly measure changes in phasic dopamine across different aspects of addiction and begin to provide some insight into the integration of these contemporary theories.

10.3.1 Longitudinal neurochemical recordings

Our lab performed longitudinal behavioral studies while recording phasic dopamine transmission with FSCV at chronic electrodes over weeks of cocaine self-administration. Rats were equipped with indwelling intravenous jugular catheters and implanted with chronic carbon-fiber microelectrodes in the NAcc. They were trained to self-administer cocaine through an operant response in which one of two nose-poke ports elicited an intravenous infusion of cocaine accompanied by an audiovisual conditioned stimulus (CS) and the other was inactive (controlling for general exploration). Animals were provided one-hour daily drug access for three weeks following training. Electrochemical recordings from this study show a phasic dopamine response in the NAcc during CS presentation following a nose poke in the active port in the first week of drug access, as previously observed with other operandum (Phillips *et al.*, 2003; Stuber *et al.*, 2005; Owesson-White *et al.*, 2009). Beyond this replication, we were able to track phasic dopamine over subsequent sessions. While the dopamine signal was present throughout our testing period, there was an incremental, but significant, decrease in its amplitude over the three weeks of cocaine use (Willuhn *et al.*, 2012). This decrease in phasic dopamine was contrary to our working hypothesis at the time, as we were preoccupied with the notion that dopamine transmission would increase during the progression of substance use. Importantly though, by capitalizing on the use of chronic electrodes, we were able to objectively track how this neurochemical signal changes over weeks of repeated drug use.

10.3.2 Neurochemical recordings in multiple brain regions

In addition to detecting dopamine in the NAcc, the chronic electrode allowed us to concurrently measure dopamine in the DLS of the same animals (Figure 10.2). Like in the NAcc, we observed phasic dopamine transmission in the DLS during CS presentation following an operant response. However, this signal did not emerge until the second week of drug access, demonstrating differences between

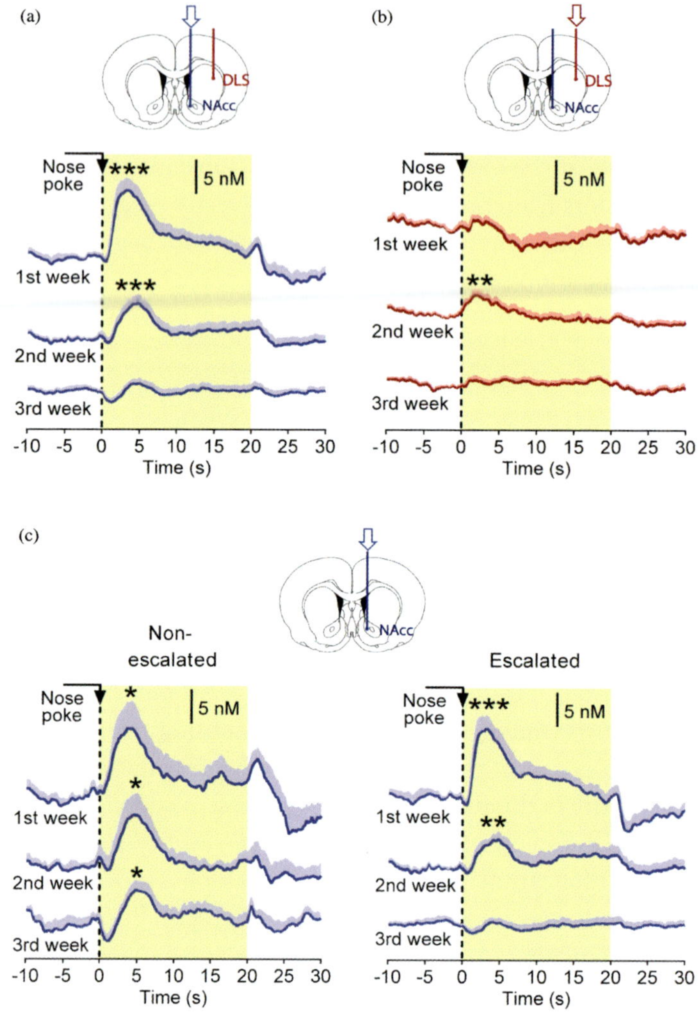

Figure 10.2 Neurochemical recording across weeks of cocaine self-administration. Active nose pokes (dashed line) elicited an infusion of cocaine and the presentation of a 20-s audiovisual stimulus (yellow box). (a) Phasic dopamine transmission in the NAcc was changed over the weeks of self-administration. During the first and second weeks of self-administration phasic dopamine signaling was significantly different from baseline, this significance was not observed in the third week of self-administration (n = 18 electrodes). (b) Phasic dopamine release in the DLS during the drug-paired audiovisual stimulus was only observed to be significantly different from base line in the second week of self-administration (n = 18 electrodes). (c) Phasic dopamine release differed among individuals. Animals were categorized by their drug-taking pattern through a linear regression between number of cocaine infusions and self-administration sessions. Non-escalated (left; n = 10) have a significant increase from baseline of phasic dopamine release in the NAcc through all three weeks of self-administration. Animals with escalated drug intake (right; n = 14) had a significant increase from baseline of phasic dopamine release only in the first two weeks of self-administration, and the peak amount of dopamine decreased across the weeks drug-taking. Data are mean + s.e.m. *$p < 0.05$, **$p < 0.01$, ***$p < 0.001$. After Willuhn *et al.*, 2014.

the long-term dynamics of these striatal systems (Willuhn *et al.*, 2012). Notably, the appearance of the phasic dopamine signal in DLS coincided with the refinement of discriminatory behavior. Specifically, as this dopamine signal emerged, animals made fewer nose-poke responses in the inactive port (i.e., actions that did not elicit drug delivery). The causality of this relationship was tested pharmacologically by administering a dopamine receptor antagonist into the DLS after the phasic dopamine signal had developed (week 3). This manipulation increased responding on the inactive lever, returning behavioral discrimination to the early-training state. Therefore, these data indicate that the evolution of the phasic dopamine signal in the DLS refines behavior by suppressing irrelevant actions. Overall, the different temporal profiles of phasic dopaminergic signaling observed between the NAcc and the DLS during cocaine self-administration support the position that there is progressive recruitment of neural activity within the striatum (Berke and Hyman, 2000; Everitt and Robbins, 2005; Kalivas and Volkow, 2005; Porrino *et al.*, 2007). This work was feasible because the use of FSCV at chronic electrodes permitted longitudinal recording of dopamine across weeks of drug access and concurrent detection of dopamine transients in distinct striatal subregions.

10.3.3 Bilateral recording to control for brain manipulations

We have discussed differences in the long-term dynamics of dopamine signaling between the NAcc and DLS as independent phenomena. However, there is evidence that the NAcc influences DLS dopamine function through serial connectivity as a sequential process (Belin and Everitt, 2008). To test the influence of this serial connectivity on the emergence of DLS phasic dopamine during self-administration, we measured dopamine in the DLS bilaterally, following a lesion of the NAcc on one side of the brain. This unilateral perturbation prevented the emergence of DLS phasic dopamine signaling on the ipsilateral, but not the contralateral side of the brain during cocaine self-administration (Willuhn *et al.*, 2012). These data demonstrate that intact NAcc circuitry is required for the development of the phasic dopamine signal in the DLS. The hemisphere selectivity of the manipulation verifies that the loss of DLS dopamine signaling is not simply due to degradation of the behavior, but a result of a specific circuit action. These data further support the conception that transitions from NAcc to DLS dopamine signaling during establishment of behavior come about through serial connectivity between these striatal regions (Belin and Everitt, 2008; Everitt and Robbins, 2016). Moreover, this finding demonstrates the application of multisite chronic electrodes to compare a control and experimental recording site within the same animal. This approach is particularly useful as within-animal comparisons appropriately control for general effects that arise when experimental manipulations compromise behavioral and/or cognitive states.

10.3.4 Tracking individual differences in neurochemical trajectories

In an extension of these experiments, our lab employed chronic carbon-fiber microelectrodes to investigate the temporal characteristics of phasic dopamine signaling during the transition from moderate to excessive drug consumption (Willuhn *et al.*, 2014). In these studies, animals were given extended six-hour drug access (commonly referred to as long access: LgA), a procedure capable of producing escalated drug intake (Ahmed and Koob, 1998). During three weeks of extended access, individual differences in behavior were observed, in which there was stable drug consumption in 42% of animals, and escalation of drug intake in the remaining 58% (Willuhn *et al.*, 2014).

Phasic dopamine release in the NAcc attenuated progressively over this period. In contrast, phasic dopamine signaling was preserved across sessions in animals with stable daily drug consumption (Figure 10.2; Willuhn *et al.*, 2014). Pharmacological manipulations were used to test if the significant correlation between neurochemistry and behavior had a causal relationship. When phasic dopamine was restored with the dopamine precursor, L-DOPA, in animals that escalated drug taking, consumption returned to pre-escalation levels. This manipulation did not have a significant effect on drug intake in the animals with stable drug consumption (i.e., those with unimpaired phasic dopamine signaling). Thus, it appears that the attenuation of phasic dopamine release in the NAcc is necessary in the development of escalated drug consumption. This work demonstrates that phasic dopamine transmission in the NAcc regulates drug intake, and it highlights the utility of being able to track trajectories of dopamine signals within animals across behavioral paradigms, in particular to identify individual differences in neurochemistry underlying behavior.

10.3.5 Measuring neurochemical profiles across multiple behaviors

Willuhn and colleagues' (2014) results provide credence for theories from Roy Wise on drug consumption (Wise, 1987) and extend these ideas on tonic dopamine levels to include phasic dopamine transmission. However, the decrease in phasic dopamine may seem at odds with the incentive sensitization hypothesis (Robinson and Berridge, 1993) where dopamine levels are posited to increase following chronic substance use. Importantly though, the incentive sensitization hypothesis demarcates craving in the absence of drug rather than rates of drug consumption. Therefore, if both of these theoretical frameworks are to bear scrutiny, then changes in dopamine transmission necessarily follow divergent trajectories for the different drug-related behaviors, craving and consumption. We have begun to test this prediction by examining dopamine across drug-related contexts throughout chronic substance use. As previously demonstrated (Phillips *et al.*, 2003; Willuhn *et al.*, 2012), we find that when drug cues are presented to rats outside a drug-taking session, they elicit dopamine release. Our recent results indicate that these signals increase with drug experience as well as with extended periods of drug withdrawal. The increase in this dopamine signal is most robust in animals that escalated cocaine consumption, the group in which we had previously observed decreased dopamine signaling (Willuhn *et al.*, 2014). Indeed, within the same individuals, dopamine release to cues presented when drug was successfully obtained during self-administration decreased whereas dopamine release to the same cues presented non-contingently increased. Therefore, changes in dopamine that take place as a result of chronic drug use in animals that escalate drug consumption are diametrically opposed depending on the drug-related context. This work again emphasizes the importance of being able to track individual differences in longitudinal neurochemical trajectories and highlights the utility of using a greater repertoire of behaviors within an individual to conduct these studies.

10.3.6 Summary of drug self-administration studies

In these studies using FSCV at chronic electrodes in rat models of substance abuse, we have observed correlations and causal relationships between neurochemistry and behavior. Our data demonstrate that phasic dopamine in the NAcc during cocaine self-administration regulates drug consumption. In a subset of animals, this signal decreases during chronic drug use producing escalation of intake. In contrast, we find augmented phasic dopamine transmission in the NAcc during non-contingent presentation of drug-associated cues in animals with a history of drug use. These augmented dopamine signals render animals reactive to drug cues when drug is not present, and they correspond with increased drug-seeking behaviors. Thus, our data demonstrate diametrically opposed changes in dopamine transmission in the NAcc between two distinct contexts which drive cardinal components of substance-use disorders: consumption and craving. Moreover, these findings have provided insight into the integration of two seemingly disparate theories of the neurobiology of substance abuse (Suto and Wise, 2011; Saunders *et al.*, 2013). We have also collected some data that speak to the role of dopamine in the DLS in addiction. Specifically, we have provided corroborative evidence that the DLS is recruited progressively during substance use under the control of the NAcc. However, our experiments have not yet addressed the crux of theories on DLS function in addiction that pertain to persistent drug seeking, maintained by the presentation of drug cues.

10.4 Further advances in chronic-electrode protocol and design

A considerable amount of information has been gained through the use FSCV at chronic electrodes. While we discussed our own studies on rat models of substance abuse in detail to illustrate some of the distinct benefits of this approach, there has also been a substantial amount of work on other topics and by other groups, of which just a subset received cursory attention in Section 10.2.2. However, despite the plethora of data that has been collected, there are still several avenues that are yet to reach fruition.

The majority of studies currently using chronic electrodes with FSCV are conducted in rats and mice. Moving beyond rodents, Schwerdt and colleagues (2017a) recently adapted the chronic-electrode approach for FSCV in non-human primates. Consequently, there are now relatively few barriers for using chronic electrodes in any mammalian laboratory model. Progress has also been made for human FSCV recordings. We collaborated with Read Montague, Kenneth Kishida and their colleagues to develop electrodes for use in humans, which were based upon the design of the fused-silica chronic electrode (Kishida *et al.*, 2011). To date, these sensors have been used in acute applications for intraoperative recordings while the subject engaged in a gambling task (Kishida *et al.*, 2016; Moran *et al.*, 2018). However, modification for chronic implantation could reap the advantages of longitudinal recordings as well as provide potential clinical benefits such as tracking progress of disease and providing closed-loop feedback for the regulation of therapeutic deep-brain stimulation (see Chang *et al.*, 2014).

Another limitation in the current use of the chronic electrodes is in the analytes that are studied, namely just dopamine. There is obvious potential for other electroactive small-molecules neurochemicals to be quantified (e.g., norepinephrine, serotonin, melatonin, adenosine) as these have been studied in acute applications. The primary barriers to date have been achieving adequate sensitivity and resolving mixtures of analytes. There are a number of potential means to overcome these obstacles, but some of the most promising are innovative approaches to data analysis that have already been successful in resolving dopamine and serotonin from intraoperative (acute) recordings in the human brain (Moran *et al.*, 2018).

While the iteration of chronic electrode that we introduced has good biocompatibility and is small enough for multiple-electrode implantation, there is always room for improvement. To this end, Schwerdt and colleagues (2017b, 2018) have developed probes using the same carbon fibers, but with an insulating layer that renders the electrode shaft an order of magnitude smaller in diameter.

In conclusion, we have argued that there are a number of tractable advantages of using chronically implanted electrodes for FSCV compared to acute recordings. Specifically, this approach enables the detection of dynamic neurochemical signatures in an individual across the course of psychological and pathological processes, and treatment regimens. These experiments permit longitudinal analysis and within-subject statistical comparisons. As the species of study and chemical analytes are expanded, and the probes are improved there is much still to be discovered using this approach.

Acknowledgments

Jeremy J. Clark, PhD, was instrumental in developing and utilizing the chronic voltammetry electrodes described in this chapter. For his pioneering work he received the Presidential Early Career Award for Scientists and Engineers from President Obama in 2014. Jeremy prematurely passed away in 2018. His scientific acumen and his friendship are sorely missed. Lauren M. Bugeno, PhD, collected the unpublished data presented in Section 10.3.5. This work was supported by NIH/NIDA grant DA039687.

References

Ahmed SH, Koob GF (1998) Transition from moderate to excessive drug intake: change in hedonic set point. *Science*, 282: 298–300.

Aitken TJ, Greenfield VY, Wassum KM (2016) Nucleus accumbens core dopamine signaling tracks the need-based motivational value of food-paired cues. *J Neurochem*, 136: 1026–1036.

Belin D, Everitt BJ (2008) Cocaine seeking habits depend upon dopamine-dependent serial connectivity linking the ventral with the dorsal striatum. *Neuron*, 57: 432–441.

Berke JD, Hyman SE (2000) Addiction, dopamine, and the molecular mechanisms of memory. *Neuron*, 25: 515–532.

Cheer JF, Wassum KM, Sombers LA, Heien ML, Ariansen JL, Aragona BJ, Phillips PEM, Wightman RM (2007) Phasic dopamine release evoked by abused substances requires cannabinoid receptor activation. *J Neurosci*, 27: 791–795.

Clark JJ, Collins AL, Sanford CA, Phillips PEM (2013) Dopamine encoding of Pavlovian incentive stimuli diminishes with extended training. *J Neurosci*, 33: 3526–3532.

Clark JJ, Hollon NG, Phillips PEM (2012) Pavlovian valuation systems in learning and decision making. *Curr Opin Neurobiol*, 22: 1054–1061.

Clark JJ, Sandberg SG, Wanat MJ, Gan JO, Horne EA, Hart AS, Akers CA, Parker JG, Willuhn I, Martinez V, Evans SB, Stella N, Phillips PEM (2010) Chronic microsensors for longitudinal, subsecond dopamine detection in behaving animals. *Nat Methods*, 7: 126–129.

Collins AL, Aitken TJ, Greenfield VY, Ostlund SB, Wassum KM (2016a) Nucleus accumbens acetylcholine receptors modulate dopamine and motivation. *Neuropsychopharmacology*, 41: 2830–2838.

Collins AL, Greenfield VY, Bye JK, Linker KE, Wang AS, Wassum KM (2016b) Dynamic mesolimbic dopamine signaling during action sequence learning and expectation violation. *Sci Rep*, 6: 20231.

Dackis CA, O'Brien CP (2001) Cocaine dependence: A disease of the brain's reward centers. *J Subst Abuse Treat*, 21: 111–117.

Day JJ, Roitman MF, Wightman RM, Carelli RM (2007) Associative learning mediates dynamic shifts in dopamine signaling in the nucleus accumbens. *Nat Neurosci*, 10: 1020–1028.

Everitt BJ, Robbins TW (2005) Neural systems of reinforcement for drug addiction: From actions to habits to compulsion. *Nat Neurosci*, 8: 1481–1489.

Everitt BJ, Robbins TW (2016) Drug addiction: Updating actions to habits to compulsions ten years on. *Annu Rev Psychol*, 67: 23–50.

Flagel SB, Clark JJ, Robinson TE, Mayo L, Czuj A, Willuhn I, Akers CA, Clinton SM, Phillips PEM, Akil H (2011) A selective role for dopamine in stimulus-reward learning. *Nature*, 469: 53–57.

Gan JO, Walton ME, Phillips PEM (2010) Dissociable cost and benefit encoding of future rewards by mesolimbic dopamine. *Nat Neurosci*, 13: 25–27.

Gentry RN, Lee B, Roesch MR (2016) Phasic dopamine release in the rat nucleus accumbens predicts approach and avoidance performance. *Nat Commun*, 7: 13154.

Hamid AA, Pettibone JR, Mabrouk OS, Hetrick VL, Schmidt R, Vander Weele CM, Kennedy RT, Aragona BJ, Berke JD (2016) Mesolimbic dopamine signals the value of work. *Nat Neurosci*, 19: 117–126.

Hart AS, Rutledge RB, Glimcher PW, Phillips PEM (2014) Phasic dopamine release in the rat nucleus accumbens symmetrically encodes a reward prediction error term. *J Neurosci*, 34: 698–704.

Hassler C, Boretius T, Stieglitz T (2011) Polymers for neural implants. *J Polym Sci B: Polym Phys*, 49: 18–33.

Hollon NG, Arnold MM, Gan JO, Walton ME, Phillips PEM (2014) Dopamine-associated cached values are not sufficient as the basis for action selection. *Proc Natl Acad Sci USA*, 111: 18357–18362.

Howe MW, Tierney PL, Sandberg SG, Phillips PEM, Graybiel AM (2013) Prolonged dopamine signalling in striatum signals proximity and value of distant rewards. *Nature*, 500: 575–579.

Kalivas PW, Volkow ND (2005) The neural basis of addiction: A pathology of motivation and choice. *Am J Psychiatry*, 162: 1403–1413.

Kishida KT, Saez I, Lohrenz T, Witcher MR, Laxton AW, Tatter SB, White JP, Ellis TL, Phillips PEM, Montague PR (2016) Subsecond dopamine fluctuations in human striatum encode superposed error signals about actual and counterfactual reward. *Proc Natl Acad Sci USA*, 113: 200–205.

Kishida KT, Sandberg SG, Lohrenz T, Comair YG, Saez I, Phillips PEM, Montague PR (2011) Sub-second dopamine detection in human striatum. *PLoS One*, 6: e23291.

Ko D, Wanat MJ (2016) Phasic dopamine transmission reflects initiation vigor and exerted effort in an action- and region-specific manner. *J Neurosci*, 36: 2202–2211.

Kruk ZL, Cheeta S, Milla J, Muscat R, Williams JE, Willner P (1998) Real time measurement of stimulated dopamine release in the conscious rat using fast cyclic voltammetry: Dopamine release is not observed during intracranial self stimulation. *J Neurosci Methods*, 79: 9–19.

Levy KA, Brodnik ZD, Shaw JK, Perrey DA, Zhang Y, Espana RA (2017) Hypocretin receptor 1 blockade produces bimodal modulation of cocaine-associated mesolimbic dopamine signaling. *Psychopharmacology (Berl)*, 234: 2761–2776.

Marsden CA, Joseph MH, Kruk ZL, Maidment NT, O'Neill RD, Schenk JO, Stamford JA (1988) *In vivo* voltammetry — Present electrodes and methods. *Neuroscience*, 25: 389–400.

Michael AC, Borland LM, eds (2007) *Electrochemical Methods for Neuroscience*. Boca Raton, FL: CRC Press/Taylor & Francis.

Moran RJ, Kishida KT, Lohrenz T, Saez I, Laxton AW, Witcher MR, Tatter SB, Ellis TL, Phillips PEM, Dayan P, Montague PR (2018) The protective action encoding of serotonin transients in the human brain. *Neuropsychopharmacology*, 43: 1425–1435.

Nasrallah NA, Clark JJ, Collins AL, Akers CA, Phillips PEM, Bernstein IL (2011) Risk preference following adolescent alcohol use is associated with corrupted encoding of costs but not rewards by mesolimbic dopamine. *Proc Natl Acad Sci USA*, 108: 5466–5471.

Oleson EB, Gentry RN, Chioma VC, Cheer JF (2012) Subsecond dopamine release in the nucleus accumbens predicts conditioned punishment and its successful avoidance. *J Neurosci*, 32: 14804–14808.

Ostlund SB, LeBlanc KH, Kosheleff AR, Wassum KM, Maidment NT (2014) Phasic mesolimbic dopamine signaling encodes the facilitation of incentive motivation produced by repeated cocaine exposure. *Neuropsychopharmacology*, 39: 2441–2449.

Owesson-White CA, Ariansen J, Stuber GD, Cleaveland NA, Cheer JF, Wightman RM, Carelli RM (2009) Neural encoding of cocaine-seeking behavior is coincident with phasic dopamine release in the accumbens core and shell. *Eur J Neurosci*, 30: 1117–1127.

Papageorgiou GK, Baudonnat M, Cucca F, Walton ME (2016) Mesolimbic Dopamine encodes prediction errors in a state-dependent manner. *Cell Rep*, 15: 221–228.

Phillips PEM, Stuber GD, Heien ML, Wightman RM, Carelli RM (2003) Subsecond dopamine release promotes cocaine seeking. *Nature*, 422: 614–618.

Phillips PEM, Wightman RM (2003) Critical guidelines for validation of the selectivity of *in-vivo* chemical microsensors. *Trends Anal Chem*, 22: 509–514.

Porrino LJ, Smith HR, Nader MA, Beveridge TJ (2007) The effects of cocaine: A shifting target over the course of addiction. *Prog Neuropsychopharmacol Biol Psychiatry*, 31: 1593–1600.

Robbins TW, Everitt BJ (1999) Drug addiction: Bad habits add up. *Nature*, 398: 567–570.

Robinson TE, Berridge KC (1993) The neural basis of drug craving: An incentive-sensitization theory of addiction. *Brain Res Brain Res Rev*, 18: 247–291.

Roitman MF, Stuber GD, Phillips PEM, Wightman RM, Carelli RM (2004) Dopamine operates as a subsecond modulator of food seeking. *J Neurosci*, 24: 1265–1271.

Saunders BT, Yager YL, Robinson TE (2013) Cue evoked cocaine 'craving': Role of dopamine in the accumbens core. *J Neurosci*, 33: 13989–14000.

Schultz W, Dayan P, Montague PR (1997) A neural substrate of prediction and reward. *Science*, 275: 1593–1599.

Schwerdt HN, Shimazu H, Amemori KI, Amemori S, Tierney PL, Gibson DJ, Hong S, Yoshida T, Langer R, Cima MJ, Graybiel AM (2017a) Long-term dopamine neurochemical monitoring in primates. *Proc Natl Acad Sci USA*, 114: 13260–13265.

Schwerdt HN, Kim MJ, Amemori S, Homma D, Yoshida T, Shimazu H, Yerramreddy H, Karasan E, Langer R, Graybiel AM, Cima MJ (2017b) Subcellular probes for neurochemical recording from multiple brain sites. *Lab Chip*, 17: 1104–1115.

Schwerdt HN, Zhang E, Kim MJ, Yoshida T, Stanwicks L, Amemori S, Dagdeviren HE, Langer R, Cima MJ, Graybiel AM (2018) Cellular-scale probes enable stable chronic subsecond monitoring of dopamine neurochemicals in a rodent model. *Commun Biol*, 1: 144.

Smith KS, Graybiel AM (2016) Habit formation coincides with shifts in reinforcement representations in the sensorimotor striatum. *J Neurophysiol*, 115: 1487–1498.

Stuber GD, Wightman RM, Carelli RM (2005) Extinction of cocaine self-administration reveals functionally and temporally distinct dopaminergic signals in the nucleus accumbens. *Neuron*, 46: 661–669.

Suto N, Wise RA (2011) Satiating effects of cocaine are controlled by dopamine actions in the nucleus accumbens core. *J Neurosci*, 31: 17917–17922.

Syed EC, Grima LL, Magill PJ, Bogacz R, Brown P, Walton ME (2016) Action initiation shapes mesolimbic dopamine encoding of future rewards. *Nat Neurosci*, 19: 34–36.

Volkow ND, Wang GJ, Fowler JS, Tomasi D (2012) Addiction circuitry in the human brain. *Annu Rev Pharmacol Toxicol*, 52: 321–336.

Wanat MJ, Bonci A, Phillips PEM (2013) CRF acts in the midbrain to attenuate accumbens dopamine release to rewards but not their predictors. *Nat Neurosci*, 16: 383–385.

Wanat MJ, Kuhnen CM, Phillips PEM (2010) Delays conferred by escalating costs modulate dopamine release to rewards but not their predictors. *J Neurosci*, 30: 12020–12027.

Wassum KM, Ostlund SB, Loewinger GC, Maidment NT (2013) Phasic mesolimbic dopamine release tracks reward seeking during expression of Pavlovian-to-instrumental transfer. *Biol Psychiatry*, 73: 747–755.

White NM (1996) Addictive drugs as reinforcers: Multiple partial actions on memory systems. *Addiction*, 91: 921–949; discussion 951–965.

Wightman RM (2006) Detection technologies. Probing cellular chemistry in biological systems with microelectrodes. *Science*, 311: 1570–1574.

Willuhn I, Burgeno LM, Everitt BJ, Phillips PEM (2012) Hierarchical recruitment of phasic dopamine signaling in the striatum during the progression of cocaine use. *Proc Natl Acad Sci USA*, 109: 20703–20708.

Willuhn I, Burgeno LM, Groblewski PA, Phillips PEM (2014) Excessive cocaine use results from decreased phasic dopamine signaling in the striatum. *Nat Neurosci*, 17: 704–709.

Wise RA, Newton P, Leeb K, Burnette B, Pocock D, Justice JB, Jr. (1995) Fluctuations in nucleus accumbens dopamine concentration during intravenous cocaine self-administration in rats. *Psychopharmacology* (*Berl*), 120: 10–20.

Wise RA (1987) The role of reward pathways in the development of drug dependence. *Pharmacol Ther*, 35: 227–263.

Yin HH, Park BS, Adermark L, Lovinger DM (2007) Ethanol reverses the direction of long-term synaptic plasticity in the dorsomedial striatum. *Eur J Neurosci*, 25: 3226–3232.

CHAPTER 11

REGIONALLY DISTINCT DOPAMINE RELEASE DYNAMICS BETWEEN STRIOSOME AND MATRIX COMPARTMENTS OF THE STRIATUM: ROLE IN BASAL GANGLIA DISORDERS

Armando G. Salinas*,† and Yolanda Mateo*,‡

*Laboratory for Integrative Neuroscience,
Division of Intramural Clinical and Biological Research,
National Institute on Alcohol Abuse and Alcoholism, USA
†Department of Bioengineering, George Mason University, USA
‡mateoy@mail.nih.gov

11.1 Introduction

Dopamine is a catecholamine neurotransmitter implicated in several physiological processes including movement, reinforcement, substance abuse, and synaptic plasticity (Graybiel *et al.*, 1990; Cenci, 2007; Pignatelli and Bonci, 2015; Volkow *et al.*, 2017). There are two primary dopaminergic pathways in the basal ganglia. Briefly, the mesocorticolimbic projection originates in dopamine neurons in the ventral tegmental area and projects to the ventral striatum, olfactory tubercle, and cortex. In contrast, the nigrostriatal pathway originates from dopamine neurons in the substantia nigra and terminates in the dorsal striatum or caudate nucleus and putamen in primates (Bjorklund and Dunnett, 2007). Abnormal function of dopamine neurons or their signaling pathways is associated with a number of neurological and psychiatric disorders including Parkinson's disease, depression, mania, and schizophrenia (Beaulieu and Gainetdinov, 2011). Thus, the study of dopamine is an area of great interest for biomedical researchers. Dopamine can be measured in several ways, each with its respective advantages and limitations. Microdialysis sample collection coupled with chromatography or other analytical assays, for example, allows for the measurement of various neurotransmitters, including dopamine, from extracellular fluid (Chefer *et al.*, 2009). This technique, however, has limited temporal resolution (Westerink *et al.*, 1987). Thus, this chapter will outline recent work examining the functional role of striosome and matrix striatal compartments in basal ganglia disorders with an emphasis on subsecond dopamine signaling measured with fast-scan cyclic voltammetry, as well as, novel transgenic models and research technologies.

11.2 Fast-Scan Cyclic Voltammetry: Monitoring dopamine dynamics in mouse striatum

Fast-scan cyclic voltammetry (FSCV) is a well-known electrochemical technique that has commonly been used to monitor dopamine release dynamics in the rodent striatum which is heavily innervated by dopamine axons (Giros *et al.*, 1996; Jones *et al.*, 1998; Budygin *et al.*, 2002; John and Jones, 2007).

In FSCV, a triangular waveform is applied to a glass- or epoxy-encased carbon fiber at 400 V/s to quickly oxidize and reduce electroactive species at the surface of the electrode. A typical waveform scan to detect dopamine goes from -0.4 V to +1.2 V and is repeated at 100 ms intervals. This rapid scan generates a large background current. To detect the smaller faradaic currents and isolate the detection of dopamine, the background current is digitally subtracted and in the resulting subtracted cyclic voltammogram, the peak oxidation and reduction potentials provide a chemical signature to identify dopamine. The oxidation peak current is proportional to the amount of dopamine at the surface of the carbon fiber and thus can be converted into concentrations using calibration factors obtained from standards of known concentration.

FSCV has frequently been coupled to carbon fiber microelectrodes to enable spatially resolved recordings of rapid dopamine dynamics. The use of transgenic mouse lines has been routinely applied to the study of dopamine neurotransmission since the first studies on the dopamine transporter knockout mouse (Giros *et al.*, 1996) and since then the use of the technique together with optogenetic tools has been a valuable asset in the characterization of striatal circuits (Cachope *et al.*, 2012; Chuhma *et al.*, 2014; Hawes *et al.*, 2017; Mateo *et al.*, 2017).

The use of acutely-prepared brain slices provides a valuable method to study intact brain regions. Using brain slices from transgenic mice expressing striosome- or matrix-enriched fluorescent proteins, electrode placement can be made considering anatomical landmarks and because the small size of the carbon fiber microelectrode (7 μm diameter, 120–200 μm in length) it is possible to differentiate structures like the striosome and matrix compartments in the mouse striatum (Salinas *et al.*, 2016).

11.3 Dopamine and basal ganglia disorders: Differences in striosome–matrix signaling

The striatum is the principal input of the basal ganglia and consequently there is abundant evidence supporting the critical function that this structure has in the control of voluntary movement and motivated behaviors (Crittenden and Graybiel, 2011). Therefore, neurons across striatum receive inputs from multiple brain regions (Smith and Bolam, 1990; Smith *et al.*, 1994; Voorn *et al.*, 2004) keeping a well-defined topographic organization that reflects different functions, with dorsolateral striatum involved in motor and action control and ventral striatum with limbic circuits (Voorn *et al.*, 2004). Moreover, the striatum is also organized in two neurochemically-distinct compartments, known as striosomes and matrix, which differ in the expression of several proteins like substance P, dopamine transporter (DAT), μ-opioid receptor (MOR), dopamine 1 (D1) receptor, met-enkephalin, Nr4a1 (nuclear receptor subfamily 4) and different cholinergic markers (Crittenden and Graybiel, 2011).

11.3.1 Anatomy and cellular composition of striosomes

Striosomes ("striatal bodies") were first identified as a series of small, complex, interconnected acetylcholinesterase-deficient regions that spanned the striatum of cats, rhesus macaques, and humans; the acetylcholinesterase rich/non-striosome regions were termed the matrix. (Graybiel and Ragsdale, 1978). Even prior to this seminal study, however, evidence for the existence of striosomes was emerging. Pert and colleagues described distinct "patches" of dense autoradiographic labeling of opiate receptors in the striatum (Pert *et al.*, 1975, 1976). Similarly, Goldman and Nauta (1977) and Ragsdale and Graybiel (1981) found intricate patterns of the prefrontal cortex to caudate projections of rhesus macaques that, in hindsight, were likely striosomes. Indeed, decades later Friedman *et al.* (2015) would provide evidence in rodents for a preferential innervation of medial prefrontal cortex corticostriatal afferents to dorsomedial striosomes.

Subsequent work found that striosomes could be neurochemically distinguished from the matrix by the differential expression of a number of additional proteins (see Crittenden and Graybiel (2011) for a comprehensive list) including

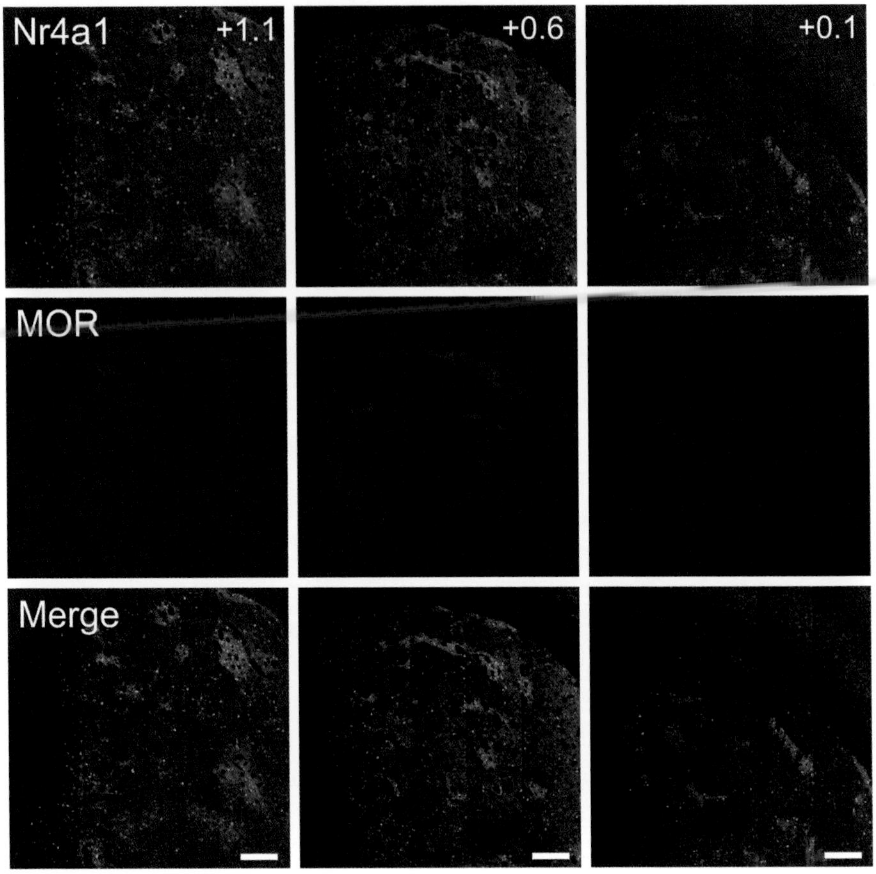

Figure 11.1 Nr4a1 is a reliable marker for striosomes throughout the rostral-caudal axis of the striatum. Nr4a1-eGFP mice were immunostained for the mu opioid receptor (MOR) and imaged at several points along the rostral-caudal axis. Note that in the caudal striatum, the MOR is no longer a suitable striosome marker. Scale bars are 200 μm. The approximate rostral-caudal Bregma coordinate is shown in the upper right for each series.

the mu and delta opioid receptors (Koizumi *et al.*, 2013; Tajima and Fukuda, 2013), enkephalin (Herkenham and Pert, 1981; Tajima and Fukuda, 2013), Nr4a1 (Davis and Puhl, 2011), calbindin (Gerfen *et al.*, 1985), aldehyde dehydrogenase 1 (Aldh1a1) (Sgobio *et al.*, 2017), and PSD95 (Morigaki and Goto, 2015). It should be noted, however, that in some cases, these striosome or matrix markers do not faithfully demarcate striosome or matrix compartments across the extent of the rostrocaudal or dorsoventral axes of the striatum (Figure 11.1). For example, Tajima and Fukuda (2013) found that the expression of three different striosome markers (the mu opioid receptor, substance P, and enkephalin) varied across the rostral caudal axis of the striatum. Furthermore, Davis *et al.* (2018) recently found that cannabinoid-1 receptors (CB1) followed an increasing ventromedial to dorsolateral distribution in mouse striatum.

The striatum is composed of two principal neuron types — direct pathway or D1 dopamine receptor-expressing and indirect pathway or D2 dopamine receptor-expressing medium spiny neurons (dMSN and iMSNs, respectively) — which combined comprise ~95% of the neurons in the striatum (Kemp and Powell, 1971; Gerfen and Surmeier, 2011). In addition, the striatum contains a number of interneuron subtypes which comprise the remaining ~5% of striatal neurons and are largely found in the matrix compartment (Figure 11.2). Striosomes are comprised of both dMSNs and iMSNs with a modest enrichment in the number of dMSNs (Fujiyama *et al.*, 2011; Watabe-Uchida *et al.*, 2012; Cui *et al.*, 2014; Banghart *et al.*, 2015; Smith *et al.*, 2016). Interestingly, the dendritic arbors of striosome or matrix MSNs remain completely within the striatal compartment they originate in. That is to say, the dendritic arbor of a striosome MSN will not cross over into proximal matrix regions and *vice versa* (Bolam *et al.*, 1988; Kawaguchi *et al.*, 1989; Walker *et al.*, 1993).

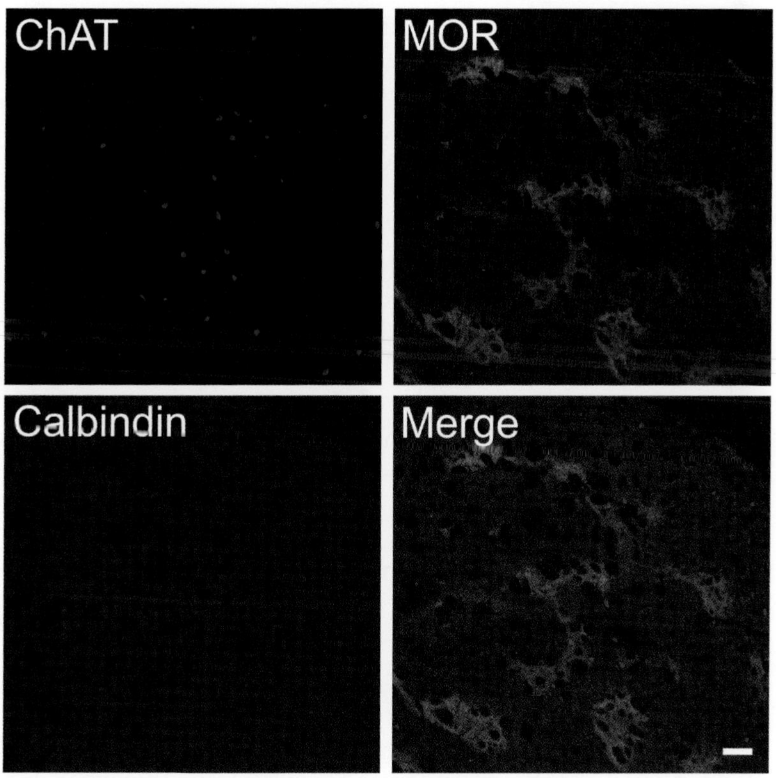

Figure 11.2 Striatal cholinergic interneurons are primarily located in the matrix compartment of the striatum. Immunohistochemical labeling of choline acetyltransferase (ChAT), the mu opioid receptor (MOR), and calbindin was carried out to label cholinergic interneurons (blue), striosomes (green), and the matrix (red), respectively. The scale bar is equal to 100 μm and applies to all panels.

11.3.2 Contrasting roles of striosome and matrix compartments in motor disorders

The differential activation of striosomes versus matrix has been best characterized for movement disorders like Huntington's disease, Parkinson's disease, dystonia, and other dyskinesias, as well as, in models of drug addiction (Crittenden and Graybiel, 2011). Huntington's disease is characterized by a progressive loss of neurons in the striatum and cerebral cortex and is caused by a CAG repeat expansion in the gene encoding huntingtin. Early studies in humans found that the disease progresses with early loss of neurons in striosomes and at a later phase neurodegeneration in both compartments (Tippett *et al.*, 2007). Several mouse models for Huntington's disease have been developed as knock-in mice with CAG repeats leading to increases in repetitive nuclear microaggregates that predominate in the striosomal compartment of the striatum, providing a possible explanation for the difference in neuronal vulnerability of striatal compartments observed in humans (Menalled *et al.*, 2002).

When looking at the different activation of striosome and matrix compartments in Parkinson's disease, the best confirmation to date comes from studies on the expression of pre-proenkephalin opioids. In postmortem striatal tissue from Parkinson's disease patients, pre-proenkephalin mRNA was enriched in striosomes relative to the matrix (Henry *et al.*, 2003). In a mouse model of dopamine deficiency, a differential loss of dopaminergic axons in striosomes relative to matrix was found (Graybiel *et al.*, 1990). More recently, a population of dopaminergic neurons in the substantia nigra pars compacta expressing aldehyde dehydrogenase 1 family member A1(Aldh1a1) was found to preferentially degenerate in Parkinson's disease. Furthermore, it has been found that Aldh1a1-positive axons project predominantly to the rostral dorsal striatum and that genetic ablation of Aldh1a1 substantially increases the dopamine release in striosomes but not in matrix (Sgobio *et al.*, 2017).

11.3.3 Striosome involvement in reinforcement and drug associated behaviors

Similar to the striosome-matrix activity imbalances in motor disorders, imbalances between striosome and matrix compartments in immediate early gene (IEG) expression have been associated with psychostimulant-induced stereotypies (Canales and Graybiel, 2000; Crittenden and Graybiel, 2011; Jedynak *et al.*, 2012). In fact, the severity of the stereotyped behavior has been correlated to the degree of IEG induction in striosomes (Saka *et al.*, 2004). Interestingly, the cooperative activation of both D1 and D2 dopamine receptor expressing medium spiny neurons (MSNs) is required for this effect (Capper-Loup *et al.*, 2002).

A role for striosomes in mediating the reinforcing or rewarding properties of natural rewards and drugs of abuse has also been investigated. The first such study used an electrical intracranial self-stimulation model in the dorsomedial striatum of rats. Briefly, White and Hiroi (1998) trained rats to operantly self-administer an electrical stimulation and, after training, they performed a detailed histological examination of the electrode placement in each subject. They found that those subjects with electrode placements in or juxtaposed to striosomes acquired the self-stimulation behavior more rapidly and more stably than those subjects with stimulating electrodes located solely in the matrix. This study provided some of the first *in vivo* evidence for a role of the striosome compartment in reinforcement. In another study, Cui *et al.* (2014) employed a novel mouse model to study the role of the mu opioid receptor (which is enriched in striosomes) in morphine-induced reward. They used a bacterial artificial chromosome driven by a prodynorphin promoter and encoding a floxed mu opioid receptor (MOR) gene to re-introduce the expression of the MOR selectively into the striatum of a global MOR knockout mouse. Cui and colleagues then found that restoring MOR expression in striosomes and nucleus accumbens MSNs rescued self-administration of an opiate drug. Importantly, they showed that analgesia in the MOR MSN rescue mouse was unaltered from the MOR knockout mouse, suggesting a specific role for striosome MORs in mediating opiate reward.

With regards to drug addiction and the differential effects of drugs of abuse on striosome and matrix signaling, there have been many studies most of them finding a preferential loss of dopamine terminals in striosomes (Adams *et al.*, 2003; Granado *et al.*, 2008, 2010) and in human studies an increase in dynorphin signaling in striosomes in the caudate (Hurd and Herkenham, 1993).

11.4 Functional physiology of striosome-matrix neurons

11.4.1 Synaptic physiology of striosome-matrix neurons

Direct and indirect medium spiny neurons in striosomes have been examined electrophysiologically and mostly found not to differ from their matrix counterparts in most basal membrane properties; however, differences in dMSN and iMSN excitability have been noted. In the matrix, iMSNs are typically more excitable than dMSNs and have a greater number of evoked action potentials in response to applied current steps, suggesting a greater capacity to spike and for sustained periods of time (Planert *et al.*, 2013). In striosomes, however, the opposite has been shown by several groups. Striosome iMSNs are less excitable than dMSNs as measured by evoked action potential spiking in response to applied current steps (Miura *et al.*, 2008; Banghart *et al.*, 2015; Crittenden *et al.*, 2017). The physiological significance of this may be tied to the more limbic nature of striosomes relative to matrix. That is to say, striosomes have been implicated in mediating reward signals from brain regions associated with reinforcement. As such, striosome dMSNs may benefit from being relatively more excitable. In this way, striosome dMSNs might be able to better filter changes in stimulus salience by feeding back to substantia nigra pars compacta dopamine neurons to dampen dopamine release, the timing of which could allow for salient stimuli to be more discreetly encoded or globally dampened (e.g. during discounting or extinction tasks). A study by Yoshizawa *et al.* (2018) attempted to address this possibility and will be discussed at a later point in the chapter.

Examination of collateral synaptic transmission between striosome and matrix compartments has failed to reveal any cross compartment synaptic communication between MSNs of either type (Banghart *et al.*, 2015; Lopez-Huerta *et al.*, 2016). However, cross compartment communication must be occurring since imbalances between striosome and matrix activity have been linked to ostensible pathological motor conditions (Crittenden and Graybiel, 2011). Cross compartment synaptic communication is thus postulated to be mediated by striatal interneurons (Koizumi *et al.*, 2013). Indeed, striatal cholinergic interneurons are poised to serve as effectors of cross compartment communication given their peri-striosomal localization and broad synaptic connectivity within the striatum (Miura *et al.*, 2008) (Figure 11.2). Crittenden *et al.* (2017) recently demonstrated that activation of cholinergic interneuron somas in the matrix or near matrix-striosome borders could modulate the spiking activity of striosomal MSNs. In this way, cholinergic interneurons may influence striosome activity, including modulation of substantia nigra pars compacta projecting MSNs. This serves as a mechanism by which the balance between striosome and matrix compartment activity can be regulated, thus avoiding many of the pathophysiological states associated with imbalances in the activity of striosome and matrix compartments (Crittenden and Graybiel, 2011). The role (if any) of other striatal interneuron subtypes in mediating cross compartment communication has yet to be examined and thus remains an area of study ripe for exploration.

11.4.2 A unique striosome striatonigral circuit

Striosome dMSNs are unique in that they have a direct projection to dopamine neurons in the substantia nigra pars compacta (Gerfen *et al.*, 1985; Fujiyama *et al.*, 2011; Watabe-Uchida *et al.*, 2012). This direct modulation of dopamine neurons places striosome dMSNs in a unique position to influence dopamine release across the striatum. These specific striatonigral projections comprise part of the recently described striosome-dendron bouquets — midbrain structures consisting of striosome dMSN axons and dendrites from substantia nigra pars compacta dopamine neurons (Crittenden *et al.*, 2016). These bouquets highly express the CB1 and are thought to originate, at least partially, from dMSN axons in the CB1 rich striosomes of the caudal portion of the dorsolateral striatum (Davis *et al.*, 2018). Thus, it appears that striosome dMSNs and the striosome-dendron bouquets are in a position to gate converging synaptic inputs into substantia nigra pars compacta dopamine neurons to control global striatal dopamine dynamics.

11.4.3 But what do striosomes do?

While a great deal of descriptive work has been done with striosomes, very little experimental work has been possible given their small size. In fact, the combined volume of striosomes comprises only ~15% of the total striatal volume (Salinas *et al.*, 2016). Furthermore, the small size of striosomes has made them very difficult to specifically manipulate without recruiting proximal matrix regions as well. Nonetheless, a few studies have examined certain aspects of striosome function in behavior. Friedman *et al.* (2015) found that stimulating or inhibiting corticostriatal projections with predominant striosome or matrix termination fields could influence choice behavior. More specifically, they found that inhibition of a corticostriatal striosome-enriched input resulted in more seeking of a high cost, high reward reinforcer. In contrast, excitation of a corticostriatal striosome-enriched input decreased seeking of the high cost, high reward reinforcer. A follow up study from the same group found that chronic stress altered corticostriatal striosome projections and that this alteration resulted in aberrant decision-making (Friedman *et al.*, 2017).

Transgenic mice expressing enhanced Green Fluorescent Protein (eGFP) or cre recombinase with promoters for a number of proteins differentially expressed between striosome and matrix compartments have been developed and exploited to begin studies examining the physiological function of striosomes (Brimblecombe and Cragg, 2017). The most common method for discerning striosome function has been the use of a striosome-enriched cre recombinase expressing mouse line with viral expression of genetically-encoded calcium indicators (e.g. GCaMP6 variants) to monitor striosome or matrix neuronal activity during behavioral tasks. One such study by Bloem *et al.* (2017) used an original approach to birthdate label striosome neurons. This strategy exploited the earlier generation time of striosome MSNs and allowed nascent striosome MSNs to express cre recombinase and activate the expression of the TdTomato reporter (Ai14) mouse line. Thus, in the adult, striosomes could be identified by the expression of TdTomato at the soma and in the neuropil. Bloem and colleagues then injected a GCaMP6 virus into the dorsal striatum and used a microendoscope with two-photon microscopy to examine neuronal activation of striosome and matrix neurons in awake, head-fixed mice during a classical conditioning task. They found that striosome neurons more strongly encoded reward-predicting cues than matrix neurons. To date, this study remains one of a handful of studies examining differences in the function of striosome and matrix compartments.

One mouse line being used increasingly in striosome research is the Sepw1_NP67 Cre mouse line (Gerfen *et al.*, 2013; Smith *et al.*, 2016). This mouse line was described as having enriched cre expression in dorsal striatal striosomes (Gerfen *et al.*, 2013) and was recently used to study the potential for encoding of reward prediction in striosome MSNs (Yoshizawa *et al.*, 2018). They used the Sepw1_NP67 cre mouse line in combination with viral transfection of a cre-dependent calcium indicator, GCaMP6, to monitor the selective activity of striosome MSNs or a combination of striosome and matrix MSNs in a classical conditioning task. Yoshizawa and colleagues found that in the initial stages of training, striosome MSNs displayed reward predictive activity. At later stages of training, however, this encoding of reward would largely dissipate. Finally, they also found that some striosome MSNs were directly activated by the reinforcer delivery.

11.5 Compartment-specific modulation of dopamine release

11.5.1 Dopamine dynamics differ between striosome and matrix compartments

During the development of the striatum, the terminal fields of midbrain dopamine neurons form a mosaic pattern that is reminiscent of striosomes. This observation prompted Graybiel (1984) to examine these "dopamine islands" and determine if they corresponded to striosomes. Indeed, Graybiel found that dopamine islands corresponded to the acetylcholinesterase poor striosome compartment of the striatum. Graybiel further demonstrated that these dopamine islands were enriched in tyrosine hydroxylase, the rate limiting enzyme in catecholamine synthesis.

Given that striosomes were enriched in dopamine synthesis enzymes and that several previous studies found preferential immediate early gene activation in striosomes after chronic psychostimulant use, Salinas and colleagues (2016) decided to compare dopamine release dynamics between the striosome and matrix compartments of the striatum. To do this, they employed a transgenic mouse model that expressed eGFP under the control of the nuclear receptor subfamily 4 group A member 1 (Nr4a1) promoter in combination with acutely-prepared brain slices and fast-scan cyclic voltammetry. The Nr4a1-eGFP mouse line was shown to have robust and consistent labeling of striosomes (Davis and Puhl, 2011) throughout the striatum (Figure 11.1) and was thus chosen for the studies by Salinas and colleagues to *a priori* delineate between the striosome and matrix compartments.

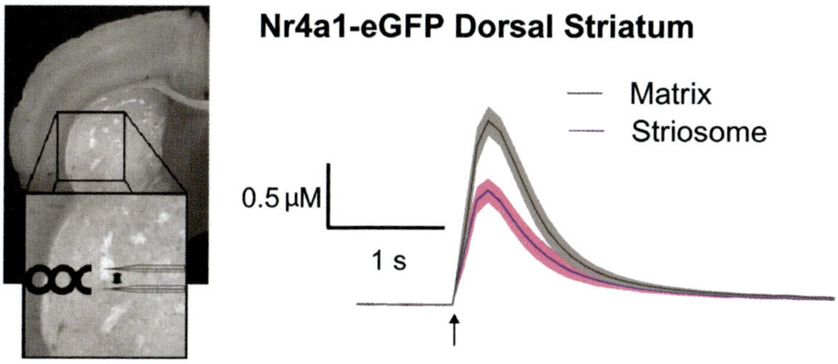

Figure 11.3 Dopamine release in the dorsal striatum differs between the matrix and striosome compartments of the striatum. The experimental configuration for recording dopamine release within the dorsal striatum with FSCV is shown on the left. The carbon fiber electrode was positioned in the matrix or striosomes compartment approximately 300 μm away from the stimulating electrode. The right panel shows average dopamine transients for matrix and striosomes compartments. The arrow represents time of electrical stimulation (1 pulse, 2 ms; 100–300 μA).

They found that evoked dopamine release was reduced in dorsal striatal striosomes relative to proximal matrix regions in the dorsal striatum (Figure 11.3). Similarly, Brimblecombe and Cragg (2015) found that dopamine release was greater in the dorsal striatal matrix relative to striosomes. However, in the ventral striatum, Salinas and colleagues found that evoked dopamine release was greater in striosomes relative to proximal matrix compartments. Given that the dorsal matrix regions had less tyrosine hydroxylase expression, it was expected that there would be less dopamine release in the dorsal matrix. The findings by Salinas *et al.* (2016), as well as Brimblecombe and Cragg (2015), were thus contrary to the initial hypothesis that evoked dopamine release would be greater in dorsal striatal striosomes. Nonetheless, to account for the observed compartment differences in dopamine release, Salinas and colleagues pharmacologically assessed the contribution of several factors known to regulate dopamine release. For example, it has been recently shown that activation of cholinergic interneurons could modulate dopamine release via nicotinic acetylcholine receptors (nAChRs) on presynaptic dopamine terminals (Zhang *et al.*, 2009; Cachope *et al.*, 2012; Threlfell *et al.*, 2012). Given that cholinergic interneurons are largely found in the matrix or matrix-striosome borders (Figure 11.2) and that the electrical stimulation used to evoke local dopamine release was also recruiting cholinergic systems to effect dopamine release, the cholinergic contribution to evoked dopamine release was assessed. Salinas *et al.* (2016) found no compartment differences in the cholinergic contribution to dopamine release in the dorsal or ventral striatum. Dopamine release was inhibited to the same degree by application of the β2-containg nAChR antagonist, DHβE. Once the dopamine terminals were pharmacologically isolated from the influence of nAChRs, the compartment difference in evoked dopamine release was reassessed and, again, found to be greater in the matrix compartment of the striatum relative to a proximal striosome region in the dorsal striatum. This suggests that the compartment differences in dopamine release are due to an intrinsic property of the dopamine terminals synapsing in each compartment. To explore this possibility, Salinas and colleagues then assessed two other parameters known to influence dopamine release and to be intrinsic to the dopamine terminal: D2 dopamine autoreceptors and the dopamine transporter (DAT).

Immunohistochemical staining for the D2 dopamine receptor reveals a modest enrichment of the D2 dopamine receptor in dorsal striatal striosomes (Salinas *et al.*, 2016). Thus, it was hypothesized that the D2 dopamine autoreceptor might account for the compartment difference in evoked dopamine release. To test this hypothesis, the function of the D2 dopamine autoreceptors was assessed pharmacologically with the D2/3 dopamine receptor agonist, quinpirole. As before, Salinas *et al.* (2016) applied quinpirole and found that dopamine release in either compartment was similarly affected, regardless of striatal dorsal or ventral subregion. Furthermore, in the presence of quinpirole, the striosome-matrix compartment difference in evoked dopamine release was still intact, suggesting that the difference was not due to differences in D2 dopamine autoreceptor activity. To reconcile the D2 dopamine receptor enrichment in immunohistochemical staining with the lack of an observed effect on evoked dopamine release, Salinas and colleagues designed an experiment to determine which population of D2 dopamine receptors in the striatum (i.e. on iMSNs, cholinergic interneurons, or on dopamine terminals) underlies the enriched D2 dopamine receptor expression in dorsal striatal striosomes. They produced a transgenic mouse in which the D2 dopamine receptor was removed from D2 dopamine receptor expressing MSNs and then repeated the immunohistochemistry experiment labeling D2 dopamine receptors. In this mouse, the immunostaining pattern of the D2 dopamine receptor was relatively homogenous across the dorsal striatum, suggesting that the "patchiness" of the D2 dopamine receptor was due to expression on iMSNs and not on dopamine terminals. Therefore, the lack of a compartment difference in evoked dopamine release modulation by the D2 dopamine autoreceptor was in accord with its neuroanatomical expression.

Immunolabeling of the DAT also revealed a slight enrichment in striosomes relative to the matrix, particularly in dorsolateral striatum (Salinas *et al.*, 2016). Therefore, Salinas and colleagues assessed the function of the DAT by modeling the uptake rates of evoked dopamine transients using a Michaelis-Menten kinetic model as well as application of the DAT inhibitor, cocaine, in a cumulative dosing experimental design. Examination of the evoked dopamine transients found no significant differences in basal DAT function (maximal uptake rate, Vmax) between the striosome and matrix compartments in either the dorsal or ventral striatum. However, application of cocaine had a greater inhibiting effect of the DAT in the dorsal striatum matrix compartment relative to striosomes. This compartment difference was not present in the ventral striatum.

11.5.2 Functional implications of compartment differences in dopamine release

Dopamine has been shown to play a critical role in corticostriatal plasticity (Calabresi *et al.*, 2007). Thus, the compartment differences in evoked dopamine release could influence the induction of synaptic plasticity. Indeed, Hawes *et al.* (2013) found that theta burst stimulation of corticostriatal afferents resulted in a D1 dopamine receptor-dependent, long-term potentiation (LTP) in the dorsomedial striatum. In a follow-up study, Hawes and colleagues attempted to augment the LTP by optogenetically stimulating dMSNs. To do this, they crossed a D1 dopamine receptor promoter-driven cre mouse line (EY217) with a floxed channelrhodopsin mouse line to express the excitatory channelrhodopsin protein selectively in dMSNs. They found that the EY217, D1 dopamine receptor cre line, is incomplete and follows an unusual expression pattern such that in the most rostral quarter of the dorsal striatum, the EY217 line appears to have a matrix-enriched distribution (unpublished observation), while in the caudal three-quarters of the dorsal striatum, the distribution is striosome-enriched (Hawes *et al.*, 2017). Interestingly, Hawes *et al.* (2017) found that light-responsive dMSNs (i.e. those dMSNs that displayed a strong channelrhodopsin-mediated current) did not show LTP under basal conditions. Given the striosome enrichment with this cre mouse line, it is likely that these dMSNs were located in striosomes and thus had a lower amount of dopamine release in response to each stimulation. Supporting this hypothesis, the same study found that in non-light responsive neurons (which were comprised of a mix of dMSNs and iMSNs), LTP was induced more readily (Hawes *et al.*, 2017).

To augment LTP, Hawes and colleagues optogenetically activated dMSNs during the induction of LTP. Rather than enhance LTP, however, optogenetic activation of dMSNs resulted in a loss of LTP. Hawes and colleagues then hypothesized that activation of dMSNs might be evoking the release of neuropeptides. Since dMSNs have been shown to express dynorphin, which acts via kappa opioid receptors (KORs), Hawes *et al.*, pharmacologically examined the dependency of the loss of LTP on KORs. Indeed, blockade of KORs during theta burst stimulation and optogenetic activation of dMSNs restored LTP. The mechanism underlying KOR activation-induced inhibition of LTP induction was unclear. Hawes and colleagues speculated that perhaps dMSN activation was inducing dynorphin release onto dopamine terminals to inhibit dopamine release. To explore this hypothesis, FSCV was employed in conjunction with the EY217, D1 Cre-channelrhodopsin mouse and pharmacology. They first applied the endogenous KOR agonist, dynorphin, to determine if this would inhibit dopamine release. They then found that optogenetic activation of dMSNs was able to reduce evoked dopamine release in a KOR-dependent manner. Altogether, the results from this study exemplify an interesting and unexpected role for striosomes and the differences in compartment dopamine release in synaptic physiology. Furthermore, these results highlight the necessity of thorough characterization of transgenic mouse lines with respect to striatal compartments.

11.5.3 Examining mouse models of Parkinson's disease with the Nr4a1-eGFP mouse

The relationship between striosomes and movement disorders, including Parkinson's disease, has been studied extensively (see Crittenden and Graybiel (2011) for review). These studies, however, have been mostly correlative and involve non-specific dopamine neuron lesions. This is an important point as not all dopamine neurons are the same (Crittenden and Graybiel, 2011; Roeper, 2013; Vogt Weisenhorn *et al.*, 2016). In fact, recent work has shown that a subtype of dopamine neurons expressing Aldh1a1 has striosome-enriched, but not exclusive, striatal projections (Cai *et al.*, 2014; Sgobio *et al.*, 2017). More specifically, the Aldh1a1-expressing dopamine neuron striosome projections are largely found in the rostral portions of the striatum and originate in the substantia nigra pars compacta. A second group of Aldh1a1-expressing dopamine neurons in the ventral tegmental area have also been described but do not display any enriched projections to striosomes. Interestingly, Aldh1a1-expressing dopamine neurons have been found to be preferentially degenerated in Parkinson's disease (Liu *et al.*, 2014).

Given the relationship between striosomes and Parkinson's disease, Sgobio and colleagues (2017) wondered if the compartment differences in dopamine release would be altered in Aldh1a1 knockout mice maintained on the Nr4a1-eGFP background mouse strain. In the Aldh1a1 knockout mouse line, Sgobio and colleagues found that the compartment difference in dopamine release was gone. In fact, the amount of dopamine released in striosomes matched the amount of dopamine release in the matrix compartment of a wild type mouse, suggesting that the level of dopamine release by striosomes in Aldh1a1 knockout mice was increased to match the matrix level. Interestingly, the compartment difference in uptake inhibition by cocaine described by Salinas and colleagues was absent in Aldh1a1 knockout mice.

To more directly explore the relationship between striosomes and Parkinson's disease in dopamine release, Sgobio *et al.* (2017) used a mouse model that overexpresses the alpha synuclein A53T mutation, which has been shown to impair dopamine release and is a widely-used model of Parkinson's disease, in the Nr4a1-eGFP mouse line. As expected, dopamine release was decreased in mice expressing the A53T mutation. In contrast to the Aldh1a1 knockout mice, however, the dopamine levels in the matrix dropped to match those of striosomes. More in-depth examination of evoked dopamine release in the A53T mouse revealed that the maximal amount of dopamine release by striosome and matrix dopamine terminals was lower than in the striosomes of wild type mice. Interestingly, the compartment difference in dopamine uptake inhibition by cocaine was gone in the A53T mouse. Altogether the study by Sgobio *et al.* (2017) was the first to directly assess the effects of a Parkinson's mouse model on differences in compartmental dopamine release with novel transgenic mouse strains.

11.6 Novel methods and future directions

In the last few years, clever use of new transgenic mouse models combined with the subsecond resolution of FSCV to measure dopamine and genetically-encoded calcium sensors to measure somatic activity has enabled researchers to address important questions about the physiological significance of striosomes. There is no doubt that as the tools and technology to investigate the function of brain circuits continues to advance so will the complexity of experiments and our knowledge of the brain.

One lingering question in the field is whether the striatal compartment differences observed *ex vivo* are present *in vivo*. This has proven a difficult question to address on account of the small size of striosomes and the expertise required to properly conduct an *in vivo* freely-moving, FSCV experiment. However, the recent advent of genetically-encoded fluorescent biosensors and microendoscopic (e.g. GRIN lens) imaging methods has made it possible to address this question. In particular, the novel dopamine sensor, dLight (Patriarchi *et al.*, 2018), has shown promise and will be used in combination with either a matrix- or striosome-enriched cre mouse lines (Plxnd1-OG1 or Sepw1_NP67, respectively) expressing TdTomato to delineate striatal compartments. By combining these transgenic mouse models with the innovative dopamine sensor, it is possible to address the *in vivo* relevance of the observed striosome-matrix difference in dopamine release. Novel genetically-encoded sensors are not limited to dopamine; indeed, novel sensors for a number of neurotransmitters are being developed and will no doubt revolutionize the study of brain circuit neuromodulation.

Acknowledgments

The authors wish to thank Dr. Margaret I. Davis for helpful discussions on the development of the striosome project and Dr. Shana M. Augustin for helpful suggestions on the manuscript.

This work was supported by the National Institutes of Health, National Institute on Alcohol Abuse and Alcoholism: K99AA025991 (AGS), Division of Intramural Clinical and Biological Research, ZIA AA000407 (YM).

References

Adams DH, Hanson GR, Keefe KA (2003) Distinct effects of methamphetamine and cocaine on preprodynorphin messenger RNA in rat striatal patch and matrix. *J Neurochem*, 84: 87–93.

Banghart MR, Neufeld SQ, Wong NC, Sabatini BL (2015) Enkephalin disinhibits mu opioid receptor-rich striatal patches via delta opioid receptors. *Neuron*, 88: 1227–1239.

Beaulieu JM, Gainetdinov RR (2011) The physiology, signaling, and pharmacology of dopamine receptors. *Pharmacol Rev*, 63: 182–217.

Bjorklund A, Dunnett SB (2007) Dopamine neuron systems in the brain: An update. *Trends Neurosci*, 30: 194–202.

Bloem B, Huda R, Sur M, Graybiel AM (2017) Two-photon imaging in mice shows striosomes and matrix have overlapping but differential reinforcement-related responses. *Elife*, 6.

Bolam JP, Izzo PN, Graybiel AM (1988) Cellular substrate of the histochemically defined striosome/matrix system of the caudate nucleus: A combined Golgi and immunocytochemical study in cat and ferret. *Neuroscience*, 24: 853–875.

Brimblecombe KR, Cragg SJ (2015) Substance P weights striatal dopamine transmission differently within the striosome-matrix axis. *J Neurosci*, 35: 9017–9023.

Brimblecombe KR, Cragg SJ (2017) The striosome and matrix compartments of the striatum: a path through the labyrinth from neurochemistry toward function. *ACS Chem Neurosci*, 8: 235–242.

Budygin EA, John CE, Mateo Y, Jones SR (2002) Lack of cocaine effect on dopamine clearance in the core and shell of the nucleus accumbens of dopamine transporter knock-out mice. *J Neurosci*, 22: RC222.

Cachope R, Mateo Y, Mathur BN, Irving J, Wang HL, Morales M, Lovinger DM, Cheer JF (2012) Selective activation of cholinergic interneurons enhances accumbal phasic dopamine release: Setting the tone for reward processing. *Cell Rep*, 2: 33–41.

Cai H, Liu G, Sun L, Ding J (2014) Aldehyde dehydrogenase 1 making molecular inroads into the differential vulnerability of nigrostriatal dopaminergic neuron subtypes in Parkinson's disease. *Transl Neurodegener*, 3: 27.

Calabresi P, Picconi B, Tozzi A, Di Filippo M (2007) Dopamine-mediated regulation of corticostriatal synaptic plasticity. *Trends Neurosci*, 30: 211–219.

Canales JJ, Graybiel AM (2000) A measure of striatal function predicts motor stereotypy. *Nat Neurosci*, 3: 377–383.

Capper-Loup C, Canales JJ, Kadaba N, Graybiel AM (2002) Concurrent activation of dopamine D1 and D2 receptors is required to evoke neural and behavioral phenotypes of cocaine sensitization. *J Neurosci*, 22: 6218–6227.

Cenci MA (2007) Dopamine dysregulation of movement control in L-DOPA-induced dyskinesia. *Trends Neurosci*, 30: 236–243.

Chefer VI, Thompson AC, Zapata A, Shippenberg TS (2009) Overview of brain microdialysis. *Curr Protoc Neurosci* Chapter 7: Unit7 1.

Chuhma N, Mingote S, Moore H, Rayport S (2014) Dopamine neurons control striatal cholinergic neurons via regionally heterogeneous dopamine and glutamate signaling. *Neuron*, 81: 901–912.

Crittenden JR, Graybiel AM (2011) Basal ganglia disorders associated with imbalances in the striatal striosome and matrix compartments. *Front Neuroanat*, 5: 59.

Crittenden JR, Lacey CJ, Weng FJ, Garrison CE, Gibson DJ, Lin Y, Graybiel AM (2017) Striatal cholinergic interneurons modulate spike-timing in striosomes and matrix by an amphetamine-sensitive mechanism. *Front Neuroanat*, 11: 20.

Crittenden JR, Tillberg PW, Riad MH, Shima Y, Gerfen CR, Curry J, Housman DE, Nelson SB, Boyden ES, Graybiel AM (2016) Striosome-dendron bouquets highlight a unique striatonigral circuit targeting dopamine-containing neurons. *Proc Natl Acad Sci USA*, 113: 11318–11323.

Cui Y, Ostlund SB, James AS, Park CS, Ge W, Roberts KW, Mittal N, Murphy NP, Cepeda C, Kieffer BL, Levine MS, Jentsch JD, Walwyn WM, Sun YE, Evans CJ, Maidment NT, Yang XW (2014) Targeted expression of mu-opioid receptors in a subset of striatal direct-pathway neurons restores opiate reward. *Nat Neurosci*, 17: 254–261.

Davis MI, Crittenden JR, Feng AY, Kupferschmidt DA, Naydenov A, Stella N, Graybiel AM, Lovinger DM (2018) The cannabinoid-1 receptor is abundantly expressed in striatal striosomes and striosome-dendron bouquets of the substantia nigra. *PLoS One*, 13: e0191436.

Davis MI, Puhl HL (2011) Nr4a1-eGFP is a marker of striosome-matrix architecture, development and activity in the extended striatum. *PLoS One*, 6: e16619.

Friedman A, Homma D, Bloem B, Gibb LG, Amemori KI, Hu D, Delcasso S, Truong TF, Yang J, Hood AS, Mikofalvy KA, Beck DW, Nguyen N, Nelson ED, Toro Arana SE, Vorder Bruegge RH, Goosens KA, Graybiel AM (2017) Chronic stress alters striosome-circuit dynamics, leading to aberrant decision-making. *Cell*, 171: 1191–1205 e1128.

Friedman A, Homma D, Gibb LG, Amemori K, Rubin SJ, Hood AS, Riad MH, Graybiel AM (2015) A corticostriatal path targeting striosomes controls decision-making under conflict. *Cell*, 161: 1320–1333.

Fujiyama F, Sohn J, Nakano T, Furuta T, Nakamura KC, Matsuda W, Kaneko T (2011) Exclusive and common targets of neostriatofugal projections of rat striosome neurons: A single neuron-tracing study using a viral vector. *Eur J Neurosci*, 33: 668–677.

Gerfen CR, Baimbridge KG, Miller JJ (1985) The neostriatal mosaic: Compartmental distribution of calcium-binding protein and parvalbumin in the basal ganglia of the rat and monkey. *Proc Natl Acad Sci USA*, 82: 8780–8784.

Gerfen CR, Surmeier DJ (2011) Modulation of striatal projection systems by dopamine. *Annu Rev Neurosci*, 34: 441–466.

Gerfen CR, Paletzki R, Heintz N (2013) GENSAT BAC cre-recombinase driver lines to study the functional organization of cerebral cortical and basal ganglia circuits. *Neuron*, 80: 1368–1383.

Giros B, Jaber M, Jones SR, Wightman RM, Caron MG (1996) Hyperlocomotion and indifference to cocaine and amphetamine in mice lacking the dopamine transporter. *Nature*, 379: 606–612.

Goldman PS, Nauta WJ (1977) An intricately patterned prefronto-caudate projection in the rhesus monkey. *J Comp Neurol*, 72: 369–386.

Granado N, Escobedo I, O'Shea E, Colado I, Moratalla R (2008) Early loss of dopaminergic terminals in striosomes after MDMA administration to mice. *Synapse*, 62: 80–84.

Granado N, Ares-Santos S, O'Shea E, Vicario-Abejon C, Colado MI, Moratalla R (2010) Selective vulnerability in striosomes and in the nigrostriatal dopaminergic pathway after methamphetamine administration: Early loss of TH in striosomes after methamphetamine. *Neurotox Res*, 18: 48–58.

Graybiel AM (1984) Correspondence between the dopamine islands and striosomes of the mammalian striatum. *Neuroscience*, 13: 1157–1187.

Graybiel AM, Hirsch EC, Agid Y (1990) The nigrostriatal system in Parkinson's disease. *Adv Neurol*, 53: 17–29.

Graybiel AM, Ragsdale CW, Jr. (1978) Histochemically distinct compartments in the striatum of human, monkeys, and cat demonstrated by acetylthiocholinesterase staining. *Proc Natl Acad Sci USA*, 75: 5723–5726.

Hawes SL, Gillani F, Evans RC, Benkert EA, Blackwell KT (2013) Sensitivity to theta-burst timing permits LTP in dorsal striatal adult brain slice. *J Neurophysiol*, 110: 2027–2036.

Hawes SL, Salinas AG, Lovinger DM, Blackwell KT (2017) Long-term plasticity of corticostriatal synapses is modulated by pathway-specific co-release of opioids through kappa-opioid receptors. *J Physiol*, 595: 5637–5652.

Henry B, Duty S, Fox SH, Crossman AR, Brotchie JM (2003) Increased striatal pre-proenkephalin B expression is associated with dyskinesia in Parkinson's disease. *Exp Neurol*, 183: 458–468.

Herkenham M, Pert CB (1981) Mosaic distribution of opiate receptors, parafascicular projections and acetylcholinesterase in rat striatum. *Nature*, 291: 415–418.

Hurd YL, Herkenham M (1993) Molecular alterations in the neostriatum of human cocaine addicts. *Synapse*, 13: 357–369.

Jedynak JP, Cameron CM, Robinson TE (2012) Repeated methamphetamine administration differentially alters fos expression in caudate-putamen patch and matrix compartments and nucleus accumbens. *PLoS One*, 7: e34227.

John CE, Jones SR (2007) Fast Scan Cyclic Voltammetry of Dopamine and Serotonin in Mouse Brain Slices. In: *Electrochemical Methods for Neuroscience* (Michael AC, Borland LM, eds). Boca Raton, FL.

Jones SR, Gainetdinov RR, Wightman RM, Caron MG (1998) Mechanisms of amphetamine action revealed in mice lacking the dopamine transporter. *J Neurosci*, 18: 1979–1986.

Kawaguchi Y, Wilson CJ, Emson PC (1989) Intracellular recording of identified neostriatal patch and matrix spiny cells in a slice preparation preserving cortical inputs. *J Neurophysiol*, 62: 1052–1068.

Kemp JM, Powell TP (1971) The structure of the caudate nucleus of the cat: Light and electron microscopy. *Philos Trans R Soc Lond B Biol Sci*, 262: 383–401.

Koizumi H, Morigaki R, Okita S, Nagahiro S, Kaji R, Nakagawa M, Goto S (2013) Response of striosomal opioid signaling to dopamine depletion in 6-hydroxydopamine-lesioned rat model of Parkinson's disease: a potential compensatory role. *Front Cell Neurosci*, 7: 74.

Liu G, Yu J, Ding J, Xie C, Sun L, Rudenko I, Zheng W, Sastry N, Luo J, Rudow G, Troncoso JC, Cai H (2014) Aldehyde dehydrogenase 1 defines and protects a nigrostriatal dopaminergic neuron subpopulation. *J Clin Invest*, 124: 3032–3046.

Lopez-Huerta VG, Nakano Y, Bausenwein J, Jaidar O, Lazarus M, Cherassse Y, Garcia-Munoz M, Arbuthnott G (2016) The neostriatum: Two entities, one structure? *Brain Struct Funct*, 221: 1737–1749.

Mateo Y, Johnson KA, Covey DP, Atwood BK, Wang HL, Zhang S, Gildish I, Cachope R, Bellocchio L, Guzman M, Morales M, Cheer JF, Lovinger DM (2017) Endocannabinoid actions on cortical terminals orchestrate local modulation of dopamine release in the nucleus accumbens. *Neuron*, 96: 1112–1126 e1115.

Menalled LB, Sison JD, Wu Y, Olivieri M, Li XJ, Li H, Zeitlin S, Chesselet MF (2002) Early motor dysfunction and striosomal distribution of huntingtin microaggregates in Huntington's disease knock-in mice. *J Neurosci*, 22: 8266–8276.

Miura M, Masuda M, Aosaki T (2008) Roles of micro-opioid receptors in GABAergic synaptic transmission in the striosome and matrix compartments of the striatum. *Mol Neurobiol*, 37: 104–115.

Morigaki R, Goto S (2015) Postsynaptic density protein 95 in the striosome and matrix compartments of the human neostriatum. *Front Neuroanat*, 9: 154.

Patriarchi T, Cho JR, Merten K, Howe MW, Marley A, Xiong WH, Folk RW, Broussard GJ, Liang R, Jang MJ, Zhong H, Dombeck D, von Zastrow M, Nimmerjahn A, Gradinaru V, Williams JT, Tian L (2018) Ultrafast neuronal imaging of dopamine dynamics with designed genetically encoded sensors. *Science*, 360, eaat4422.

Pert CB, Kuhar MJ, Snyder SH (1975) Autoradiograhic localization of the opiate receptor in rat brain. *Life Sci*, 16: 1849–1853.

Pert CB, Kuhar MJ, Snyder SH (1976) Opiate receptor: Autoradiographic localization in rat brain. *Proc Natl Acad Sci USA*, 73: 3729–3733.

Pignatelli M, Bonci A (2015) Role of dopamine neurons in reward and aversion: A synaptic plasticity perspective. *Neuron*, 86: 1145–1157.

Planert H, Berger TK, Silberberg G (2013) Membrane properties of striatal direct and indirect pathway neurons in mouse and rat slices and their modulation by dopamine. *PLoS One*, 8: e57054.

Ragsdale CW, Jr., Graybiel AM (1981) The fronto-striatal projection in the cat and monkey and its relationship to inhomogeneities established by acetylcholinesterase histochemistry. *Brain Res*, 208: 259–266.

Roeper J (2013) Dissecting the diversity of midbrain dopamine neurons. *Trends Neurosci*, 36: 336–342.

Saka E, Goodrich C, Harlan P, Madras BK, Graybiel AM (2004) Repetitive behaviors in monkeys are linked to specific striatal activation patterns. *J Neurosci*, 24: 7557–7565.

Salinas AG, Davis MI, Lovinger DM, Mateo Y (2016) Dopamine dynamics and cocaine sensitivity differ between striosome and matrix compartments of the striatum. *Neuropharmacology*, 108: 275–283.

Sgobio C, Wu J, Zheng W, Chen X, Pan J, Salinas AG, Davis MI, Lovinger DM, Cai H (2017) Aldehyde dehydrogenase 1-positive nigrostriatal dopaminergic fibers exhibit distinct projection pattern and dopamine release dynamics at mouse dorsal striatum. *Sci Rep*, 7: 5283.

Smith Y, Bolam JP (1990) The output neurones and the dopaminergic neurones of the substantia nigra receive a GABA-containing input from the globus pallidus in the rat. *J Comp Neurol*, 296: 47–64.

Smith Y, Bennett BD, Bolam JP, Parent A, Sadikot AF (1994) Synaptic relationships between dopaminergic afferents and cortical or thalamic input in the sensorimotor territory of the striatum in monkey. *J Comp Neurol*, 344: 1–19.

Smith JB, Klug JR, Ross DL, Howard CD, Hollon NG, Ko VI, Hoffman H, Callaway EM, Gerfen CR, Jin X (2016) Genetic-based dissection unveils the inputs and outputs of striatal patch and matrix compartments. *Neuron*, 91: 1069–1084.

Tajima K, Fukuda T (2013) Region-specific diversity of striosomes in the mouse striatum revealed by the differential immunoreactivities for mu-opioid receptor, substance P, and enkephalin. *Neuroscience*, 241: 215–228.

Threlfell S, Lalic T, Platt NJ, Jennings KA, Deisseroth K, Cragg SJ (2012) Striatal dopamine release is triggered by synchronized activity in cholinergic interneurons. *Neuron*, 75: 58–64.

Tippett LJ, Waldvogel HJ, Thomas SJ, Hogg VM, van Roon-Mom W, Synek BJ, Graybiel AM, Faull RL (2007) Striosomes and mood dysfunction in Huntington's disease. *Brain*, 130: 206–221.

Vogt Weisenhorn DM, Giesert F, Wurst W (2016) Diversity matters — heterogeneity of dopaminergic neurons in the ventral mesencephalon and its relation to Parkinson's Disease. *J Neurochem*, 139 Suppl 1: 8–26.

Volkow ND, Wise RA, Baler R (2017) The dopamine motive system: Implications for drug and food addiction. *Nat Rev Neurosci*, 18: 741–752.

Voorn P, Vanderschuren LJ, Groenewegen HJ, Robbins TW, Pennartz CM (2004) Putting a spin on the dorsal-ventral divide of the striatum. *Trends Neurosci*, 27: 468–474.

Walker RH, Arbuthnott GW, Baughman RW, Graybiel AM (1993) Dendritic domains of medium spiny neurons in the primate striatum: relationships to striosomal borders. *J Comp Neurol*, 337: 614–628.

Watabe-Uchida M, Zhu L, Ogawa SK, Vamanrao A, Uchida N (2012) Whole-brain mapping of direct inputs to midbrain dopamine neurons. *Neuron*, 74: 858–873.

Westerink BH, Damsma G, Rollema H, De Vries JB, Horn AS (1987) Scope and limitations of in vivo brain dialysis: A comparison of its application to various neurotransmitter systems. *Life Sci*, 41: 1763–1776.

White NM, Hiroi N (1998) Preferential localization of self-stimulation sites in striosomes/patches in the rat striatum. *Proc Natl Acad Sci USA*, 95: 6486–6491.

Yoshizawa T, Ito M, Doya K (2018) Reward-predictive neural activities in striatal striosome compartments. *eNeuro*, 5.

Zhang L, Doyon WM, Clark JJ, Phillips PE, Dani JA (2009) Controls of tonic and phasic dopamine transmission in the dorsal and ventral striatum. *Mol Pharmacol*, 76: 396–404.

CHAPTER 12

PHYSIOLOGICAL AND DRUG-INDUCED FLUCTUATIONS IN BRAIN OXYGEN AND GLUCOSE ASSESSED BY SUBSTRATE-SELECTIVE SENSORS COUPLED WITH HIGH-SPEED AMPEROMETRY

Eugene A. Kiyatkin

In Vivo Electrophysiology Unit,
Behavioral Neuroscience Branch,
National Institute on Drug Abuse — Intramural Research Program,
National Institutes of Health, Baltimore, MD, USA
ekiyatki@intra.nida.nih.gov

12.1 Introduction

Metabolic activity of brain cells essentially depends on proper delivery of oxygen and glucose, which arrive into brain tissue from arterial blood, where their concentrations are higher than in the brain's extracellular space. Therefore, brain levels of oxygen and glucose directly depend on their blood levels, which are tightly regulated and remain relatively stable in healthy organisms under quiet resting conditions. While blood glucose is an important homeostatic parameter, its levels can be strongly increased (hyperglycemia) following consumption of glucose-containing or glucose-producing products, thus elevating glucose levels in brain tissue. Blood glucose never falls below its normal concentrations under physiological conditions, and decreases in blood glucose (hypoglycemia) are typically associated with certain diseases or side effects of certain drugs. Oxygenation of arterial blood is also maintained at relatively stable levels by respiration, but it depends on oxygen content in the inspired air and the efficiency of respiratory activity, which determines the oxygen transfer from the external medium to the lung's vessels. Respiration is a sensitive parameter that can be strongly increased during physiological and behavioral activation and decreased following exposure to certain drugs. While the respiration-dependent changes in blood oxygen content will definitely modulate oxygen entry into brain tissue, it is quite difficult to distinguish the contribution of this factor due to simultaneous changes in neuronal activity, brain metabolism, and the tone of cerebral vessels, which all are affected by physiological and drug stimuli.

Along with the gradient-dependent entry of oxygen and glucose as the basic mechanism determining their levels in the brain's extracellular space, numerous data suggest that this entry could be actively modulated by neural activity via changes in the tone of cerebral vessels and fluctuations in local cerebral blood flow (Roy and Sherrington, 1890; Fox and Raichle, 1986; Fellows and Boutelle, 1993; Attwell *et al.*, 2010; Lecrux and Hamel, 2011; Muoio *et al.*, 2014). This gradient-independent neural mechanism, usually considered in terms of neurovascular coupling, allows the brain to receive more glucose and oxygen in advance of their enhanced use during functional neural activation, thus preventing any possible metabolic deficit. However, the enhanced inflow of oxygen and glucose into the brain's extracellular space is opposed by continuous metabolic consumption of these substances tending to decrease their levels in brain tissue. Moreover, both oxygen consumption and glucose utilization are increased during functional neural activation (Siesjo, 1978; Sokoloff, 1999; Mergenthaler *et al.*, 2013), also affecting the levels of these substances in brain tissue. While it is well-established that the extracellular levels of oxygen and glucose reflect a balance between highly dynamic and opposing influences between their entry from arterial blood and consumption in brain tissue, our knowledge of physiological and drug-induced fluctuations in these homeostatic parameters is very limited.

In this review paper, I will present and discuss the results of our recent studies, in which substrate-specific sensors were coupled with amperometry to examine physiological and drug-induced fluctuations in brain oxygen and glucose in awake, freely-moving rats. Since intra-brain entry of oxygen and glucose is governed by common mechanisms, I will consider the relationships between changes in these parameters under physiological conditions and will discuss possible mechanisms underlying these changes. Finally, I will present data on changes in brain oxygen and glucose induced by widely used drugs of abuse that affect brain metabolism and induce multiple physiological and behavioral effects. My focus here will be on two highly popular drugs: cocaine, a psychostimulant, and heroin, an opioid that is usually viewed as a CNS depressant.

Since changes in brain glucose and oxygen are tightly related to cerebral metabolism, our neurochemical data will be considered in conjunction with brain thermorecording data, which allowed us to assess the effects of physiological and drug stimuli on brain metabolic activity and vascular response. Similar to glucose and oxygen, brain temperature is an important homeostatic parameter related to brain metabolism. Also similar to oxygen and glucose, our knowledge of brain temperature and its physiological and drug-induced fluctuations is limited, and the existing data are generally ignored. Since all the energy used for brain metabolism is finally transformed into heat (Hodgkin, 1967; Ritchie, 1973; Siesjo, 1978; Schmidt-Nielsen, 1987; Sokoloff, 1999), temperature monitoring provides a valid tool to assess changes in metabolic activity (Kiyatkin, 2010). Using thermorecording data, we were able to examine how the changes in brain oxygen and glucose are related to changes in brain metabolic activity.

12.2 Substrate-selective sensors coupled with amperometry as a tool to evaluate fluctuations in brain oxygen and glucose under physiologically relevant conditions

Electrochemical oxygen detection has a long history (Davies and Brink, 1942; Davies and Bronk, 1957; Clark *et al.*, 1958; Clark and Lyons, 1962; Kahn, 1964). The study conducted by Travis and Clark (1965) was the first, in which oxygen sensors were used in awake, freely-moving rats to examine physiological fluctuations in brain oxygen levels. Important studies employing oxygen electrochemistry have been conducted in Oxford, UK in the laboratory of Marianne Fillenz (Lowry *et al.*, 1996, 1997; Lowry and Fillenz, 2001; Fillenz, 2005) and have been continued in the laboratory of John Lowry in the National University of Ireland (Bolger *et al.*, 2011; Bolger and Lowry, 2005; Lowry *et al.*, 2010; Kealy *et al.*, 2013, 2017).

The classic study by Travis and Clark revealed that oxygen levels in brain tissue, as examined in several brain structures in freely-moving rats, are maintained at relatively low and stable levels with rapid "spontaneous" fluctuations that appear to correlate with phasic fluctuations in arterial blood pressure (Travis and Clark, 1965). In this study, it was demonstrated that oxygen levels phasically increased following exposure to arousing stimuli of different modalities. While differing in each of the several brain structures examined, these changes showed between-structure correlations, suggesting their generalized nature. Due to short latencies of observed responses, it was suggested that rapid changes in brain oxygen levels result from rapid changes in cerebral blood flow due to neuronal activation induced by arousing stimuli. The importance of neural activity in regulating blood supply to the brain has originally been proposed by Charles S. Sherrington in the very early stages of neuroscience (Roy and Sherrington, 1890). This basic idea, in different variants, modifications and complications, still dominates in current studies of cerebral blood flow and brain metabolism.

While electrochemistry was widely used for the monitoring of oxidizable brain substances, such as monoamines (Robinson *et al.*, 2008; Bucher and Wightman, 2015; Fox and Wightman, 2017), and oxygen which is reduced by applied potentials (Travis and Clark, 1965; Lowry *et al.*, 1996, 1998), glucose is not electroactive and *in vivo* microdialysis appears to be the method of choice to examine basal levels and physiological changes in this parameter in brain tissue (Fellows *et al.*, 1992; Fellows and Boutelle, 1993; McNay and Gold, 1999; McNay *et al.*, 2001). However, slow temporal resolution is perhaps the most serious limitation of microdialysis that made it impossible to learn about the dynamic aspects of this parameter. This problem was resolved with the development of enzyme-based glucose biosensors — a discovery also originating from the laboratory of Leland Clark (Clark and Lyons, 1962). This technology was significantly improved and sophisticated in later years (Boutelle *et al.*, 1986; Silver and Erecinska, 1994; Hu and Wilson, 1997; Lowry *et al.*, 1998), and the study by Martyn Boutelle and collaborators was the first which described methodology of an amperometric enzyme-based biosensor for assessing brain glucose levels in freely-moving rats (Boutelle *et al.*, 1986).

To examine physiological and drug-induced fluctuations in brain oxygen and glucose, we used commercially produced oxygen and glucose sensors (Pinnacle Technology Inc., Lawrence, Kansas, USA) coupled with fixed-potential amperometry in awake, freely-moving rats. The technical aspects of these sensors, their advantages and disadvantages, as well as multiple tests and controls necessary to determine their sensitivity and selectivity were described in detail elsewhere (Kiyatkin and Lenoir, 2012; Kiyatkin and Wakabayashi, 2015; Solis *et al.*, 2017a). Below, I will present some basic information about these sensors and their advantages over other techniques to learn about the dynamic aspects of changes in brain oxygen and glucose.

Pinnacle oxygen sensors consist of an epoxy-sheathed disc electrode that is ground to a fine surface using a diamond-lapping disc. These sensors are prepared from Pt-Ir wire, 180 μm in diameter, with a relatively small sensing area of 0.025 mm^2 at the sensor's tip. In contrast to other sensor designs, an active electrode in the Pinnacle oxygen sensors is incorporated with an integrated Ag/AgCl reference electrode. Dissolved oxygen is reduced on the active surface of these sensors, which is held at a stable potential of –0.6 V vs. the reference electrode, producing an amperometric current. Pinnacle glucose sensors are prepared from the same-diameter Pt-Ir wire, but the sensing cavity is ~1 mm in length on its tip and the sensing area (~0.56 mm^2) is much larger than the sensing area of the oxygen sensors. Similar to the oxygen sensor, the active electrode is incorporated with an integrated Ag/AgCl reference electrode. On the active surface of these sensors, that is held at a potential of +0.6 V vs. the reference electrode, glucose oxidase enzyme converts glucose to α-ketoglutarate and hydrogen peroxide, which is detected as an amperometric oxidation current (Hu and Wilson, 1997). The potential contribution of ascorbic acid to the measured current is competitively reduced by co-localizing ascorbic acid oxidase enzymes on the active surface of the sensor. This enzyme converts ascorbic acid to non-electroactive dehydroascorbate and water. In addition, a negatively charged Nafion polymer layer under the enzyme layer serves to exclude the contribution of endogenous anionic compounds. An outer polyurethane membrane is cast over the entire sensor surface, serving to limit diffusion of glucose to the enzymatic layer as well as to provide additional screening of interferents. The calibration of oxygen sensors was performed by the manufacturer (Pinnacle Technology) according to a standard protocol described elsewhere (Bolger *et al.*, 2011). Glucose biosensors were calibrated *in vitro* in-house before and after each recording session. We also determined responses of glucose sensors to other possible chemical interferents, including ascorbate and dopamine. In our work with glucose biosensors, we utilized acute recording, in which a sensor was inserted into brain tissue via a previously implanted cannula and removed after the recording session. While acute recording was also used in most of our experiments with oxygen sensors, currently we are also utilizing chronic sensor implantation into target brain structure, thus allowing electrochemical recordings to be continued during several daily sessions.

Measurement selectivity has been a long-standing troublesome issue of voltammetric techniques, which were used to assess fluctuations in oxidizable brain substances such as monoamines (Bucher and Wightman, 2015). The development of enzyme-based biosensors has made it possible to use electrochemistry for assessing brain fluctuations of non-oxidizable substances such as glutamate and glucose. However, careful calibrations and multiple selectivity controls are still necessary due to possible contamination of specific currents by other chemical and physical contributions (Kiyatkin *et al.*, 2013; Kiyatkin and Wakabayashi, 2015). In contrast to glutamate, which is contained in the brain's extracellular space at low levels (~1 μM) and shows physiological fluctuations within 20–200 nM (Wakabayashi and Kiyatkin, 2012), basal concentrations of glucose assessed in brain tissue by *in vivo* microdialysis in freely-moving rats are much larger (0.5–1.0 mM; Boutelle *et al.*, 1986; Lowry *et al.*, 1998; McNey and Gold, 1999; Dunn-Maynell *et al.*, 2009), thus making any non-specific contributions much smaller.

While the performance of any electrochemical sensor is temperature-dependent and brain temperature fluctuates within 2–4°C within the physiological continuum (Kiyatkin, 2010), temperature contribution to "glucose" currents as tested in our experiments is very small and estimated as 2–3% of baseline concentration. While naturally occurring fluctuations in brain temperature are unlikely to affect physiological changes in oxygen and glucose, measurement accuracy could be seriously affected if sensors are calibrated at room temperature. While the best strategy is to calibrate sensors at physiological, 37°C temperature, mimicking the physical environment of the awake brain, it is not always done due to technical complications. While our calibrations of glucose sensors were conducted at room temperature (22–23°C), the values obtained were adjusted by using a temperature coefficient determined in separate experiments. To validate the reliability of glucose measurements *in vivo*, we also used glucose-null sensors (Pinnacle Technology), which were identical to the glucose sensors, except the absence of glucose oxidase. Therefore, when used under identical *in vivo* conditions, these electrodes are exposed to the same chemical and physical factors (i.e., various electroactive interferents, temperature, pH, etc.) as glucose sensors, but are insensitive to glucose. Therefore, the difference in currents detected by glucose and glucose-null sensors ("current differentials") in the same brain areas under identical conditions is the best approach to reveal actual changes in extracellular glucose levels. Both types of sensors were tested *in vitro* using the same protocol before and after *in vivo* recording and they were used in identical *in vivo* experiments.

Since oxygen is reduced by applied potentials, thus excluding possible contamination by multiple oxidizable substances, and its brain levels are relatively large (10–30 μM; Feng *et al.*, 1988; Zimmerman and Wightman, 1991; Bazzu *et al.*, 2009; Calia *et al.*, 2009), measurement selectivity is a lesser problem with oxygen sensors. Furthermore, oxygen sensors were calibrated at 37°C, thus excluding a possible contribution of temperature to oxygen currents and increasing the accuracy of our measurements.

In our studies, currents generated by glucose and oxygen sensors were recorded with 1-s time resolution and then transformed into concentration values, which were averaged for different time intervals. This high temporal resolution of the primary recording allowed us to accurately assess the dynamic aspects of fluctuations in these parameters. Typically, 1-min resolution was used to represent slow changes in oxygen and glucose, whereas 4- or 10-s resolution was used to reveal rapid changes in these substances. All of our experiments were conducted in freely-moving rats, which had a previously implanted intravenous (i.v.) catheter and which were extensively habituated to the recording environment. Moreover, the use of catheter extensions and an electrical swivel allowed us to administer drugs when the rat was in quiet resting conditions, without any sensory and arousing influences typical to standard subcutaneous or intraperitoneal drug administration.

12.3 Physiological fluctuations in brain oxygen and glucose: Underlying mechanisms

Before we initiated our studies with drugs, we examined how brain oxygen and glucose levels fluctuate in brain tissue under physiological conditions. Most of our brain recordings were conducted in the nucleus accumbens (NAc), an important brain structure involved in sensorimotor integration and mechanisms of natural and drug reinforcement (Mogenson *et al.*, 1980; Wise and Bozarth, 1987; Di Chiara, 2002). To determine the structural specificity of oxygen and glucose responses, we also conducted electrochemical measurements in other brain structures, particularly in the substantia nigra pars reticulata (Kiyatkin and Lenoir, 2013) and amygdala (Solis *et al.*, 2017a), distinctly located brain structures that differ from the NAc in basal neuronal activity and responses to environmental challenges. Our experiments were conducted in awake, freely-moving male rats that were extensively habituated to the recording environment and exposed to a number of arousing stimuli from a simple auditory tone to a tail-pinch and social interaction with a male conspecific.

Consistent with early work (Travis and Clark, 1965; Lowry *et al.*, 1997), we found that NAc oxygen levels rapidly increase following exposure to all of the arousing stimuli (Figure 12.1, top graphs). These increases were relatively small in magnitude (1–3 μM or ~5–15% over baseline) with the most rapid acceleration occurring immediately after the stimulus onset. The responses also differed depending upon the stimulus duration and its presumed biological significance. A short auditory stimulus induced the weakest change, which was evident within ~2–3 min, but both the 3-min tail-pinch and 3-min social interaction induced stronger and more prolonged oxygen increases that peaked at the end of or after the offset of stimulus exposure. These phasic responses elicited by arousing stimuli were superimposed over spontaneous fluctuations in basal oxygen levels, which were often equally rapid and strong as those induced by arousing stimuli.

The NAc glucose levels also phasically increased following exposure to arousing stimuli and these increases also differed depending on the stimulus nature (Figure 12.1, bottom graphs). These increases were even more rapid than those of oxygen, with significant increases at the first data point (0–4 s) following the stimulus onset. Similar to that with oxygen, the response was weakest and most transient after auditory stimulus and stronger and more prolonged with tail-pinch and social interaction.

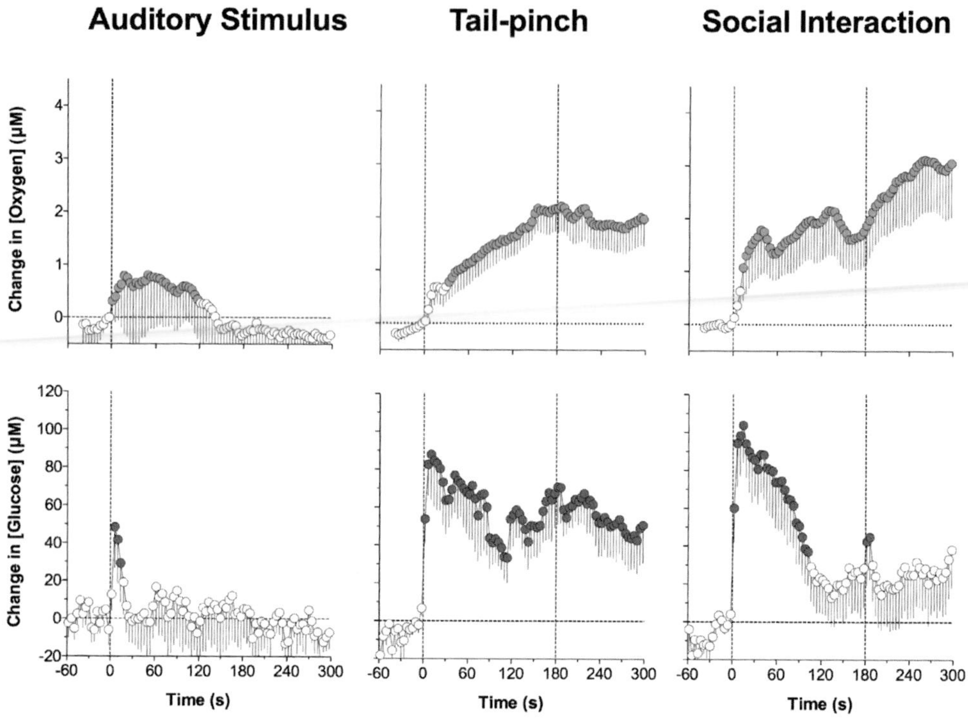

Figure 12.1 Mean (±SEM) changes in NAc oxygen and glucose induced by short auditory stimulus, tail-pinch, and social interaction in awake, freely-moving rats. Filled symbols show values significantly different from pre-stimulus baseline. Dotted lines on X-axis show onset (0 s) and offset (180 s) of stimulation and horizontal dashed lines show basal values.

Although both oxygen and glucose levels phasically increased following exposure to arousing stimuli, there were important differences in their time-course as evidenced by time-dependent correlative analyses that show how changes in one parameter depend on changes in another parameter (Figure 12.2). As can be seen, increases in glucose levels were much more rapid than increases in oxygen levels, showing maximal acceleration within the first three data points (0–12 s) while oxygen levels remained at levels close to baseline. Then, oxygen levels began to increase when glucose levels decreased towards baseline. As calculated to the peak of oxygen responses, there was no significant correlation between these two parameters for all three stimuli. Therefore, although the intra-brain entry of both oxygen and glucose are governed by similar mechanisms, the time-dependent correlative analysis reveals important differences in the dynamics of these parameters. While speculative at this stage, these differences could be related to distinct mechanisms involved in the transfer of these two substances via the blood-brain barrier. While oxygen passively diffuses into the brain's extracellular space, glucose is transported into the brain environment via a specialized type of GLUT-1

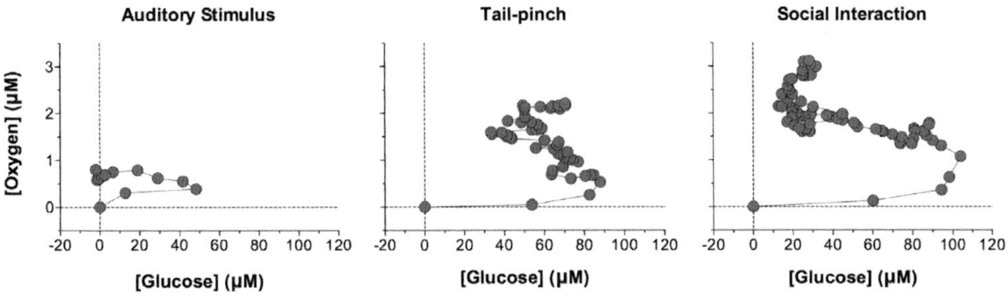

Figure 12.2 Time-dependent correlative relationships between changes in NAc oxygen and glucose induced by auditory stimulus, tail-pinch, and social interaction. Basal value in each graph is set at 0 μM for both parameters.

transporter, densely expressed on the endothelial cells of the blood-brain barrier (Simpson *et al.*, 2007). Thus, a transporter-mediated mechanism may determine higher efficiency of glucose transfer from arterial blood into the brain's extracellular environment.

Data shown above clearly demonstrate that both oxygen and glucose in the NAc rapidly increase following arousing stimulation. Since blood levels of these substances are maintained at relatively stable levels, the rapidity of changes in NAc oxygen and glucose points at neural activation as the triggering force for their increases. This mechanism is consistent with electrophysiological data obtained in dorsal and ventral striatal neurons in awake, freely-moving rats. These cells have low, sporadic activity during quiet rest and are phasically excited by various arousing stimuli (Carelli and West, 1991; Kiyatkin and Rebec, 1996, 1999; Rebec, 2006). Additional proof for neuronal activation as the initial trigger for rapid entry of oxygen and glucose was obtained by comparing changes in these parameters with cortical EEG. As shown in Figure 12.3, a brief auditory stimulus induced a rapid drop in cortical EEG total power, suggesting desynchronization, a classic measure of neural activation. The time-course of changes in this parameter was virtually congruent to the time-course of changes in NAc glucose. Among different EEG bands, high-frequency gamma activity, which is maintained at low levels at quiet resting conditions, showed the largest increase, which was very similar to NAc glucose increase.

A tight correlation between stimulus-induced increases in NAc glucose and indices of neuronal activation as assessed by EEG may suggest causal relationships between these parameters. To directly test this mechanism, we monitored oxygen and glucose levels during intra-NAc micro-injections of glutamate nearby the electrode's sensing area. Accumbal neurons are universally sensitive to both locally applied or iontophoretically delivered glutamate (Kiyatkin and Rebec, 1996, 1999). When injected at the optimal concentration, glutamate microinjections increased

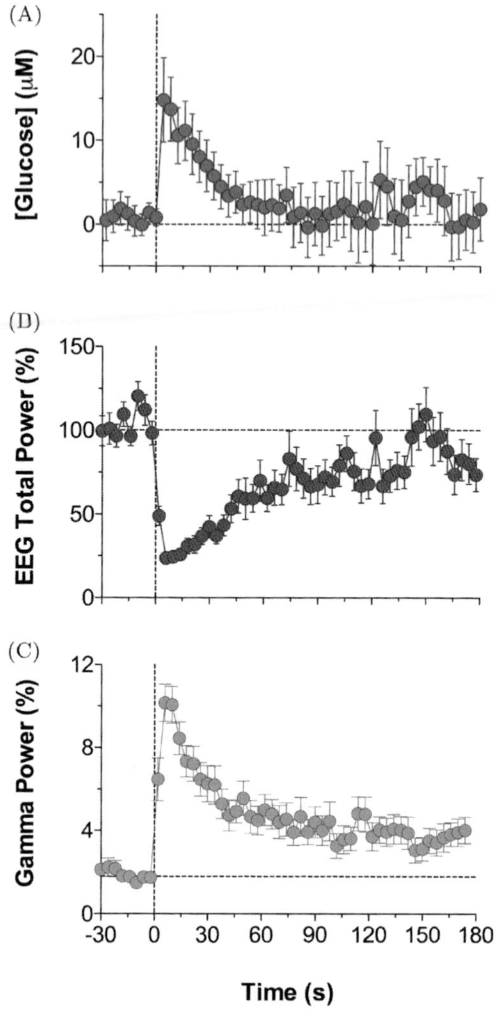

Figure 12.3 Mean (±SEM) changes in NAc glucose (A), EEG total power (B), and gamma EEG activity (C) elicited by an auditory stimulus presented to awake, freely-moving rats during quiet resting conditions. Vertical dotted lines show the moment of stimulus presentation.

NAc levels of both oxygen and glucose (Figure 12.4). In contrast to responses elicited by natural arousing stimuli, glutamate-induced increases in oxygen and glucose exhibited longer and more variable onset latencies with relatively small magnitudes, which were similar or lower than those seen with the arousing stimuli. Therefore, activation of accumbal neurons induced by arousing stimuli appears to trigger local vasodilation that enhances gradient-dependent entry of oxygen and glucose into the brain's extracellular space.

Most of our electrochemical recordings were conducted in the NAc, the structure tightly involved in sensorimotor integration. In this structure, arousing stimuli

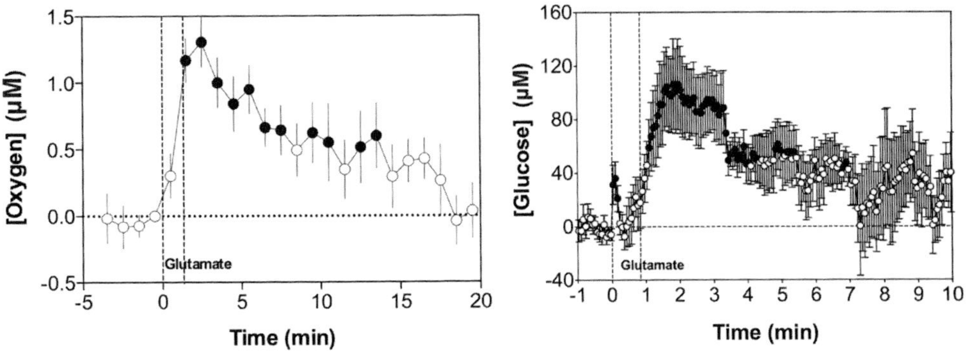

Figure 12.4 Mean (±SEM) changes in NAc oxygen and glucose induced by intra-NAc microinjections of glutamate. Vertical dotted lines show duration of glutamate microinjection.

induced rapid increases in oxygen and glucose, and activation of accumbal cells appear to be the cause inducing these increases. Since neurons in other brain structures have different impulse activity and show distinct responses to arousing stimuli, it is possible that changes in oxygen and glucose will show structural specificity. To explore this issue, we conducted glucose recordings in the substantia nigra, pars retuculata (SNr). In contrast to the slow, sporadic activity of accumbal neurons, most SNr neurons are autoactive, having high rates of basal impulse activity and typically showing transient inhibitions following sensory stimulation and movement activation (DeLong *et al.*, 1984; Windels and Kiyatkin 2004; Deniau *et al.*, 2007). Consistent with our expectations, glucose responses in the SNr differed from those seen in the NAc. In contrast to increases seen in the NAc, glucose levels slowly decreased with all strong arousing stimuli (Kiyatkin and Lenoir, 2013). Figure 12.5 shows between-structure differences in glucose responses induced by 3-min social interaction. While a slow decrease in glucose levels was the basic effect elicited by this stimulation in the SNr, a weak, but significant increase was observed immediately after the start of social interaction. While this finding is primarily phenomenological, it suggests that the pattern of physiological fluctuations in glucose has significant structural specificity.

Glucose and oxygen are critical parameters of the brain's energy metabolism. Therefore, it was important to evaluate how changes in oxygen and glucose in the extracellular space are related to the brain's metabolic activity. In contrast to energy metabolism of the whole organism that can be assessed by either gas analyses through direct or indirect calorimetry or by measurement of heat production (Schmidt-Nielsen, 1987), it is quite difficult to examine metabolic activity of the brain and its physiological fluctuations. In contrast to a significant energy expenditure associated with muscular activity in the whole organism (Rowell, 1983; Donaldson *et al.*, 2003), the brain is constantly active and a significant amount of

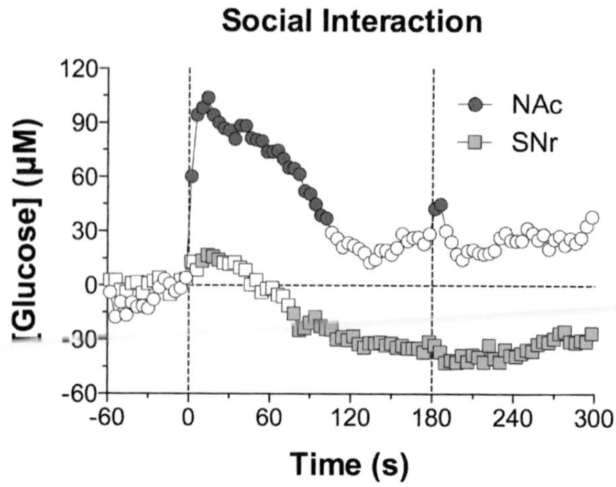

Figure 12.5 Mean changes in glucose levels in the nucleus accumbens (NAc) and substantia nigra pars reticulata (SNr) induced in awake, freely-moving rats by three-minute social interaction with a male conspecific. Filled symbols show values significantly different from baseline. Standard errors are not shown for visual clarity.

energy is used for normal functioning of brain cells under quiet resting conditions. It is well-known that the brain, which represents ~2% of the human's body weight, accounts for ~20% of the organism's total oxygen consumption at rest (Siesjo, 1978; Schmidt-Nielsen, 1997). Traditionally, two approaches, measurements of glucose utilization and oxygen consumption, are used to estimate brain metabolic activity (Clarke and Sokoloff, 1999; Mergenthaler *et al.*, 2013; Rodgers, 2016). However, these approaches have significant limitations in assessing dynamic aspects of metabolic activity in awake, behaving animals. As a way to resolve this issue, we used a multi-site thermorecording technique, which allowed us to examine the effects of physiological and drug stimuli on intra-brain heat production and peripheral vascular response (Kiyatkin, 2010).

It is general knowledge that neural activity is highly energy-consuming and all of the energy used for brain metabolism is finally transformed into heat (Hodgkin, 1967; Ritchie, 1973; Siesjo, 1978; Schmidt-Nielsen, 1987; Sokoloff, 1999). Similar to brain oxygen and glucose levels, which reflect the balance between their intra-brain consumption and inflow from arterial blood, brain temperature depends on changes in intra-brain heat production, a direct result of metabolic brain activation, and the efficiency of heat loss from brain tissue to the rest of the body and then to the external environment. This latter variable mainly depends on the tone of peripheral blood vessels, which could constrict, decreasing heat outflow, or dilate, increasing heat outflow.

To reveal the component of the temperature response that is related to intra-brain heat production, we simultaneously recorded temperature in the brain and temporal muscle, the latter being a non-locomotor muscle that receives its arterial blood supply from the same carotid arteries as the brain. Therefore, by analyzing the difference between brain and muscle temperature (or the brain-muscle differential) it is possible to eliminate the influence of arterial blood temperature on brain temperature, thus revealing the component of the temperature response that is related to intra-brain heat production resulting from metabolic brain activation (Kiyatkin, 2010). To assess the contribution of peripheral heat loss to physiological fluctuations in brain temperature, brain and muscle temperature recordings were supplemented by recordings of skin temperature. Skin temperature primarily depends on the tone of skin vessels, but it is also modulated by temperature of the arterial blood supply. The difference between these two measures, or the skin-muscle differential, eliminates the latter influence, providing an accurate measure of changes in peripheral vascular tone.

Our thermorecording studies revealed that brain temperature as assessed in multiple brain structures is not stable, fluctuating within 3–4°C within a normal physiological continuum (Kiyatkin, 2010). While temperature in deep brain structures in freely-moving rats is at its lowest levels (35–36°C) during natural sleep (Smirnov and Kiyatkin, 2008), it transiently peaks at ~39.0–39.5°C at ejaculation points during male copulatory behavior (Kiyatkin and Mitchum, 2003). These temperature increases are derived from two simultaneously acting variables: more transient and weak increases in brain-muscle differentials resulting from brain metabolic activity, and stronger and more prolonged decreases in skin-muscle differentials resulting from peripheral vasoconstriction, a known adaptive response that is triggered by various stressful stimuli and aimed at heat retention under conditions of potential danger (Altschule, 1951; Solomon *et al.*, 1964; Baker *et al.*, 1976).

Figure 12.6 shows the relationships between changes in NAc oxygen and glucose with changes in NAc temperature and NAc-muscle differentials during 3-min tail-pinch. As can be seen, increases in oxygen and glucose were more rapid than increases in NAc temperature (A and D). The differences were especially pronounced with glucose, which peaked in the first minute of stimulation, whereas brain temperature in the same location peaked only at ~10–12 minutes. More tight relations with significant correlations were found when changes in oxygen and glucose were assessed in relation to changes in brain-muscle differentials (B and E). Correlation was especially strong for oxygen ($r = 0.90, p < 0.001$), but even in this case oxygen increased more rapidly than brain-muscle differentials (C). Correlation was weaker for glucose ($r = 0.51, p < 0.01$), which showed more rapid dynamics than increases in brain-muscle differentials (F). Glucose rapidly peaked at the first minute of stimulation when brain-muscle differentials were still close to baseline. Then, brain-muscle differentials slowly increased while glucose levels were maintained near their peak levels.

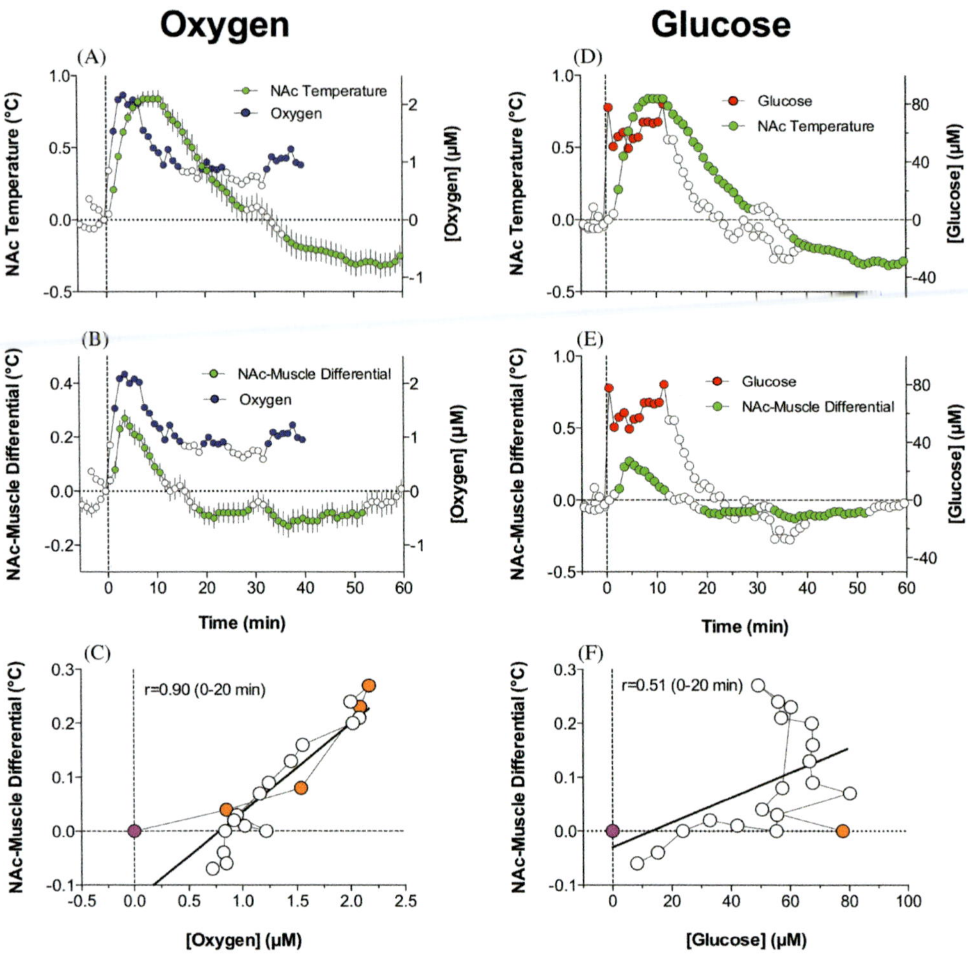

Figure 12.6 Relationships between changes in NAc oxygen (A–C) and glucose (D–F) and temperature parameters induced by 3-min tail-pinch in awake, freely-moving rats. Top graphs show mean changes in NAc temperature and oxygen (A) and glucose (D). Middle graphs show changes in NAc-muscle differentials and oxygen (B) and glucose (E). Filled symbols show values significantly different from baseline. Bottom graphs show correlative relationships between changes in oxygen and glucose and NAc-muscle differentials. Basal values are shown as magenta symbols and orange symbols show values until the peak of glucose or oxygen increase. For more explanation see the text.

Therefore, while increases in brain oxygen and glucose are related to brain metabolic activation, there are important differences in their time-course. Metabolic activation, as assessed by intra-brain heat production, occurs gradually, reflecting enhanced energy use following neuronal activation. In contrast, both oxygen and glucose enter the brain environment preceding metabolic activation, thus preventing any possible metabolic deficit.

12.4 Changes in brain oxygen and glucose induced by drugs of abuse

In contrast to therapeutic psychoactive drugs, which are taken either voluntarily or given by a medical professional to alleviate specific pathological symptoms, people also take drugs recreationally to induce desired psychoactive effects. The variety and popularity of these drugs differ greatly, ranging from the very "safe" consumption of caffeine-containing drinks to the compulsive intravenous injections of heroin and cocaine. While the desire to experience new, unusual or pleasurable psychic effects is the initial drive for the recreational use of most drugs, experience (i.e., repeated drug intake) slowly adds another powerful contributor, a subjective or physical discomfort of being without the drug, that promotes further drug intake.

Each psychoactive drug induces specific behavioral, physiological and psycho-emotional effects, which vary significantly depending upon the drug dose, the individual's activity state preceding drug intake, and the environmental conditions under which the drug is taken. While some physiological effects of the drugs at low doses mimic those induced by natural arousing stimuli, at higher doses these effects could exceed the physiological range, resulting in acute drug-induced intoxication and posing a significant risk for human health. Despite differences in behavioral and physiological effects, all drugs of abuse induce multiple neural effects and affect brain metabolism. Therefore, we conducted a serious of studies, in which we examined changes in oxygen and glucose induced by major drugs of abuse. We also conducted parallel thermorecording experiments, which allowed us to establish relationships between drug-induced changes in brain oxygen and glucose, metabolic activity, and the vascular response. Below, I will consider some of the data obtained from experiments with cocaine and heroin, two popular drugs of abuse with opposite patterns of their initial effects.

12.4.1 Heroin

Heroin is perhaps the most known drug of abuse, which has rapid and exceptionally strong psychoactive and physiological effects. Heroin belongs to the opioid class of drugs, which are widely used therapeutically to alleviate pain of different origins. However, all opioids have strong addictive potential, predisposing individuals for their abuse, which typically occurs with gradually increasing drug doses. Similar to other opioid drugs, heroin has a number of side effects, including hypoactivity, sedation, inhibition of gastro-intestinal activity, and respiratory depression (Yeadon and Kitchen, 1990; Jaffe *et al.*, 1997; Simon, 1997; Baud, 2009). The latter effect is especially dangerous, being responsible for the development of acute brain hypoxia, coma and lethality following overdose of these drugs. Due to the critical role of respiratory depression in mediating acute life-threatening effects of opioid drugs, we examined changes in brain oxygen and glucose induced by heroin (Solis *et al.*, 2017a, 2017c). The effects of heroin were examined in awake, freely-moving rats, using i.v. administration within the wide range of doses from low ones maintaining self-administration to large doses likely to be abused by humans.

In contrast to oxygen increases elicited by natural arousing stimuli, i.v. heroin decreased NAc oxygen levels, suggesting acute brain hypoxia (Figure 12.7). At a low dose (0.1 mg/kg) that is optimal for self-administration in rats, this effect was very rapid, strong (~20–40% drop from pre-injection baseline), but relatively transient in its duration. When the dose was doubled, this effect was greatly increased in its magnitude (~50% drop from baseline) and duration (~20 min), suggesting robust brain hypoxia. At this dose, the effect of heroin was biphasic, with a slight increase in NAc oxygen levels after a strong decrease seen immediately after i.v. injection.

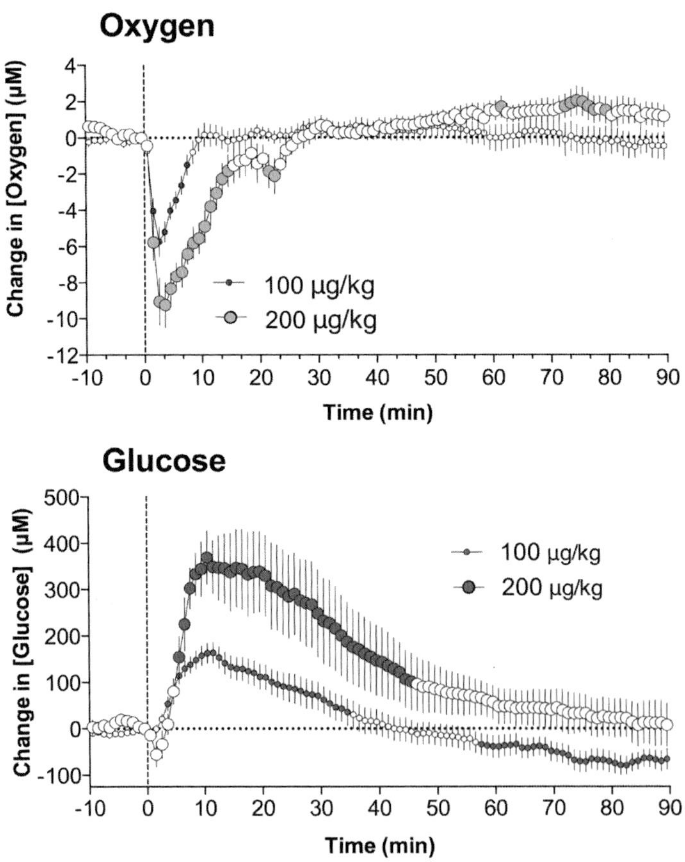

Figure 12.7 Mean (±SEM) changes in NAc oxygen and glucose induced by i.v. administration of heroin at two behaviorally relevant doses. Filled symbols show values significantly different from baseline.

In contrast to oxygen, heroin dose-dependently increased NAc glucose levels. These effects were more prolonged than those of oxygen and larger in their absolute amplitude (increases to 180 and 350 μM for 0.1 and 0.2 mg/kg, respectively), but comparable in terms of percent change. Importantly, the heroin-induced oxygen decrease was rapid (onset latency ~40 s) and peaked at 120–180 s, but the glucose rise developed with a definite latency (~3–4 min), suggesting that this change can be dependent on and secondary to brain hypoxia that results from respiratory depression. Since hypoxia consistently results in accumulation of CO_2, a strong vasodilator (Schmidt and Kety, 1947; Kontos, 1981; Battisti-Charbonney *et al.*, 2011), a heroin-induced rise in CO_2 levels in the brain is a likely cause for dilation of cerebral vessels, which results in increased cerebral blood flow and enhanced glucose entry into the brain tissue.

While physiological increases in NAc glucose were more rapid than increases in oxygen (see above), opposite changes were found with heroin (Figure 12.8). In this case, as demonstrated with heroin at a 0.1 mg/kg dose, NAc oxygen rapidly dropped with ~30 s latency after injection onset, but glucose remained unchanged until ~140–150 s when it began to increase, strongly correlating with oxygen's return towards baseline. It is only when oxygen slowly returned to baseline that glucose reaches its peak. This dissociation of changes suggests that the decrease in oxygen is the initial effect of heroin that causes the more delayed but larger rise in glucose, due to cerebral vasodilation.

While we observed that heroin induces shallow breathing, suggesting respiratory depression, this mechanism was confirmed by obtaining oxygen measurements in the subcutaneous space, a densely-vascularized location with no metabolic activity of its own. As shown in Figure 12.9, heroin significantly decreased oxygen levels in the subcutaneous space; this effect was dose-dependent and larger in terms of absolute concentration change than that in the NAc. In contrast to the NAc, where the oxygen decrease was transient, in the subcutaneous space it was much longer in duration. Therefore, respiratory depression and a decrease in blood oxygen levels is the primary cause of the oxygen drop in brain tissue.

Finally, we examined how heroin-induced changes in NAc oxygen and glucose are related to changes in NAc temperature and changes in NAc-muscle differentials that reflect drug-induced brain metabolic activity (Figure 12.10). As can be seen, rapid changes in NAc oxygen and glucose induced by heroin did not correlate with NAc temperature, which increased slowly, peaked at ~60 min post-injection, and returned to baseline at ~120 min (A). Rapid heroin-induced changes in NAc oxygen and glucose also did not correlate with changes in NAc-muscle temperature differentials, which increased much more slowly than the neurochemical parameters (B). Heroin also induced a robust but relatively short-term decrease in locomotor activity (freezing); this effect occurred when NAc oxygen dropped and NAc glucose increased above baseline (C).

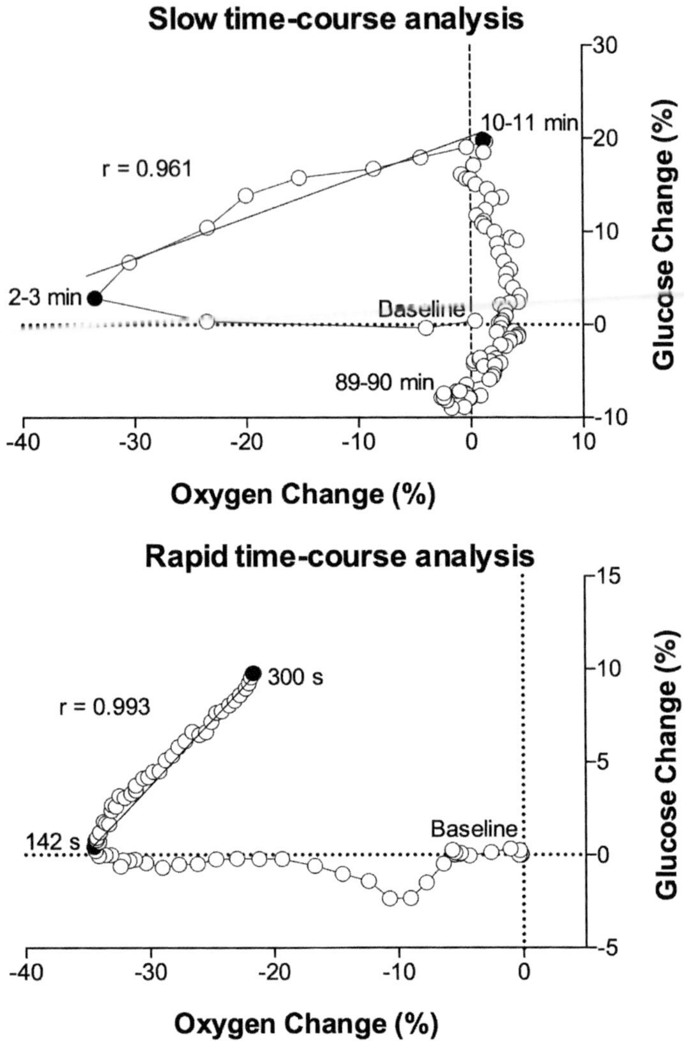

Figure 12.8 Time-dependent correlative relationships between changes in NAc glucose and oxygen (in percent *versus* baseline = 0%) induced by i.v. heroin (100 μg/kg) and assessed with slow (1-min) and rapid (4-s) time resolution. As can be seen, oxygen levels rapidly decreased after the injection onset, while glucose levels remained at baseline for ~2–3 min. Then, oxygen levels slowly returned to baseline, while glucose levels slowly increased up to 10–11 min after the injection.

Therefore, i.v. heroin delivered to awake, drug-naive rats at low, behaviorally-relevant doses (0.1–0.2 mg/kg) dose-dependently decreases NAc oxygen levels due to respiratory depression and a subsequent decrease in blood oxygen levels. Due to this gradient-dependent mechanism, heroin-induced brain hypoxia appears to be generalized in the brain as a whole as shown by oxygen measurements in the

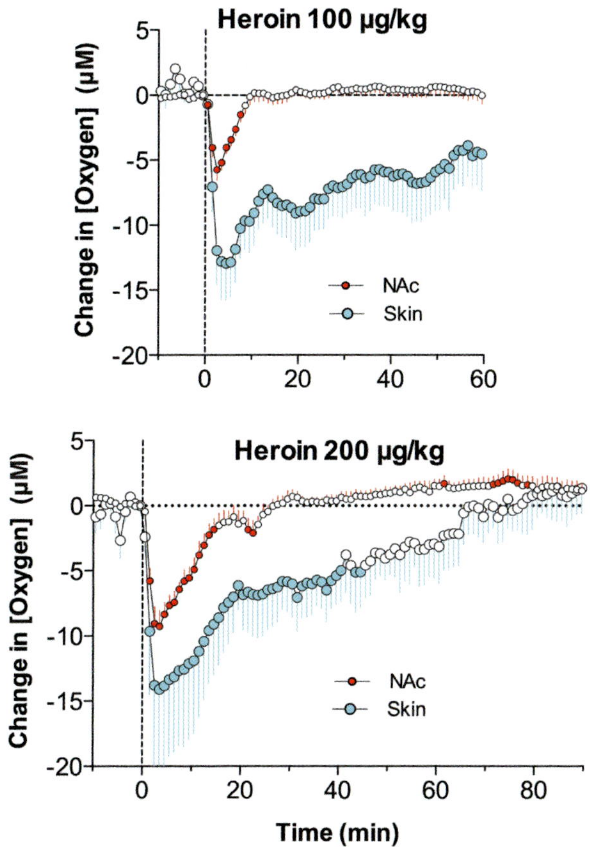

Figure 12.9 Mean (±SEM) changes in oxygen levels in the NAc and subcutaneous space (Skin) induced by i.v. heroin at two doses. Filled symbols show values significantly different from baseline. Original data for this figure were published in Solis *et al.*, 2017c.

basolateral amygdala (Solis *et al.*, 2017a), a distinct brain structure involved in different neural functions. Heroin also induces relatively strong brain hyperglycemia. This effect occurs with a definite latency following oxygen decreases, and it could result from cerebral vasodilation induced by CO_2 that is accumulated in brain tissue during hypoxia.

12.4.2 Cocaine

Cocaine is a widely used drug of abuse that acts on multiple neural substrates localized in both the central and peripheral nervous systems. Cocaine is a non-selective inhibitor of monoamine reuptake and cocaine's action on monoamine transporters, particularly the dopamine transporter, is usually viewed as the critical

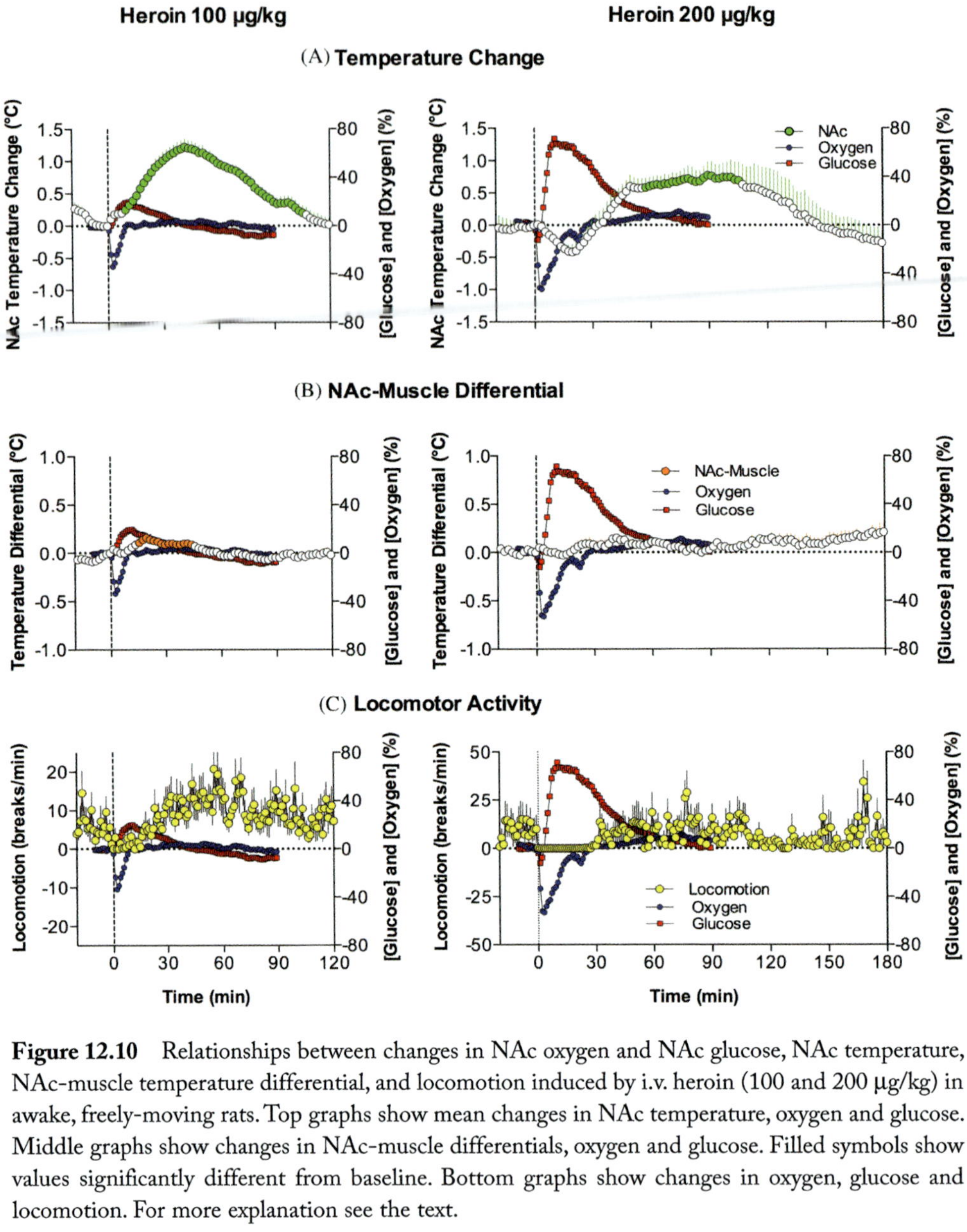

Figure 12.10 Relationships between changes in NAc oxygen and NAc glucose, NAc temperature, NAc-muscle temperature differential, and locomotion induced by i.v. heroin (100 and 200 µg/kg) in awake, freely-moving rats. Top graphs show mean changes in NAc temperature, oxygen and glucose. Middle graphs show changes in NAc-muscle differentials, oxygen and glucose. Filled symbols show values significantly different from baseline. Bottom graphs show changes in oxygen, glucose and locomotion. For more explanation see the text.

action in mediating its reinforcing properties (Kuhar *et al.*, 1991; Wise and Bozarth, 1987). Interaction with these central targets is also often viewed as the cause of the multiple neural, physiological and behavioral changes induced by cocaine.

Through our thermorecording studies, we found that i.v. cocaine administered to awake, quietly resting rats at a low (1 mg/kg) dose that is optimal for drug

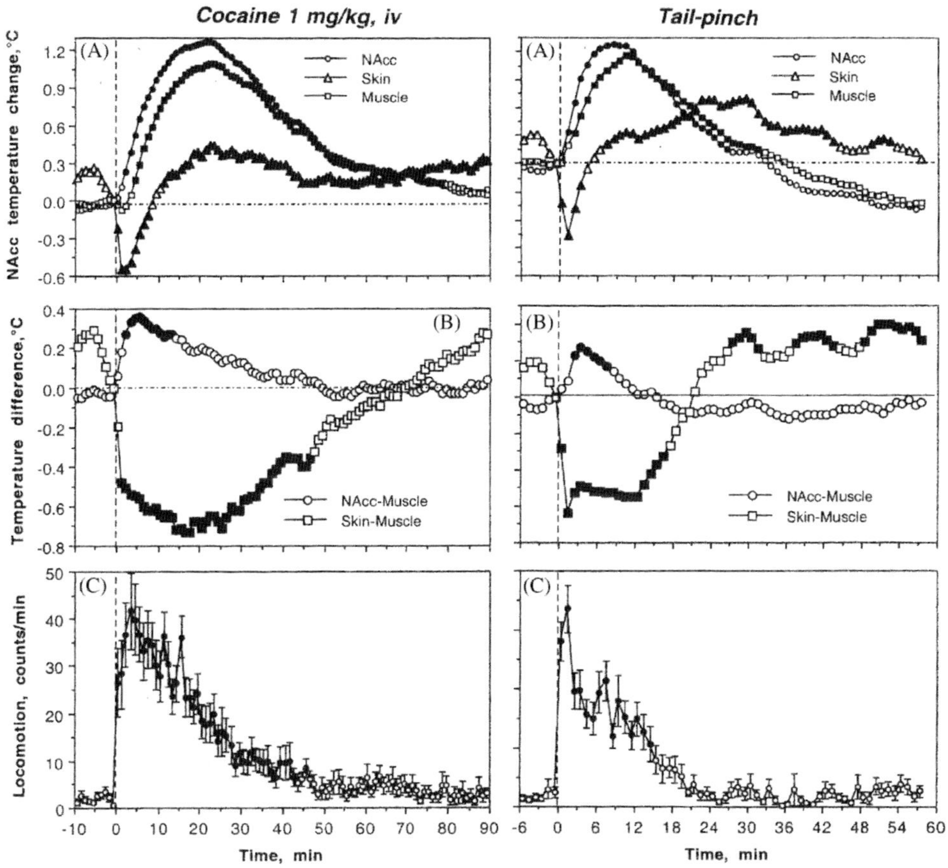

Figure 12.11 Mean changes in NAc temperature, temperature differentials, and locomotion induced by i.v. cocaine in awake, freely-moving rats. This picture has been originally published in *Temperature*, 1 (3), 201–203. This Open Access article is not subject to US copyright law and it is distributed under the terms of the Creative Commons Attribution-Non-Commercial License (http://creativecommons. org/licenses/by-nc/3.0/).

self-administration, induces strong locomotor activation and a modest increase in NAc temperature (Figures 12.11(A)–12.11(C), left panel). Importantly, the temperature increase in brain tissue was more rapid and stronger than in the temporal muscle, resulting in a transient increase in the NAc-muscle differential, reflecting brain metabolic activation. While this metabolic activation is a definite contributor to cocaine-induced brain hyperthermia, cocaine also strongly decreased skin temperature and the skin-muscle differential, indicating skin vasoconstriction. Therefore, heat dissipation from the body surfaces to the external environment is diminished in response to cocaine. In terms of strength, this vascular effect is the more significant contributor to cocaine-induced brain hyperthermia. Importantly,

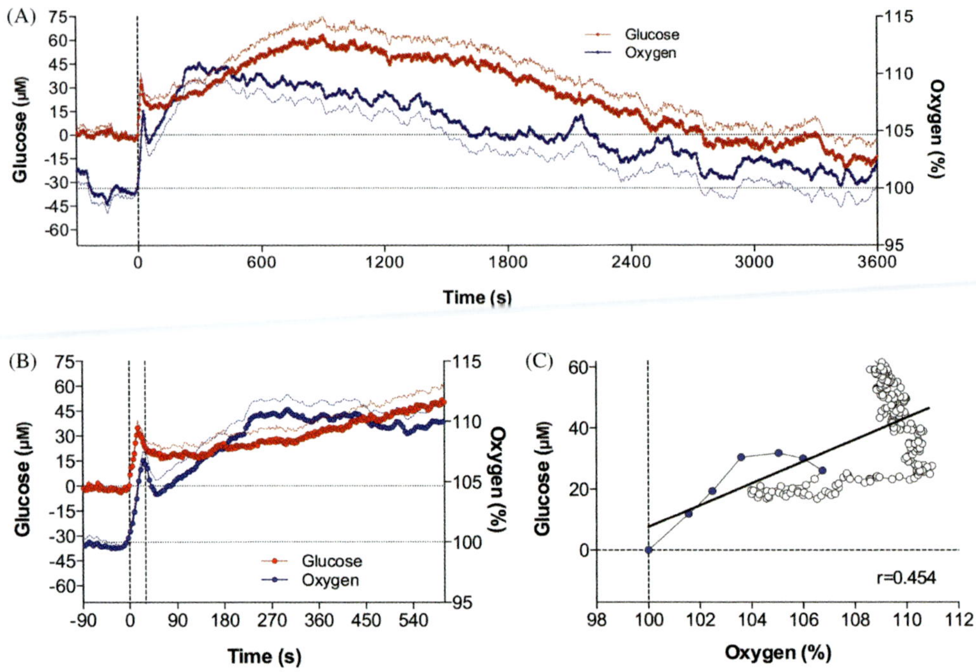

Figure 12.12 Changes in NAc oxygen and glucose induced by i.v. cocaine (1 mg/kg) delivered passively to awake, freely-moving rats. (A) shows mean (±SEM) changes for one hour after the injection. (B) shows changes for the initial time interval following the injection. (C) shows time-dependent correlative relationships between cocaine-induced changes in oxygen and glucose. Values to oxygen peak are shown as blue symbols.

the pattern of temperature response induced by i.v. cocaine was very similar to that induced by natural arousing stimuli such as a 3-min tail-pinch (see Figures 12.11(A)–12.11(C), right panel). Similar to cocaine, tail-pinch induced locomotor activation, increased brain and muscle temperatures, and induced a biphasic fluctuation in skin temperature. Also similar to cocaine, tail-pinch induced increases in the NAc-muscle differential and rapidly decreased the skin-muscle differential. These changes, however, were less robust and more transient than those seen with cocaine. Therefore, based on changes in locomotor activity and physiological parameters, cocaine mimics the effects of arousing stimuli, transiently increasing intra-brain heat production and decreasing heat loss to the external environment.

Intravenous cocaine also significantly affected NAc oxygen and glucose levels (Figure 12.12). In this case, both parameters increased, showing a biphasic response. Initially, both oxygen and glucose rapidly increased during the injection duration, but then they both showed a second elevation, which was clearly stronger and more prolonged for glucose than oxygen (A). Using high-speed resolution analysis, we observed a rapid and almost identical increase in glucose and oxygen that peaked at the offset of cocaine injection at ~30 s from its onset (B). Then both parameters slightly decreased but began accelerating again from ~60–90 s after the injection onset. Using time-dependent correlative analysis, we showed that cocaine-induced changes in NAc oxygen and glucose significantly correlate ($r = 0.454$, $p < 0.05$) and revealed differences in the time-course of their changes (C). Importantly, both parameters similarly increased for ~20–25 s after the injection onset, suggesting a common trigger as driving this synchronous response.

Our previous electrophysiological data suggest that i.v. cocaine induces short-latency EEG desynchronization, a classic feature of neural activation (Kiyatkin and Smirnov, 2010). This effect was very similar in both the cortex and VTA, suggesting its generalized nature. Therefore, neuronal activation appears to be the cause of rapid changes in NAc oxygen and glucose induced by i.v. cocaine. While cocaine injection resulted in changes in all of the EEG frequencies, the changes in high-frequency gamma power were maximal. As shown in Figures 12.13A and 12.13(C), i.v. cocaine induced a robust increase in cortical gamma power and this

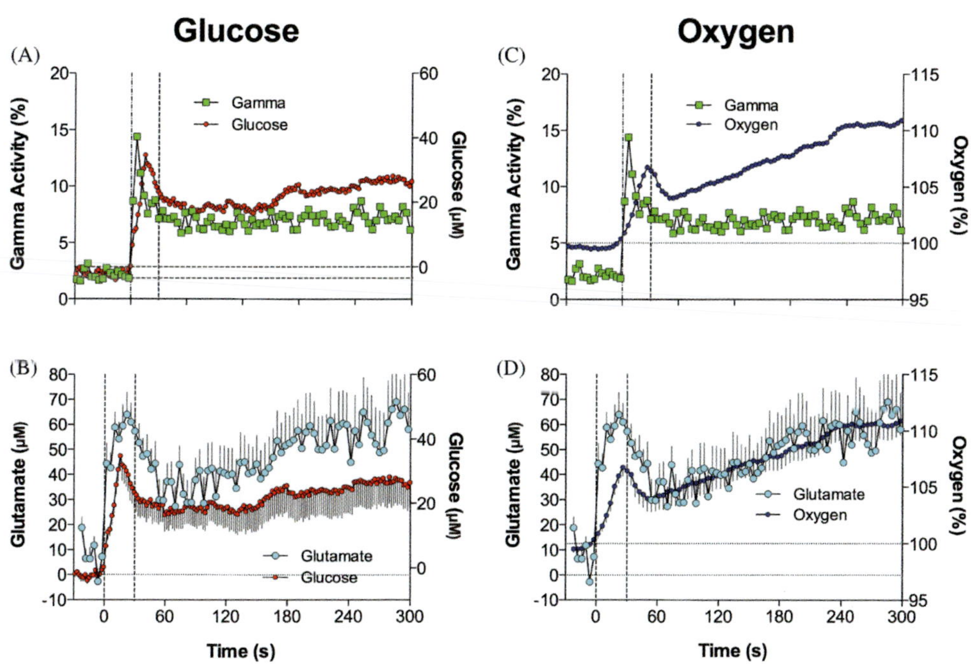

Figure 12.13 A–B: Relationships between cocaine-induced changes in NAc glucose and (A) EEG activation as assessed by changes in gamma power and (B) NAc glutamate levels. C–D: Relationships between changes in NAc oxygen and (C) EEG activation as assessed by changes in gamma power and (D) NAc glutamate levels. Drug-induced changes in glutamate levels were assessed by using Pinnacle glutamate biosensors; these data were described in detail in Wakabayashi *et al.*, 2015 and Wakabayashi and Kiyatkin, 2015.

change, reflecting neuronal activation, preceded increases in NAc oxygen and glucose (Wakabayashi and Kiyatkin, 2015).

Since accumbal neurons receive dense glutamate inputs and are very sensitive to glutamate, cocaine-induced release of this neurotransmitter appears to be the cause of excitation of accumbal neurons. This was confirmed by our comparison of changes in NAc oxygen and glucose with changes in glutamate as assessed by biosensors (Figures 12.13(B) and 12.13(D); Wakabayashi and Kiyatkin, 2015). In this case, NAc glutamate levels rapidly increased, preceding slower changes in glucose and oxygen. Therefore, similar to excitations of accumbal neurons elicited by natural arousing stimuli, cocaine-induced excitation of these cells triggers local cerebral vasodilation that enhances entry of oxygen and glucose into brain tissue. Similar to that occurring with natural stimuli, increases in oxygen and glucose elicited by i.v. cocaine followed neuronal activation but preceded slower and more prolonged changes in drug-induced metabolic brain activity as assessed by intra-brain heat production (see Figure 12.11).

While it is generally believed that neural effects of cocaine are mediated via direct action of this drug on centrally located neural substrates, some of these effects are exceptionally rapid, suggesting the possible involvement of peripherally located substrates as their trigger. For example, i.v. cocaine at behaviorally relevant doses induces a rapid increase in arterial blood pressure, which has second-scale latencies and peaks within one minute from the start of injection (Poon and van den Buuse 1998). Such a rapid time-course is difficult to reconcile with cocaine's pharmacokinetics and the time necessary for the drug to reach brain blood vessels, cross the blood-brain barrier (BBB), diffuse to reach and bind to centrally located neural substrates, and affect neurotransmission. Furthermore, this effect is resistant to dopamine receptor blockade (Kiritsy-Roy *et al.*, 1990; Poon and van den Buuse 1998) and is mimicked by cocaine's analog cocaine methiodide, which does not cross the blood-brain barrier (Dickerson *et al.*, 1999). Both cocaine analogs injected to freely-moving rats at low, behaviorally relevant doses (0.25–1.0 mg/kg) also induce rapid and strong cortical EEG desynchronization, suggesting generalized neural activation (Kiyatkin and Smirnov 2010), an effect which is also resistant to full dopamine receptor blockade (Kiyatkin and Smirnov 2010). Finally, a critical role of peripheral actions of cocaine in mediating its acute neural effects was demonstrated by electrochemical assessment of NAc glucose (Wakabayashi and Kiyatkin 2015). In this study, we found that i.v. cocaine induces a biphasic increase in NAc glucose, with the initial prominent rise occurring with 4–6 s latency and peaking 20–25 s after the injection onset. Importantly, this initial effect of cocaine was only slightly attenuated by full dopamine receptor blockade and was mimicked by cocaine methiodide.

To provide additional support for the role of peripherally driven mechanisms in mediating rapid effects of cocaine, we recently examined changes in NAc oxygen elicited by BBB-permeable cocaine HCl and BBB-impermeable cocaine

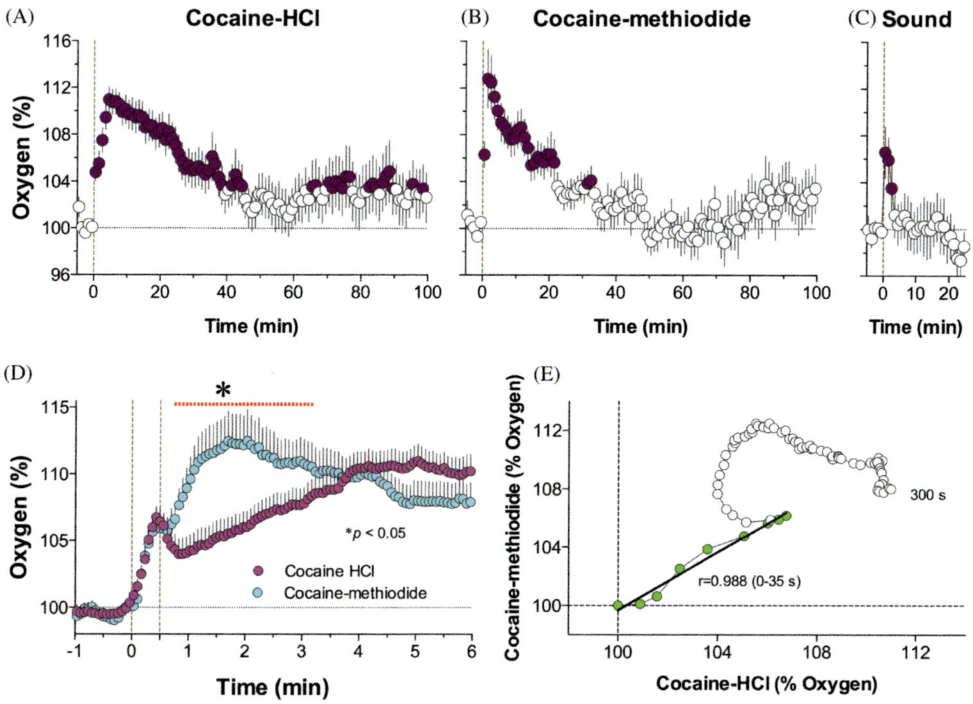

Figure 12.14 Mean (±SEM) changes in NAc oxygen induced by cocaine HCl (A), cocaine methiodide (B) and an auditory stimulus in awake, freely-moving rats. Filled symbols show values significantly different from pre-injection baseline. (D) shows rapid time-course analysis of immediate effects of two cocaine analogs. Red line with asterisk shows between-group differences. (E) shows time-dependent correlation between NAc oxygen changes induced by two cocaine analogs. The initial effects of both drugs are virtually identical.

methiodide (Solis *et al.*, 2018). As can be seen in Figure 12.14, both cocaine analogs induced rapid increases in NAc oxygen (A and B). Despite their larger magnitude and longer duration, these effects were similar to those induced by a short auditory stimulus (C). The initial, most rapid oxygen increase was virtually identical for both drugs, but the oxygen increase elicited by cocaine methiodide was monotonic while the increase elicited by cocaine HCl was biphasic, showing a second, much slower acceleration from ~45–50 s following injection onset (D). Importantly, this difference was evident only with rapid-time course analyses. As shown by time-dependent correlative analysis (E), the initial effects of both cocaine analogs on NAc oxygen were virtually identical, showing an exceptionally high correlation ($r = 0.988$, $p < 0.0001$). However, curves diverged from ~35 s after the injection onset, possibly reflecting the timing when cocaine HCl reaches centrally located neural substrates and directly induces changes in neural activity.

While the immediate effects of both cocaine analogs on NAc oxygen and glucose were surprisingly similar, there were important differences in the delayed effects of these drugs and in their behavioral outcomes. In contrast to cocaine HCl, cocaine methiodide did not induce notable locomotor activation (Brown and Kiyatkin, 2006) and its effects on NAc glutamate were much weaker, showing within-session tolerance (Wakabayashi and Kiyatkin 2012). These differences obviously arise due to the direct central action of cocaine, whereas such action is absent with cocaine methiodide. The similarity of the rapid neural effects induced by somatosensory stimuli and both cocaine analogs suggests that they share a common underlying mechanism involving activation of peripherally located neural substrates and rapid neural transmission. However, the exact nature of these substrates and pathways transmitting this signal to the CNS remains unclear. One possibility is the interaction of cocaine with multiple ionic channels (i.e., voltage-gated Na^+, K^+, Ca^{2+}, TRP channels) that are densely expressed on terminals of sensory nerves that innervate peripheral vessels (Lee *et al.*, 2005; Premkumar 2005; Wu *et al.*, 2006; Chen *et al.*, 2006). These channels are activated by multiple types of "pathological" shifts in blood chemicals and often viewed as visceral nociceptors. These substrates could be directly activated by cocaine when its levels rapidly and strongly increase after i.v. administration, thus generating an ascending excitatory drive to the CNS using neural pathways which transmit the effects of visceral somatosensory stimuli. While analytical research employing other techniques is necessary to determine the nature of peripheral neural substrates activated following i.v. cocaine administration, it is a difficult challenge to undertake because of the inability to pharmacologically block multiple neural targets in the periphery and the complexity of neural pathways which transmit visceral signals to the CNS.

12.5 Summary and conclusions

Proper entry of oxygen and glucose into the brain's internal environment is essential for normal functioning of the CNS and maintaining life and behavioral activity of living beings. While it is well known that these critical metabolic substances arrive to the brain from arterial blood and they are continuously used for metabolic activity of brain cells, this work was focused on physiological fluctuations of oxygen and glucose in the brain's extracellular space. In contrast to different indirect measures, the development of oxygen and glucose microsensors made it possible to directly examine natural fluctuations in these homeostatic parameters in awake, freely-moving rats. The major advantage of such electrochemical technologies is the ability to study dynamic aspects of changes in brain oxygen and glucose and examine their relationships with neuronal activity and changes in metabolic brain activity as assessed by multi-site temperature recordings. While neuronal responses elicited by natural arousing stimuli occur with millisecond latencies and usually are short in duration, metabolic activity as assessed by intra-brain heat production always follows neuronal activity and has definite latencies and longer duration. Our studies have clearly demonstrated that both oxygen and glucose rapidly enter the brain's extracellular environment following neuronal activation and preceding increases in metabolic activity, thus preventing any possible deficit of these substances actively used during functional activation. We also considered how brain levels of oxygen and glucose are affected during exposure to two highly popular drugs of abuse with either stimulant (cocaine) or depressant (heroin) modes of initial action. While cocaine mimics the effects of natural arousing stimuli, rapidly increasing brain oxygen and glucose, heroin induced distinct effects on these parameters. Specifically, heroin decreased entry of oxygen into brain tissue due to respiratory depression and a subsequent decrease in blood oxygen levels induced by this drug. In addition, glucose levels in response to heroin strongly increased due to cerebral vasodilation, which occurs as an adaptive response aimed to minimize oxygen drops in blood inflow. Importantly, heroin at low doses maintaining drug self-administration induced relatively strong brain hypoxia, underscoring its danger in terms of inducing serious health complications, including coma and death.

Acknowledgments

The original research reviewed in this manuscript was supported by the Intramural Research Program of the NIH, NIDA. The Author wants to thank Anum Afzal for valuable help in preparing figures for this manuscript, editorial assistance, and fruitful discussions on the matters discussed in this work.

References

Altschule MD (1951) Emotion and circulation. *Circulation*, 3: 444–454.

Attwell D, Buchan AM, Charpak S, Lauritzen M, Macvicar BA, Newman EA (2010) Glial and neuronal control of brain blood flow. *Nature*, 468: 232–243.

Battisti-Charbonney A, Fisher J, Duffin J (2011) The cerebrovascular response to carbon dioxide in humans. *J Physiol*, 589: 3039–3048.

Baker MA, Cronin MJ, Mountjoy DG (1976) Variability of skin temperature in the waking monkey. *Am J Physiol*, 230: 449–455.

Baud FJ (2009) Mechanisms of opioid-induced overdose: Experimental approach to clinical concerns. *Ann Parm Fr*, 67: 353–359 (in French).

Bazzu G, Puggioni GGM, Dedola S, Calla G, Rocchitta G, Mgheli R, Desole MS, Lowry JP, O'Neill RD, Serra PA (2009) Real-time monitoring of brain tissue oxygen using a miniaturized biotelemetric device implanted in freely moving rats. *Anal Chem*, 81: 2235–2241.

Bolger FB, Lowry JP (2005) Brain tissue oxygen: *In vivo* monitoring with carbon paste electrodes. *Sensors*, 5: 473–487.

Bolger FB, Bennett R, Lowry JP (2011) An *in vitro* characterization comparing carbon paste and Pt microelectrodes for real-time detection of brain tissue oxygen. *Analyst*, 136: 4028–4035.

Boutelle MG, Stanford C, Fillenz M, Albery WJ, Bartlett PN (1986) An amperometric enzyme electrode for monitoring brain glucose in freely moving rat. *Neurosci Lett*, 72: 283–288.

Brown PL, Kiyatkin EA (2006) The role of peripheral Na(+) channels in triggering the central excitatory effects of intravenous cocaine. *Eur J Neurosci*, 24: 1182–1192.

Bucher ES, Wightman RM (2015) Electrochemical analysis of neurotransmitters. *Annu Rev Anal Chem (Palo Alto Calif)*, 8: 239–261.

Calia G, Rocchita G, Megheli R, Puggioni G, Spussu Y, Bazzu G (2009) Biotelemetric monitoring of brain neurochemistry in conscious rats using microsensors and biosensors. *Sensors*, 9: 2511–2523.

Carelli RM, West MO (1991) Representation of the body by single neurons in the dorsolateral striatum of the awake, unrestrained rats. *J Comp Neurol*, 309: 231–249.

Chen YH, Lin CH, Lin PL, Tsai MC (2006) Cocaine elicits action potential bursts in a central snail neuron: The role of delayed rectifying K^+ current. *Neuroscience*, 138: 257–280.

Clark LC, Misrahy G, Fox RP (1958) Chronically implanted polarographic electrodes. *J Appl Physiol*, 13: 85–91.

Clark LC, Lyons C (1962) Electrode systems for continuous monitoring in cardiovascular surgery. *Ann NY Acad Sci*, 102: 29–45.

Clarke DD, Sokoloff L (1999) Circulation and energy metabolism. In: *Basic Neurochemistry: Molecular, Cellular and Medical Aspects* (Siegel GJ, Agranoff BW, Alberts RW, eds.), pp. 638–669. Philadelphia: Lippincott-Raven.

Davies PW, Brink FJ (1942) Microelectrodes for measuring local oxygen tension in animal tissues. *Rev Sci Instrum*, 13: 524–533.

Davies PW, Bronk DW (1957) Oxygen tension in mammalian brain. *Fed Proc*, 16: 689–692.

DeLong MR, Alexander GE, Georgopoulos AP, Crutcher MD, Mitchell SJ, Richardson RT (1984) Role of basal ganglia in limb movements. *Hum Neurobiol*, 2: 235–244.

Deniau JM, Mailly P, Maurice N, Charpier S (2007) The pars reticulata of the substantia nigra: A window to basal ganglia output. *Progr Brain Res*, 160: 151–172.

Dickerson LW, Rodak DJ, Kuhn FE, Wahlstrom SK, Tessel RE, Visner MS, Schaer GL, Gillis RA (1999) Cocaine-induced cardiovascular effects: Lack of evidence for a central nervous system site of action based on hemodynamic studies with cocaine methiodide. *J Cardiovasc Pharmacol*, 33: 36–42.

Di Chiara G (2002) Nucleus accumbens shell and core dopamine: Differential role in behavior and addiction. *Behav Brain Res*, 137: 75–114.

Donaldson GC, Keatinge WR, Saunders RD (2003) Cardiovascular responses to heat stress and their adverse consequences in healthy and vulnerable human populations. *Int J Hyperthermia*, 19: 225–235.

Duelli R, Kuschinsky W (2001) Brain glucose transporters: Relationship to local energy demand. *News Physiol Sci*, 16: 71–76.

Dunn-Meynell AA, Sanders NM, Compton D, Becker TC, Eiki J, Zhang BB, Levin BE (2009) Relationship among brain and blood glucose levels spontaneous and glucoprivic feeding. *J Neurosci*, 29: 7015–7022.

Fellows LK, Boutelle MG, Fillenz M (1992) Extracellular brain glucose levels reflect local neuronal activity: A microdialysis study in awake, freely moving rats. *J Neurochem*, 59: 2141–2147.

Fellows LK, Boutelle MG (1993) Rapid changes in extracellular glucose levels and blood flow in the striatum of the freely moving rat. *Brain Res*, 604: 225–231.

Feng ZC, Roberts EL, Sick TJ, Rosenthal M (1988) Depth profile of local oxygen tension and blood flow in rat cerebral cortex, white matter and hippocampus. *Brain Res*, 445: 280–288.

Fillenz M (2005) *In vivo* neurochemical monitoring and the study of behavior. *Neurosci Biobehav Rev*, 29: 949–962.

Fox PT, Raichle ME (1986) Focal physiological uncoupling of cerebral blood flow and oxidative metabolism during somatosensory stimulation in human subjects. *Proc Natl Acad Sci USA*, 83: 1140–1144.

Fox ME, Wightman RM (2017) Contrasting regulation of catecholamine neurotransmission in the behaving brain: Pharmacological insights from an electrochemical perspective. *Pharmacol Rev*, 69: 12–32.

Hodgkin AL (1967) *The Conduction of the Nervous Impulse.* Liverpool: Liverpool University Press.

Hu Y, Wilson GS (1997) Rapid changes in local extracellular rat brain glucose observed with an *in vivo* glucose sensor. *J Neurochem*, 68: 1745–1752.

Jaffe JH, Knapp CM, Ciraulo DA (1997) Opiates: Clinical aspects. In: *Substance Abuse. A Comprehensive Textbook* (3rd ed.) (Lowinson JH, Ruis P., Millman RB Langrod JG, eds), pp.158–165. Baltimore *et al.*: Williams and Wilkins.

Kahn JS (1964) Construction of a simple and sensitive Clark-type electrode for measurement of oxygen concentrations. *Anal Biochem*, 9: 389–391.

Kealy J, Bennett R, Lowry JP (2013) Simultaneous recording of hippocampal oxygen and glucose in real time using constant potential amperometry in freely-moving rat. *J Neurosci Meth*, 215: 110–120.

Kealy J, Bennett R, Woods B, Lowry JP (2017) Real-time changes in hippocampal energy demands during a spatial working memory task. *Behav Brain Res*, 326: 59–68.

Kiritsy-Roy JA, Halter JB, Gordon SM, Smith MJ, Terry LC (1990) Role of the central nervous system in hemodynamic and sympathoadrenal responses to cocaine in rats. *J Pharmacol Exp Ther*, 255: 154–160.

Kiyatkin EA (2010). Brain temperature homeostasis: Physiological fluctuations and pathological shifts. *Front Biosci*, 15: 73–92.

Kiyatkin EA, Lenoir M (2012) Rapid fluctuations in extracellular brain glucose levels induced by natural arousing stimuli and intravenous cocaine: Fueling the brain during neural activation. *J Neurophysiol*, 108: 1669–1684.

Kiyatkin EA, Mitchum R (2003) Fluctuations in brain temperatures during sexual behavior in male rats: An approach for evaluating neural activity underlying motivated behavior. *Neuroscience*, 119: 1169–1183.

Kiyatkin EA, Rebec GV (1996) Dopaminergic modulation of glutamate-induced excitations of neurons in the neostriatum and nucleus accumbens of awake, unrestrained rats. *J Neurophysiol*, 75: 142–153.

Kiyatkin EA, Rebec GV (1999) Modulation of striatal neuronal activity by glutamate and GABA: Iontophoresis in awake, unrestrained rats. *Brain Res*, 822: 88–106.

Kiyatkin EA, Smirnov MS (2010) Rapid EEG desynchronization and EMG activation induced by intravenous cocaine in freely moving rats: A peripheral, nondopamine neural triggering. *Am J Physiol Regul Integr Comp Physiol*, 298: R285–R300.

Kiyatkin EA, Wakabayashi KT, Lenoir M (2013) Physiological fluctuations in brain temperature as a factor affecting electrochemical evaluations of extracellular glutamate and glucose in behavioral experiment. *ACS Chem Neurosci*, 4: 652–665.

Kiyatkin EA, Wakabayashi KT (2015) Parsing glucose entry into the brain: Novel findings obtained with enzyme-based glucose biosensors. *ACS Chem Neurosci*, 6: 108–116.

Kontos HA (1981) Regulation of the cerebral circulation. *Annu Rev Physiol*, 43: 397–407.

Kuhar MJ, Ritz MC, Boja JM (1991) The dopamine hypothesis of the reinforcing properties of cocaine. *Trends Neurosci*, 14: 299–302.

Lecrux C, Hamel E (2011) The neurovascular unit in brain function and disease. *Acta Physiol (Oxf)*, 203: 47–59.

Lee Y, Lee CH, Oh U (2005) Painful channels in sensory neurons. *Mol Cells*, 20: 315–324.

Lowry JP, Boutelle MG, O'Neill RD, Fillenz M (1996) Characterization of carbon paste electrodes *in vitro* for simultaneous amperometric measurements of changes in oxygen and ascorbic acid *in vivo*. *Analyst*, 121: 761–766.

Lowry JP, Boutelle MG, Fillenz M (1997) Measurement of brain tissue oxygen at a carbon paste electrode can serve as an index of increases in regional cerebral blood flow. *J Neurosci Methods*, 71: 177–182.

Lowry JP, O'Neill RD, Boutelle MG, Fillenz M (1998) Continuous monitoring of extracellular glucose concentrations in the striatm of freely moving rats with implanted glucose biosensor. *J Neurochem*, 70: 391–396.

Lowry JP, Griffin K, McHugh SB, Lowe AS, Trickebank M, Sibson NR (2010) Real-time electrochemical monitoring of brain tissue oxygen: A surrogate for functional magnetic resonance imaging in rodents. *Neuroimage*, 52: 549–555.

McNay EC, McCarty RC, Gold PE (2001) Fluctuations in brain glucose concentration during behavioral testing: Dissociations between brain areas and between brain and blood. Neurobiol Lern Mem, 75: 325–337.

McNay EC, Gold PE (1999) Extracellular glucose concentrations in the rat hippocampus measured by zero-net-flux: Effects of microdialysis flow rate, strain, and age. *J Neurochem*, 72: 785–790.

Mergenthaler P, Lindauer U, Dienel G, Meisel A (2013) Sugar for the brain: The role of glucose in physiological and pathological brain function. *Trends Neurosci*, 36: 587–597.

Mogenson GJ, Jones DL, Yim CY (1980) From motivation to action: Functional interface between the limbic system and the motor system. *Prog Neurobiol*, 14: 69–97.

Muoio V, Persson PB, Sendeski MM (2014) The neurovascular unit — concept review. *Acta Physiol (Oxf)*, 210: 790–798.

Poon J, van den Buuse M (1998) Autonomic mechanisms in the acute cardiovascular effects of cocaine in conscious rats. *Eur J Pharmacol*, 363: 147–152.

Premkumar LS (2005) Block of a Ca(2+)-activated potassium channel by cocaine. *J Membr Biol*, 204: 129–136.

Rebec GV (2006) Behavioral electrophysiology of psychostimulants. *Neuropsychopharmacology*, 31: 2341–2348.

Ritchie JM (1973) Energetic aspects of nerve conductance: The relationships between heat production, electrical activity and metabolism. *Progr Biophys Molec Biol*, 26: 147–187.

Robinson DL, Hermans A, Seipel AT, Wightman RM (2008) Monitoring rapid chemical communication in the brain. *Chem Rev*, 108: 2554–2584.

Rodgers ZB, Defte JA, Wehrli FM (2016) MRI-based methods for quantification of the cerebral metabolic rate of oxygen. *J Cereb Blood Flow Metab*, 36: 1165–1185.

Roy CS, Sherrington CS (1890) On the regulation of the blood-supply of the brain. *J Physiol*, 11: 85–108.

Rowell LB (1983) Cardiovascular aspects of human thermoregulation. *Circ Res*, 52: 367–376.

Schmidt CF, Kety SS (1947) Recent studies of cerebral blood flow and cerebral metabolism in man. *Trans Assoc Am Physicians*, 60: 52–58.

Schmidt-Nielsen K (1997) *Animal Physiology. Adaptation and Environment* (5th ed.). Cambridge: Cambridge University Press.

Silver IA, Erecinska M (1994) Extracellular glucose concentration in mammalian brain: Continuous monitoring of changes during increased neuronal activity and upon limitation in oxygen supply in normo-, hypo-, and hyperglycemic animals. *J Neurosci*, 14: 5068–5076.

Simpson IA, Carrutheers A, Vannucci SJ (2007) Supply and demand in cerebral energy metabolism: The role of nutrient transporters. *J Cereb Blood Flow Metab*, 27: 1766–1791.

Siesjo B (1978) Brain Energy Metabolism. New York: Wiley.

Simon EJ (1997) Opiates: Neurobiology. In: *Substance Abuse* (3rd ed.), (Lowinson JH, Ruiz P, Millman RB, Langrod JG, eds.), pp.148–158. Baltimore: Williams & Wilkins.

Smirnov MS, Kiyatkin EA (2008) Fluctuations in central and peripheral temperatures associated with feeding behavior in rats. *Am J Physiol*, 295: R1414–R1424.

Sokoloff L (1999) Energetics of functional activation in neural tissues. *Neurochem Res*, 24: 321–329.

Solis E, Bola RA, Fasulo BJ, Kiyatkin EA (2017a). Brain hyperglycemia induced by heroin: Association with metabolic neural activation. *ACS Chem Neurosci*, 8: 265–271.

Solis E, Cameron-Burr KT, Kiyatkin EA (2017b) Rapid physiological fluctuations in nucleus accumbens oxygen levels induced by arousing stimuli: Relationships with changes in brain glucose and metabolic neural activation. *Front Integr Neurosci*, 11: 9.

Solis E, Cameron-Burr KT, Shaham Y, Kiyatkin EA (2017c) Intravenous heroin induces rapid brain hypoxia and hyperglycemia that precede brain metabolic response. *eNeuro*, 7 Jun 2017; 4(3). Pii.

Solis E, Afzal A, Kiyatkin EA (2019) Intravenous cocaine increases oxygen entry into brain tissue: Critical role of peripheral drug actions. *ACS Chem Neurosci*, 10: 1923–1928.

Solomon GF, Moos RH, Stone GC, Fessel WJ (1964) Peripheral vasoconstriction induced by emotional stress in rats. *Angiology*, 15: 362–365.

Travis RP, Clark LC (1965) Changes in evoked brain oxygen during sensory stimulation and conditioning. *EEG & Clin Neurophysiol*, 19: 484–491.

Wakabayashi KT, Myal S, Kiyatkin EA (2015). Fluctuations in nucleus accumbens extracellular glutamate and glucose during motivated glucose-drinking behavior. *J Neurochem*, 132: 327–341.

Wakabayashi KT, Kiyatkin EA (2015) Central and peripheral contributions to dynamic changes in nucleus accumbens glucose induced by intravenous cocaine. *Front Neurosci*, 9: 42.

Windels F, Kiyatkin EA (2004) GABA, not glutamate, controls the activity of substantia nigra reticulata neurons in awake, unrestrained rats. *J Neurosci*, 24: 6751–6754.

Wise RA, Bozarth MA (1987) A psychomotor stimulant theory of addiction. *Psychol Rev*, 94: 469–492.

Wu SN, Chang HD, Sung RJ (2006) Cocaine-induced inhibition of ATP-sensitive K+ channels in rat ventricular myocytes and in heart-derived H9c2 cells. *Basic Clin Pharmacol Toxicol*, 98: 510–517.

Yeadon M, Kitchen I (1990) Multiple opioid receptors mediate the respiratory depressant effects of fentanyl-like drugs in the rat. *Gen Pharmacol*, 21: 655–664.

Zimmerman JB, Wightman RM (1991) Simultaneous electrochemical measurements of oxygen and dopamine *in vivo*. *Anal Chem*, 63: 24–28.

CHAPTER 13

MICRODIALYSIS OF CATECHOLAMINES REVEALS ETHANOL ACTIONS IN THE RODENT BRAIN

Saul Jaime and Rueben A. Gonzales*

Division of Pharmacology and Toxicology, College of Pharmacy, The University of Texas at Austin, USA

***rgonzales@austin.utexas.edu**

13.1 Introduction

Ethanol has long been known to alter various neurotransmitter systems in the brain, and catecholamines have been a focus of ethanol research for many decades (reviewed in Di Chiara, 1999; Gonzales *et al.*, 2004). The major focus of most of the previous research on ethanol and catecholamines has been on dopamine, and some early work was also done on norepinephrine. Both major catecholamine neurotransmitters in the brain are implicated in numerous behaviors, and some of these will be briefly reviewed in this chapter. While some of the behaviors mediated by these two catecholamines overlap, there are several differences and behavioral features that are unique to each one.

The major goals of this chapter are to review and summarize some of the recent advances in our understanding of the behavioral and neurochemical responses of ethanol focusing on both major central nervous system catecholamines: dopamine and norepinephrine. First, some of the major behavioral effects of ethanol and the potential roles that the catecholamines may play in these behaviors will be examined. Second, we will emphasize the brain regions that have been studied that exhibit a catecholamine response to ethanol using either microdialysis or voltammetric techniques. Third, the unique methodological advantage of microdialysis with respect to concurrent analysis of the ethanol and the catecholamine content from the same sample will be discussed. Finally, we provide recent data that increase our insight into how catecholamines are involved in the mechanisms of action of ethanol on selected behaviors.

13.2 Behavioral effects of ethanol

Ethanol, also known as alcohol, is a psychoactive substance that affects the entire body including the brain of the person who consumes it (National Institute on Alcohol Abuse and Alcoholism, *https://www.niaaa.nih.gov/alcohol-health/alcohols-effects-body*). Reports have demonstrated that ethanol has both stimulating and depressing effects on numerous behaviors that are dose dependent. Many of these have been investigated with respect to the neurochemical mechanisms that contribute to these various behaviors. It is worth noting that the doses of ethanol that produce these various behavioral effects are much larger than most centrally acting drugs with similar effects. In other words, ethanol is a very low potency drug with behaviorally effective doses in the g/kg range instead of the typical range of mg/kg for most medications.

For example, a commonly recognized reference point for ethanol intoxication is the legal driving limit in terms of blood alcohol concentration (BAC). This is commonly expressed as mg/dl or mg% and in the United States this limit is 80 mg/dl or 0.08 mg%. In terms of standard chemical terminology, this concentration is equivalent to 17.4 mM. Although there is much variability in the amount of ethanol consumed that produces this concentration in the blood for humans, a general rule of thumb is that this is achieved one hour after consuming 3–4 standard drinks (total dose of 0.7–1.0 g/kg). We now briefly review the major behavioral effects of ethanol intoxication to lay the foundation for discussion of neurochemical studies which implicate the catecholamines in subsequent sections.

13.2.1 Motor effects

One of the most common and recognized behavioral effects of ethanol is loss of motor coordination. The ethanol-induced loss of motor coordination is dose-dependent, but it occurs at relatively low doses (below the legal driving limit). The amount of incoordination increases with greater doses. With very large doses, motor coordination is completely lost. These motor effects are virtually universally seen in all animal models ranging from invertebrate to rodents to humans. In humans this can be observed as changes in reaction time, changes in gait (stumbling), or slurred speech. At the legal driving limit, the loss of motor coordination inhibits the safe use of motor vehicles.

In contrast to the loss of motor coordination, ethanol also produces stimulating effects on locomotor activity in some animal models, often in combination with the loss of motor coordination. In an early study by Waller *et al.* (1986), the authors showed that ethanol at low doses of 0.12 and 0.25 g/kg increased spontaneous motor activity of alcohol-preferring rats (Li *et al.*, 1993) and Maudsley Reactive rats (Adams *et al.*, 2002) in comparison to alcohol non-preferring control rats. In contrast, when the ethanol dose was increased to 1.0 g/kg and 1.5 g/kg, the spontaneous motor activity was suppressed in all strains of rats tested. In a more recent study, the authors tested the effects of alcohol on behavior of outbred adult Wistar rats (Karlsson and Roman, 2016). They found that a low dose of ethanol (0.5 g/kg) initially increased locomotor activity, which then decreased over time. These authors also replicated the suppression of locomotor activity at higher doses of 1.0 and 1.5 g/kg. Taken together with numerous other studies using different animal models, it is clear that ethanol has both a stimulating and depressing effect on locomotor behavior that is dose dependent (for review see Brabant *et al.*, 2014).

13.2.2 Cognitive effects

Ethanol also has well-established deleterious effects on cognitive function with both acute (Popke *et al.*, 2000) and chronic use (Oscar-Berman *et al.*, 1997). While these cognitive functions are carried out by numerous brain regions, one of the prominent and most well studied is the prefrontal cortex. Some examples of these findings will be briefly summarized in this section.

The prefrontal cortex is known to be involved in various aspects of executive function, including decision making, and impairment of prefrontal function clearly produces cognitive dysfunction (Bechara and Van Der Linden, 2005). In a similar manner to the lesion studies, chronic ethanol consumption in humans is known to reduce cognitive functions associated with prefrontal cortex activity (Goldstein *et al.*, 2004). Moreover, chronic alcoholism has also been found to produce brain damage in prefrontal cortical areas (Pfefferbaum *et al.*, 1997). Animal models have also suggested that ethanol exposure alters prefrontal cortical activity along with behaviors that are mediated, at least in part, by prefrontal cortical function (Badanich *et al.*, 2011; Trantham-Davidson *et al.*, 2017). Other studies have established that catecholamine signaling in the prefrontal cortex is critical for optimal cognitive function (Arnsten and Pliszka, 2011).

Taken together, the body of work on the role of the prefrontal cortex on cognition, in combination with the demonstrated reduction in prefrontal activity upon ethanol exposure, makes it plausible to suggest that ethanol may produce deficits in cognitive function by altering catecholamine activity in the prefrontal cortex. Subsequent sections will review the recent data from neurochemical studies that provide some insight to this question.

13.2.3 Reward and motivation

With regard to alcohol use disorders, one of the most critical effects of ethanol is the promotion of a positive hedonic state also referred to as drug reward. Many theories of the development of addiction to ethanol and other drugs propose that these rewarding effects occur with initial self-administration (Koob and Volkow, 2016). This action, in turn, promotes future drug-seeking and self-administration in a process known as reinforcement. Furthermore, it is well-accepted in the field of drug abuse that drugs, including ethanol, produce these behaviors by acting on neural circuits that exist to help individuals survive and maintain critical homeostatic behaviors, such as eating when hungry and drinking when thirsty. Thus, these neurobiological circuits mediate motivation for natural rewards but abused drugs, including ethanol, also activate these circuits in the absence of hunger and thirst. Some of these ideas have been recently discussed in previously published chapters of this book series (Belle and Wightman, 2015; Wise and You, 2017).

Dopamine has long been suggested to be involved in the rewarding and motivational effects of natural and drug reward, including ethanol reward (Di Chiara, 1999; Gonzales *et al.*, 2004). Sections 13.3 and 13.4 will review some examples of the studies that support the role of dopamine in ethanol-induced reward and motivation. Unlike the major focus of the field on dopamine, the other catecholamine, norepinephrine, has also been suggested to be involved in some of the rewarding and motivational properties of ethanol, but less data exist on role of norepinephrine. However, in the past few years there has been a resurgence of neurochemical research on the potential effects of ethanol on norepinephrine (Haass-Koffler *et al.*, 2018) and these recent studies will also be examined.

13.2.4 Aversive responses

In addition to the well-known and well-studied positive hedonic effects of ethanol, it is also clear that ethanol possesses aversive properties. In this way, ethanol is like other drugs of abuse that have both positive and negative motivational properties (Risinger and Boyce, 2002). In general, the neurochemical mechanisms that underlie the negative or aversive responses to ethanol are less understood compared with the rewarding effects of ethanol. As mentioned in the previous section, dopamine circuits are implicated in the positive, rewarding, and motivational approach behavior. On the other hand, the role of norepinephrine in either the positive or aversive effects of acute ethanol is unclear.

One way that the aversive effects of acutely administered ethanol have been studied is the use of conditioning procedures. For example, in the conditioned taste aversion paradigm, administration of ethanol to a mouse after the mouse has access to a palatable solution will promote the avoidance of the palatable solution in future trials (Chester *et al.*, 1998; Barkley-Levenson *et al.*, 2015). The aversive effects of ethanol can be demonstrated in naïve mice or after a few pairings in which the mouse has limited exposure to ethanol.

In addition to these relatively acute aversive effects of ethanol, the chronic effects of ethanol after high dose exposure over several weeks can lead to physical dependence on ethanol (Gilpin *et al.*, 2008). This can be demonstrated experimentally after withdrawal of the ethanol from the experimental subject. Within a few hours after the ethanol has been metabolized and the BAC drops to near zero, the subjects typically begin showing somatic signs that include hyperactivity, tremor, increased muscle tone and changes in posture, and possibly seizure activity (Roberts *et al.*, 1996). These signs of physical withdrawal from ethanol may last for up to 24 hours with continued abstinence (Metten *et al.*, 2010). During this acute withdrawal period, the experimental subject is clearly experiencing an aversive and stressful state. It has been proposed that relief of the ethanol withdrawal-induced aversive state by ethanol consumption is part of the mechanism for the escalation of consumption of ethanol with prolonged use (Roberts *et al.*, 2000). Also, norepinephrine is known to be involved in stress responses (Morilak *et al.*, 2005) and central norepinephrine circuits may contribute to the behavioral effects of ethanol withdrawal (Becker, 2012). During this acute withdrawal period, several studies have shown that extracellular dopamine decreases in the nucleus accumbens (Weiss *et al.*, 1996; Rossetti *et al.*, 1992). Overall, it is plausible that both dopamine and norepinephrine are involved in some of the behavioral effects of ethanol withdrawal after induction of physical dependence.

13.3 Brain regions that show a neurochemical response to ethanol: Microdialysis and voltammetry

13.3.1 Dopamine

We, and others, have investigated the dopaminergic response to ethanol in the medial prefrontal cortex (mPFC) and in the ventral and dorsal striatum using microdialysis in various models of alcohol administration to study the rewarding and reinforcing effects of ethanol.

The mesolimbic dopaminergic system has been shown to be involved in natural and drug reward. The ventral tegmental area (VTA) sends dopaminergic projections to the nucleus accumbens, prefrontal cortex, amygdala, and hippocampus (Ikemoto, 2007). While not originally recognized, recent work suggests that additional projections to the dorsal striatum also exist (Ferenczi *et al.*, 2016; Decot *et al.*, 2017; Lohani *et al.*, 2017). The following sections will summarize data on whether ethanol stimulates dopamine release in these major projections from the VTA.

Ding *et al.* (2011) reported a 150% increase over baseline in mPFC extracellular dopamine after a microinjection of 200 mg% ethanol into the posterior VTA whereas microinjections into the anterior VTA caused no difference from controls. Similarly, Schier *et al.* (2013) measured a 160% increase over baseline in mPFC extracellular dopamine after both a slow and fast i.v. infusion of 10% ethanol. More recently, Doherty *et al.* (2016) extended these studies using a model of operant ethanol self-administration and reported a significant increase in dopamine during the consumption phase of the paradigm. Collectively, these studies show that dopamine terminals in the mPFC respond to ethanol administration and that the stimulatory effect of ethanol on mPFC dopamine may, in part, be driven by the posterior VTA.

The nucleus accumbens is one of the critical regions of the reward system and a major target of the mesolimbic dopaminergic system. Previous studies show that the nucleus accumbens responds to many substances of abuse including ethanol (reviewed in Gonzales *et al.*, 2004; Howard *et al.*, 2008, 2009; Ding *et al.*, 2009). For example, Robinson *et al.* (2009) used microdialysis and fast scan voltammetry to measure phasic and tonic dopamine levels in the nucleus accumbens after ethanol administration. The authors reported a dose-dependent tonic increase in dopamine levels that was 140% above baseline at doses of 1.0 and 2.0 g/kg ethanol. Fast scan cyclic voltammetry data showed that the effect of ethanol on dopamine release is selective across regions of the nucleus accumbens and not uniform in comparison to the microdialysis data. The effects of ethanol on extracellular dopamine levels extends to a self-administration paradigm demonstrating the reinforcing effects of ethanol. Carrillo and Gonzales (2011) showed that a 10% sucrose plus 10% ethanol solution enhanced the levels of dopamine 140% above baseline in the nucleus accumbens after a single voluntary exposure.

In contrast, less is known about the potential dopamine response to ethanol in the dorsal striatum. According to the ventral to dorsal striatum transition hypothesis of drug addiction, it would be of interest to understand how the dorsal striatum dopaminergic system responds to acute and chronic ethanol administration in both rodents and humans (Everitt and Robbins, 2013). In a recent fMRI study, using a cue-reactivity paradigm, Vollstädt-Klein *et al.* (2010) showed an increase in the blood oxygen level dependent (BOLD) signal in the dorsal striatum of heavy drinkers in comparison to light drinkers. Moreover, light drinkers showed a higher activation of the ventral striatum and mPFC than heavy drinkers. Although major advances have been made in understanding the neural mechanisms of the BOLD signal (Jaime *et al.*, 2018) a more invasive approach will be needed to measure the neurochemical signatures of the BOLD increases measured in this study.

An early microdialysis study indicated that ethanol administration stimulated dopamine release in the dorsal striatum at moderate to high doses (1–5 g/kg, i.p.) (Imperato and Di Chiara, 1986). More recently, Vena *et al.* (2016) examined subregions of the dorsal striatum (dorsomedial and dorsolateral) with microdialysis and measured extracellular dopamine after i.v. administration of 10% ethanol. The authors reported no significant changes in dopamine levels in either region after the 1 g/kg ethanol infusion. However, a cumulative ethanol dosing approach, in which the highest dose was 2.25 g/kg, showed a significant change in both the dorsomedial and dorsolateral striatum with the dorsomedial striatum showing the most robust increase in dopamine levels. Overall it appears that the dorsomedial striatum is more involved in the effects of severe ethanol intoxication compared with the dorsolateral striatum. More studies will be needed to confidently conclude the role of the dorsal striatum to ethanol exposure in a chronic model of alcohol use disorders.

13.3.2 Norepinephrine

In contrast to the large body of work on the effects of ethanol on dopamine systems, much less work is available on the effects of ethanol on norepinephrine. However, in recent years, there has been a resurgence of interest in the ethanol field on the possible roles of noradrenergic signaling in ethanol behaviors and as a possible therapeutic target for the treatment of alcohol use disorders (Haass-Koffler *et al.*, 2018). We start this section with a brief review of the anatomy of noradrenergic systems, and this is followed by the data available to date on norepinephrine responses to ethanol.

Noradrenergic cell bodies expressing dopamine beta hydroxylase (DBH) are found in seven nuclei, A1–A7, in the medulla and pons region of the hindbrain. The nucleus of the solitary tract (NTS) and the locus coeruleus (LC) contain the highest number of DBH+ cell bodies (Berridge *et al.*, 1997; Delfs *et al.*, 1998; Rinaman 2011). The LC has been shown to project to the cortex, hippocampus, amygdala, cerebellum, and thalamus via the dorsal noradrenergic bundle, while the NTS projects to the medial and lateral septum, nucleus accumbens shell, hypothalamus, and the bed nucleus of the stria terminalis (BNST) via the ventral noradrenergic bundle (Moore and Bloom, 1979; Grzanna and Molliver, 1980; Robertson *et al.*, 2013; Schwarz and Luo, 2015).

Stress, substance abuse, aversive stimuli, and reinstatement have been suggested to modulate norepinephrine levels in many of these noradrenergic target regions (Forray and Gysling 2004; Park *et al.*, 2012; McElligot *et al.*, 2013; Fox *et al.*, 2015; Chandler, 2016; Espana *et al.*, 2016). Though the underlying mechanisms are unknown, there is some evidence that ethanol also alters the function of brain regions that modulate noradrenergic signaling (Chang *et al.*, 1995; Rybinin *et al.* 1997; Thiele *et al.*, 2000; Covarrubias *et al.*, 2005; Vilpoux *et al.*, 2009; Lee *et al.*, 2011; Aimino *et al.*, 2018).

Unlike dopamine, research identifying brain regions that show a noradrenergic response to ethanol are sparse but growing. Using microdialysis, Rossetti *et al.* (1992) demonstrated that in awake and freely behaving Sprague-Dawley rats, 10% ethanol stimulated extracellular levels of norepinephrine in the medial PFC (mPFC) at low doses (0.2 g/kg) to 160% compared with baseline within 60 minutes of i.p. administration. At a higher dose of 10% ethanol (2.0 g/kg) there was also a reported increase in extracellular norepinephrine initially that was followed by a decrease below baseline 60 minutes after ethanol i.p. administration. More recently, we have found similar results to Rossetti *et al.* (1992) (data submitted for publication). When we placed a 3.0 mm active area microdialysis probe into the mPFC, we observed a 77% increase above baseline after i.v. administration of 10% ethanol (1 g/kg). The major difference between our work and previous work was the route of administration. Intravenous injections allowed us to remove the handling aspect of the experimental design. This issue is important to tackle as norepinephrine is the main neurotransmitter involved in the stress response and there is the possibility that the response reported in studies involving i.p. administration could be confounded by the stress of the i.p. injection that involves physical handling and restraint of the experimental subject (Ihalainen *et al.*, 1999; Kawahara *et al.*, 2000; Marsteller *et al.*, 2002). Moreover, in our study we did not observe reduction of extracellular norepinephrine below the baseline that was previously reported by Rossetti *et al.* (1992). This could be due to the moderate dose of ethanol (1 g/kg) we infused in comparison to previous work (2 g/kg).

In addition to the mPFC, the basal forebrain is a region of interest when considering alcohol use disorders and whether these regions respond to ethanol is not well studied. Previous preclinical work had suggested a potential noradrenergic response to ethanol in the nucleus accumbens. Marinelli *et al.* (2003) studied the noradrenergic response to 0.8, 1.6, 2.4 g/kg of either 7, 14, or 21% (v/v) ethanol. No dose or concentration of ethanol caused a statistically significant change in extracellular norepinephrine levels in the nucleus accumbens. The authors reported a p-value of 0.067 for the main effect of ethanol injection, and after examination of the table it can be concluded that the lack of significance was due to the large between-subject variability that may have been from the handling of the animal for the ethanol i.p. administration. The authors pointed out a potential pitfall to their study was the lack of HPLC optimization for norepinephrine measurements, leading them to suggest that further studies of the norepinephrine and ethanol relationship were warranted. More recently, Karkhanis *et al.* (2014) tackled a similar question using a social isolation induced stress model. The authors administered a 1.0 or 2.0 g/kg dose of ethanol i.p. to rats that were group-housed or isolated during adolescence. Using microdialysis, this study also found no effect of ethanol, at either dose, on extracellular norepinephrine levels in the nucleus accumbens of group-housed rats. In contrast, in the socially isolated rats both doses of ethanol stimulated norepinephrine release. Another possible experimental factor that could be involved in the lack of stimulatory effects of ethanol on extracellular norepinephrine levels in the nucleus accumbens might be the temporal resolution of microdialysis sampling. Many of the studies measuring extracellular norepinephrine employ 20–30 minute sampling intervals to account for the expected low dialysate concentrations of norepinephrine. In our most recent work (submitted for publication), we improved the temporal resolution by using a 10-minute sampling interval. We found that 10% ethanol stimulated extracellular norepinephrine to 74% and 106% above baseline after a 1 g/kg dose (i.v.) at 1.0 ml/min and 0.25 ml/min infusion rates, respectively. Once again, our i.v. route of administration was used to remove any potential confounding effects that may come from the handling or restraint during ethanol administration.

Collectively, findings from our laboratory and others demonstrate that the mesocorticolimbic noradrenergic system also responds to ethanol administration as previously established for the dopaminergic systems. Further studies are necessary to fully understand if and how dopaminergic and noradrenergic systems are altered by ethanol synergistically and whether such synergistic alterations might contribute to the development of alcohol use disorders.

13.4 Monitoring dialysate ethanol along with dialysate catecholamines

One of the unique features of monitoring the effects of ethanol on the extracellular concentration of neurotransmitters using microdialysis is that the ethanol reaching the brain can also be measured. Capturing both the ethanol that is producing the neurochemical response along with the neurotransmitter provides additional insight into the pharmacodynamics of the response *in vivo*. This section will briefly review some of the original studies in which ethanol was measured using intracerebral microdialysis, and more recent work will be covered as well.

Original work in the 1990s illustrated that ethanol could be measured in microdialysis samples taken from probes placed in various brain areas (Ferraro *et al.*, 1990; Yoshimoto and Komura, 1993; Nurmi *et al.*, 1994, 1999). These early studies were important for establishing the time course of changes in dialysate ethanol concentration sampled from various brain regions after i.p. or self-administration. These studies measured the concentration of dialysate ethanol with a higher temporal resolution than previously possible from analysis of post-mortem brain tissue. However, one of the limitations of conventional microdialysis, as performed under typical experimental conditions, is that the concentration of ethanol in the dialysate is only a fraction of that in the tissue surrounding the microdialysis probe. Thus, a limitation of these early studies was that the true concentration in the brain was not known.

This limitation was overcome in a study which incorporated quantitative microdialysis techniques. One of the basic principles underlying the use of quantitative microdialysis is that a calibration method must be used to account for the fraction of the analyte that diffuses from the tissue into the lumen of the probe. Robinson *et al.* (2000) examined this issue in two ways. First, the probes were calibrated by delivering a constant concentration of ethanol through the microdialysis probe and measuring the dialysate concentrations leaving the probe. In this way, a calibration can be done for delivery of ethanol to the tissue. If the diffusion of ethanol from the tissue into the probe is equivalent to diffusion of ethanol in the opposite direction, then the calibration factor can be used to calculate tissue concentrations of ethanol from dialysate concentrations after ethanol administration. Unfortunately, *in vitro* studies found that the direction of diffusion affected the diffusion rate of ethanol across the probe such that diffusion out of the probe was higher than diffusion into the probe. This led to an underestimation of the brain tissue concentrations.

Robinson *et al.* (2000) also used a second method for calibration of probes for ethanol diffusion *in vivo*, i.e., the no-net-flux method (Justice, 1993). This method also led to an underestimation of the brain ethanol concentrations which was also presumably due to the directional difference in diffusion of ethanol across the probe. To account for this difference, Robinson *et al.* (2000) suggested a calibration method in which the delivery of ethanol through the probe into the tissue is first used, and then this calibration factor is adjusted for the directional difference. Although these findings were unexpected, the study suggested that it is possible to successfully use quantitative methods to measure ethanol concentrations in brain tissue using microdialysis.

More recent work has focused on comparison of the time courses of ethanol and dopamine when both analytes are measured from the same sample. An earlier study in which both ethanol and dopamine were concurrently analyzed from the same dialysis samples taken from the nucleus accumbens after i.p. administration of ethanol had clearly shown that the dopamine response declines faster than the ethanol (Yim *et al.*, 2000). Vena and Gonzales (2015) followed this initial finding by analyzing the time courses of both ethanol and dopamine across several brain regions (nucleus accumbens, medial prefrontal cortex, and dorsomedial striatum). This analysis indicated that the relative change of dopamine to ethanol across time declined in the medial prefrontal cortex compared with the other brain regions. This brain regional difference in pharmacodynamics could reflect the differences in clearance mechanisms for dopamine in the brain regions. Dopamine clearance from the extracellular space in the medial prefrontal cortex relies more on the enzyme catechol O-methyl transferase than the striatum (Yavich *et al.*, 2007). Collectively, these studies illustrate that new insights into pharmacodynamics of catecholamine responses to ethanol can be gained by concurrent analysis of ethanol along with the catecholamine across time using microdialysis.

13.5 Role of catecholamines in ethanol-induced behaviors

13.5.1 Dopamine

As mentioned in the introduction, a large body of work has examined the effect of ethanol on dopamine using microdialysis methods (reviewed in Gonzales *et al.*, 2004). Research has continued along these lines with over 200 papers published between 2008–2018. A large majority of this work has used forced, noncontingent administration of ethanol to examine possible mechanisms of the ethanol-induced dopamine release. However, this section will focus on the handful of recent papers that have looked at behavioral models of ethanol self-administration using microdialysis and voltammetry.

The pioneering work of Weiss and colleagues (Weiss *et al.*, 1993, 1996; Gonzales and Weiss, 1998) established that operant self-administration of ethanol after lever press training produced an increase in extracellular dopamine in the nucleus accumbens. Subsequently, another lab confirmed (Melendez *et al.*, 2002) and extended these findings to the ventral pallidum (Melendez *et al.*, 2004). Work in our lab followed up with a new model of operant self-administration of ethanol in which the lever press period was separated from the consumption period of the operant session. This work found that the increase in dopamine in the nucleus accumbens occurred during the initial consumption of ethanol but not later in the session when ethanol concentrations were highest in the brain (Doyon *et al.*, 2003, Carrillo and Gonzales, 2011; Howard *et al.*, 2009). Bassareo *et al.* (2017) recently confirmed that operant responding for ethanol using a nose-poke training also increases dopamine in the nucleus accumbens. In addition, operant self-administration also promotes dopamine release in the mPFC during the initial consumption of ethanol (Doherty *et al.*, 2016). Together, the studies using microdialysis in conjunction with models of the reinforcing and rewarding effects of ethanol clearly show that the mesocorticolimbic dopamine system is activated during ethanol seeking behavior.

Voltammetric techniques to monitor dopamine in the dopaminergic mesolimbic terminal regions during ethanol self-administration have also been used. While fast scan cyclic voltammetry has high temporal resolution, the detection limits are not quite as good as those possible with microdialysis. Bass *et al.* (2013) expressed channelrhodopsin-2 selectively in VTA dopamine neurons in rats by injecting a virus that infects neurons and allows the expression of channelrhodopsin-2 only in cells that express tyrosine hydroxylase. The investigators then were able to implant a carbon fiber electrode into the nucleus accumbens while also placing an optical fiber above the VTA. This enabled the selective activation of the dopamine neurons in a temporally controlled manner by delivering blue light pulses to the VTA. The investigators found that patterns of stimulation of VTA dopamine neurons that produce an increase in extracellular dopamine that mimics an increase in tonic dopamine concentrations diminished ethanol consumption when the rats had access to the ethanol drinking solution. These results illustrate the promise of combining optogenetic technology to questions about selective dopamine circuits and how they relate to ethanol-induced behaviors.

Fast scan cyclic voltammetry has also been used to examine dopamine changes in both ventral and dorsal striatum during operant self-administration of ethanol (Schnitko and Robinson, 2015). The nucleus accumbens showed dopamine release that was cue-induced as expected. Dopamine release after the lever press was also found in the dorsolateral striatum, but little effect was seen in the dorsomedial striatum. These investigators did not report an increase in dopamine release after ethanol consumption. The difference between the fast scan cyclic voltammetry results reported in this paper and the microdialysis studies are interesting and may be due in part to the much higher temporal resolution of the dopamine signals in the voltammetry studies relative to the microdialysis work.

13.5.2 Norepinephrine

In a previous section we presented evidence that various regions of the rodent brain neurochemically respond to ethanol administration using different routes of administration. The previous discussion did not focus on ethanol-induced behaviors and the role that noradrenergic pathways played in those behaviors. To explore this relationship, studies have targeted the noradrenergic system by agonizing or antagonizing the noradrenergic receptors or by genetically manipulating norepinephrine levels before, during, or after ethanol exposure.

A preclinical study showed that norepinephrine plays a major role in alcohol drinking behavior. When dopamine beta-hydroxylase (DBH), the enzyme (EC 1.14.17.1) that converts dopamine into norepinephrine was knocked out in mice, Weinshenker *et al.* (2000) reported a reduction in alcohol drinking and a reduction in ethanol preference in a two-bottle choice design compared to a wild-type control group. This study also reported that the reduced drinking may be due to a delay in extinction of taste aversion that may promote an aversive response for ethanol. Depletion of norepinephrine in the prefrontal cortex also was found to reduce ethanol consumption and preference (Ventura *et al.*, 2006).

Other animal studies have shown that by systematically targeting adrenergic receptors, ethanol-induced behavior can be modified. These results suggest that adrenergic receptors may serve as a therapeutic target for alcohol use disorders. For example, after the administration of the α_1 antagonist, prazosin, Walker *et al.* (2008) measured a reduction in ethanol-seeking behavior in dependent and non-dependent Wistar rats, but only at 1.5 and 2.0 mg/kg, the highest tested doses in their study. Similarly, two independent studies reported a reduction of ethanol consumption and seeking behavior in alcohol-preferring P rats after varying doses of prazosin treatment (Rasmussen *et al.*, 2009; Verplaetse *et al.*, 2012). Since prazosin requires multiple administrations to obtain maximum efficacy, O'Neil *et al.* (2013) tested whether doxazosin, an α_1 antagonist and a structurally related compound to prazosin, had similar effects on ethanol induced behaviors with a single daily administration. In this study the authors reported that alcohol-preferring P rats also demonstrated a reduction in alcohol intake, in a two-bottle choice of 15% ethanol *versus* water, after all tested doses of doxazosin. Prazosin has also been shown to attenuate stress-induced reinstatement of ethanol seeking behavior (Le *et al.*, 2011).

Similar results have been reported when the α_2 noradrenergic receptor was targeted. For example, after clonidine administration, an α_2 agonist, at 75 and 175 μg/kg (p.o.) or 75 and 150 μg/kg (s.c.), alcohol preferring rats demonstrated a reduction in alcohol intake in a free choice between 10% ethanol and drinking water paradigm (Optiz, 1990). More recently, Rasmussen *et al.* (2014) also reported a reduction in ethanol-drinking behavior in alcohol preferring P rats after i.p. doses of clonidine, 40 and 80 μg/kg, on two consecutive days prior to 2-hour ethanol access. In conjunction, the authors also reported that two weeks after the initiation of ethanol drinking, five consecutive days of 40 or 80 μg/kg i.p. clonidine also reduced ethanol intake each day. Guanfacine is also effective at reducing cue-induced relapse for ethanol in a model of comorbidity of depression and alcohol use disorder (Riga *et al.*, 2014).

These observations have also been extended in a clinical setting. In a study involving an alcohol dependent group, patients reported fewer drinking days and fewer drinks per week after prazosin treatment in comparison to the placebo group (Simpson *et al.*, 2009). But in a separate study, doxazosin administration resulted in no change in drinks per week or in heavy drinking days per week (Kenna *et al.*, 2016). In this study, a difference was only observed in either measure when family history of alcoholism was used as a moderator. In fact, doxazosin only reduced alcohol drinking in patients with a high family history of alcoholism. While doxazosin did not ameliorate drinking behavior in the present ethanol dependent individuals, this study suggests the possibility of personalized medicine and the administration of this drug to a patient population at a higher risk of alcohol dependence before the development of alcohol dependence.

Together the studies presented in this section support the idea that noradrenergic neurotransmission is an important regulator of alcohol drinking behavior and that noradrenergic receptors can serve as a therapeutic target to minimize the severity of ethanol drinking behavior in an ethanol dependent population. However, little is known about whether norepinephrine release is increased in the brain during the induction of physical dependence on ethanol, withdrawal from ethanol dependence, or after prolonged abstinence following dependence. Thus, the mechanisms by which ethanol might affect noradrenergic systems is not fully understood and further studies are necessary.

13.6 Conclusions

Although ethanol is one of the oldest drugs known to man, the neurobiological mechanisms responsible for its pharmacological effects are still not completely understood. Even more so, the neurobiological roots that underlie alcohol use disorders are less well known. The acute effects of ethanol on behavior are presumably due to changes in multiple neurotransmitter systems in the brain. Among these numerous neurotransmitter systems in the brain that are affected by ethanol, the catecholamines have been firmly established as contributing to some of the critical behavioral effects of ethanol. This chapter has reviewed some of the work that implicates dopamine as a critical mediator of reward and reinforcement for ethanol. Models of physical dependence on ethanol have also implicated catecholamines in some of the chronic effects of ethanol. While most of the work to date has focused on dopamine because of the critical role it plays in motivated behavior, the field has begun to swing towards examination of norepinephrine in many of the acute and chronic effects of ethanol. More work is clearly needed in this area, and in particular, the potential role of norepinephrine in ethanol-induced behaviors after long term ethanol use should be studied.

Soon after the introduction of the microdialysis technique for the study of brain neurochemistry, this technique has proven to be extremely helpful in furthering our understanding of how ethanol affects catecholamine function in the central nervous system. Looking toward the future, it seems clear that microdialysis will continue to be one of the techniques that will provide new insights into the role of catecholamines in ethanol's mechanisms of action.

Acknowledgments

Research in the authors' lab has been supported by NIH (R37AA011852 and (R01AA014874). Dr. Jaime is supported by an NIH training grant (T32AA007471). The authors thank Dr. Vorani Ramachandra for editorial assistance.

References

Adams N, Mitchell PS, Campbell SD, Samson HH (2002) Ethanol self-administration in Maudsley reactive and Maudsley nonreactive inbred rats. *Alcohol*, 26: 155–161.

Aimino MA, Coker CR, Silberman Y (2018) Acute ethanol modulation of neurocircuit function in the nucleus of the tractus solitarius. *Brain Res Bull*, 138: 5–11.

Arnsten AF, Pliszka SR (2011) Catecholamine influences on prefrontal cortical function: relevance to treatment of attention deficit/hyperactivity disorder and related disorders. *Pharmacol Biochem Behav*, 99: 211–216.

Badanich KA, Becker HC, Woodward JJ (2011) Effects of chronic intermittent ethanol exposure on orbitofrontal and medial prefrontal cortex-dependent behaviors in mice. *Behav Neurosci*, 125: 879–891.

Barkley-Levenson AM, Cunningham CL, Smitasin PJ, Crabbe JC (2015) Rewarding and aversive effects of ethanol in high drinking in the dark selectively bred mice. *Addict Biol*, 20: 80–90.

Bass CE, Grinevich VP, Gioia D, Day-Brown JD, Bonin KD, Stuber GD, Weiner JL, Budygin EA (2013) Optogenetic stimulation of VTA dopamine neurons reveals that tonic but not phasic patterns of dopamine transmission reduce ethanol self-administration. *Front Behav Neurosci*, 7: 173.

Bassareo V, Cucca F, Frau R, Di Chiara G (2017) Changes in dopamine transmission in the nucleus accumbens shell and core during ethanol and sucrose self-administration. *Front Behav Neurosci*, 11: 71.

Bechara A, Van Der Linden M (2005) Decision-making and impulse control after frontal lobe injuries. *Curr Opin Neurol*, 18: 734–739.

Becker HC (2012) Effects of alcohol dependence and withdrawal on stress responsiveness and alcohol consumption. *Alcohol Res*, 34: 448–458.

Belle AM, Wightman RM (2015) Monitoring and modulating dopamine release and unit activity in real-time. In: *Compendium of In Vivo Monitoring in Real-time Molecular Neuroscience*. Volume 1: *Fundamentals and Applications*. World Scientific Publishing, Singapore, pp.161–180.

Berridge CW, Stratford TL, Foote SL, Kelley AE (1997) Distribution of dopamine beta-hydroxylase-like immunoreactive fibers within the shell subregion of the nucleus accumbens. *Synapse*, 27: 230–241.

Brabant C, Guarnieri DJ, Quertemont E (2014) Stimulant and motivational effects of alcohol: lessons from rodent and primate models. *Pharmacol Biochem Behav*, 122: 37–52.

Carrillo J, Gonzales RA (2011) A single exposure to voluntary ethanol self-administration produces adaptations in ethanol consumption and accumbal dopamine signaling. *Alcohol*, 45: 559–566.

Chandler DJ (2016) Evidence for a specialized role of the locus coeruleus noradrenergic system in cortical circuitries and behavioral operations. *Brain Res*, 1641: 197–206.

Chang SL, Patel NA, Romero AA (1995) Activation and desensitization of Fos immuno-reactivity in the rat brain following ethanol administration. *Brain Res*, 679: 89–98.

Chester JA, Risinger FO, Cunningham CL (1998) Ethanol reward and aversion in mice bred for sensitivity to ethanol withdrawal. *Alcohol Clin Exp Res*, 22: 468–473.

Covarrubias MY, Khan RL, Vadigepalli R, Hoek JB, Schwaber JS (2005) Chronic alcohol exposure alters transcription broadly in a key integrative brain nucleus for homeostasis: the nucleus tractus solitarius. *Physiol Genomics*, 24: 45–58.

Decot HK, Namboodiri VM, Gao W, McHenry JA, Jennings JH, Lee SH, Kantak PA, Jill Kao YC, Das M, Witten IB, Deisseroth K, Shih YI, Stuber GD (2017) Coordination of brain-wide activity dynamics by dopaminergic neurons. *Neuropsychopharmacology*, 42: 615–627.

Delfs JM, Zhu Y, Druhan JP, Aston-Jones GS (1998) Origin of noradrenergic afferents to the shell subregion of the nucleus accumbens: anterograde and retrograde tract-tracing studies in the rat. *Brain Res*, 806: 127–140.

Di Chiara G (1999) Drug addiction as dopamine-dependent associative learning disorder. *Eur J Pharmacol*, 375: 13–30.

Ding ZM, Rodd ZA, Engleman EA, McBride WJ (2009) Sensitization of ventral tegmental area dopamine neurons to the stimulating effects of ethanol. *Alcohol Clin Exp Res*, 33: 1571–1581.

Ding ZM, Oster SM, Hall SR, Engleman EA, Hauser SR, McBride WJ, Rodd ZA (2011) The stimulating effects of ethanol on ventral tegmental area dopamine neurons projecting to the ventral pallidum and medial prefrontal cortex in female Wistar rats: regional difference and involvement of serotonin-3 receptors. *Psychopharmacology (Berl)*, 216: 245–255.

Doherty JM, Gonzales RA (2015) Operant self-administration of sweetened ethanol and time course of blood ethanol levels in adolescent and adult male Long-Evans rats. *Alcohol Clin Exp Res*, 39: 485–495.

Doherty JM, Schier CJ, Vena AA, Dilly GA, Gonzales RA (2016) Medial prefrontal cortical dopamine responses during operant self-administration of sweetened ethanol. *Alcohol Clin Exp Res*, 40: 1662–1670.

Doyon WM, Anders SK, Ramachandra VS, Czachowski CL, Gonzales RA (2005) Effect of operant self-administration of 10% ethanol plus 10% sucrose on dopamine and ethanol concentrations in the nucleus accumbens. *J Neurochem*, 93: 1469–1481.

Doyon WM, York JL, Diaz LM, Samson HH, Czachowski CL, Gonzales RA (2003) Dopamine activity in the nucleus accumbens during consummatory phases of oral ethanol self-administration. *Alcohol Clin Exp Res*, 27: 1573–1582.

Espana RA, Schmeichel BE, Berridge CW (2016) Norepinephrine at the nexus of arousal, motivation and relapse. *Brain Res*, 1641: 207–216.

Everitt BJ, Robbins TW (2013) From the ventral to the dorsal striatum: Devolving views of their roles in drug addiction. *Neurosci Biobehav Rev*, 37: 1946–1954.

Ferenczi EA, Zalocusky KA, Liston C, Grosenick L, Warden MR, Amatya D, Katovich K, Mehta H, Patenaude B, Ramakrishnan C, Kalanithi P, Etkin A, Knutson B, Glover GH, Deisseroth K (2016) Prefrontal cortical regulation of brainwide circuit dynamics and reward-related behavior. *Science*, 351: aac9698.

Ferraro TN, Weyers P, Carrozza DP, Vogel WH (1990) Continuous monitoring of brain ethanol levels by intracerebral microdialysis. *Alcohol*, 7: 129–132.

Forray MI, Gysling K (2004) Role of noradrenergic projections to the bed nucleus of the stria terminalis in the regulation of the hypothalamic-pituitary-adrenal axis. *Brain Res Brain Res Rev*, 47: 145–160.

Fox ME, Studebaker RI, Swofford NJ, Wightman RM (2015) Stress and drug dependence differentially modulate norepinephrine signaling in animals with varied HPA axis function. *Neuropsychopharmacology*, 40: 1752–1761.

Gilpin NW, Richardson HN, Cole M, Koob GF (2008) Vapor inhalation of alcohol in rats. *Curr Protoc Neurosci*, Chapter 9: Unit 9 29.

Goldstein RZ, Leskovjan AC, Hoff AL, Hitzemann R, Bashan F, Khalsa SS, Wang GJ, Fowler JS, Volkow ND (2004) Severity of neuropsychological impairment in cocaine and alcohol addiction: association with metabolism in the prefrontal cortex. *Neuropsychologia*, 42: 1447–1458.

Gonzales RA, Weiss F (1998) Suppression of ethanol-reinforced behavior by naltrexone is associated with attenuation of the ethanol-induced increase in dialysate dopamine levels in the nucleus accumbens. *J Neurosci*, 18: 10663–10671.

Gonzales RA, Job MO, Doyon WM (2004) The role of mesolimbic dopamine in the development and maintenance of ethanol reinforcement. *Pharmacol Ther*, 103: 121–146.

Grzanna R, Molliver ME (1980) The locus coeruleus in the rat: an immunohistochemical delineation. *Neuroscience*, 5: 21–40.

Haass-Koffler CL, Swift RM, Leggio L (2018) Noradrenergic targets for the treatment of alcohol use disorder. *Psychopharmacology (Berl)*, 235: 1625–1634.

Howard EC, Schier CJ, Wetzel JS, Duvauchelle CL, Gonzales RA (2008) The shell of the nucleus accumbens has a higher dopamine response compared with the core after non-contingent intravenous ethanol administration. *Neuroscience*, 154: 1042–1053.

Howard EC, Schier CJ, Wetzel JS, Gonzales RA (2009) The dopamine response in the nucleus accumbens core-shell border differs from that in the core and shell during operant ethanol self-administration. *Alcohol Clin Exp Res*, 33: 1355–1365.

Ihalainen JA, Riekkinen P, Jr., Feenstra MG (1999) Comparison of dopamine and noradrenaline release in mouse prefrontal cortex, striatum and hippocampus using microdialysis. *Neurosci Lett*, 277: 71–74.

Ikemoto S (2007) Dopamine reward circuitry: Two projection systems from the ventral midbrain to the nucleus accumbens-olfactory tubercle complex. *Brain Res Rev*, 56: 27–78.

Imperato A, Di Chiara G (1986) Preferential stimulation of dopamine release in the nucleus accumbens of freely moving rats by ethanol. *J Pharmacol Exp Ther*, 239: 219–228.

Jaime S, Cavazos JE, Yang Y, Lu H (2018) Longitudinal observations using simultaneous fMRI, multiple channel electrophysiology recording, and chemical microiontophoresis in the rat brain. *J Neurosci Methods*, 306: 68–76.

Justice JB, Jr. (1993) Quantitative microdialysis of neurotransmitters. *J Neurosci Methods*, 48: 263–276.

Karkhanis AN, Locke JL, McCool BA, Weiner JL, Jones SR (2014) Social isolation rearing increases nucleus accumbens dopamine and norepinephrine responses to acute ethanol in adulthood. *Alcohol Clin Exp Res*, 38: 2770–2779.

Karlsson O, Roman E (2016) Dose-dependent effects of alcohol administration on behavioral profiles in the MCSF test. *Alcohol*, 50: 51–56.

Kawahara H, Kawahara Y, Westerink BH (2000) The role of afferents to the locus coeruleus in the handling stress-induced increase in the release of noradrenaline in the medial prefrontal cortex: a dual-probe microdialysis study in the rat brain. *Eur J Pharmacol*, 387: 279–286.

Kenna GA, Haass-Koffler CL, Zywiak WH, Edwards SM, Brickley MB, Swift RM, Leggio L (2016) Role of the alpha1 blocker doxazosin in alcoholism: A proof-of-concept randomized controlled trial. *Addict Biol*, 21: 904–914.

Koob GF, Volkow ND (2016) Neurobiology of addiction: A neurocircuitry analysis. *Lancet Psychiat*, 3: 760–773.

Le AD, Funk D, Juzytsch W, Coen K, Navarre BM, Cifani C, Shaham Y (2011) Effect of prazosin and guanfacine on stress-induced reinstatement of alcohol and food seeking in rats. *Psychopharmacology* (*Berl*), 218: 89–99.

Lee S, Craddock Z, Rivier C (2011) Brain stem catecholamines circuitry: Activation by alcohol and role in the hypothalamic-pituitary-adrenal response to this drug. *J Neuroendocrinol*, 23: 531–541.

Li TK, Lumeng L, Doolittle DP (1993) Selective breeding for alcohol preference and associated responses. *Behav Genet*, 23: 163–170.

Lohani S, Poplawsky AJ, Kim SG, Moghaddam B (2017) Unexpected global impact of VTA dopamine neuron activation as measured by opto-fMRI. *Mol Psychiatry*, 22: 585–594.

Marinelli PW, Quirion R, Gianoulakis C (2003) A microdialysis profile of beta-endorphin and catecholamines in the rat nucleus accumbens following alcohol administration. *Psychopharmacology* (*Berl*), 169: 60–67.

Marsteller DA, Gerasimov MR, Schiffer WK, Geiger JM, Barnett CR, Schaich Borg J, Scott S, Ceccarelli J, Volkow ND, Molina PE, Alexoff DL, Dewey SL (2002) Acute handling stress modulates methylphenidate-induced catecholamine overflow in the medial prefrontal cortex. *Neuropsychopharmacology*, 27: 163–170.

McElligott ZA, Fox ME, Walsh PL, Urban DJ, Ferrel MS, Roth BL, Wightman RM (2013) Noradrenergic synaptic function in the bed nucleus of the stria terminalis varies in animal models of anxiety and addiction. *Neuropsychopharmacology*, 38: 1665–1673.

Melendez RI, Rodd ZA, McBride WJ, Murphy JM (2004) Involvement of the mesopallidal dopamine system in ethanol reinforcement. *Alcohol*, 32: 137–144.

Melendez RI, Rodd-Henricks ZA, Engleman EA, Li TK, McBride WJ, Murphy JM (2002) Microdialysis of dopamine in the nucleus accumbens of alcohol-preferring (P) rats during anticipation and operant self-administration of ethanol. *Alcohol Clin Exp Res*, 26: 318–325.

Metten P, Sorensen ML, Cameron AJ, Yu CH, Crabbe JC (2010) Withdrawal severity after chronic intermittent ethanol in inbred mouse strains. *Alcohol Clin Exp Res*, 34: 1552–1564.

Moore RY, Bloom FE (1979) Central catecholamine neuron systems: anatomy and physiology of the norepinephrine and epinephrine systems. *Annu Rev Neurosci*, 2: 113–168.

Morilak DA, Barrera G, Echevarria DJ, Garcia AS, Hernandez A, Ma S, Petre CO (2005) Role of brain norepinephrine in the behavioral response to stress. *Prog Neuropsychopharmacol Biol Psychiatry*, 29: 1214–1224.

Nurmi M, Kiianmaa K, Sinclair JD (1994) Brain ethanol in AA, ANA, and Wistar rats monitored with one-minute microdialysis. *Alcohol*, 11: 315–321.

Nurmi M, Kiianmaa K, Sinclair JD (1999) Brain ethanol levels after voluntary ethanol drinking in AA and Wistar rats. *Alcohol*, 19: 113–118.

O'Neil ML, Beckwith LE, Kincaid CL, Rasmussen DD (2013) The alpha1-adrenergic receptor antagonist, doxazosin, reduces alcohol drinking in alcohol-preferring (P) Rats. *Alcohol Clin Exp Res*, 37: 202–212.

Opitz K (1990) The effect of clonidine and related substances on voluntary ethanol consumption in rats. *Drug Alcohol Depend*, 25: 43–48.

Oscar-Berman M, Shagrin B, Evert DL, Epstein C (1997) Impairments of brain and behavior: the neurological effects of alcohol. *Alcohol Health Res World*, 21: 65–75.

Park J, Wheeler RA, Fontillas K, Keithley RB, Carelli RM, Wightman RM (2012) Catecholamines in the bed nucleus of the stria terminalis reciprocally respond to reward and aversion. *Biol Psychiatry*, 71: 327–334.

Pfefferbaum A, Sullivan EV, Mathalon DH, Lim KO (1997) Frontal lobe volume loss observed with magnetic resonance imaging in older chronic alcoholics. *Alcohol Clin Exp Res*, 21: 521–529.

Popke EJ, Allen SR, Paule MG (2000) Effects of acute ethanol on indices of cognitive-behavioral performance in rats. *Alcohol*, 20: 187–192.

Rasmussen DD, Alexander LL, Raskind MA, Froehlich JC (2009) The alpha1-adrenergic receptor antagonist, prazosin, reduces alcohol drinking in alcohol-preferring (P) rats. *Alcohol Clin Exp Res*, 33: 264–272.

Rasmussen DD, Alexander L, Malone J, Federoff D, Froehlich JC (2014) The alpha2-adrenergic receptor agonist, clonidine, reduces alcohol drinking in alcohol-preferring (P) rats. *Alcohol*, 48: 543–549.

Riga D, Schmitz LJ, van der Harst JE, van Mourik Y, Hoogendijk WJ, Smit AB, De Vries TJ, Spijker S (2014) A sustained depressive state promotes a guanfacine reversible susceptibility to alcohol seeking in rats. *Neuropsychopharmacology*, 39: 1115–1124.

Rinaman L (2011) Hindbrain noradrenergic A2 neurons: diverse roles in autonomic, endocrine, cognitive, and behavioral functions. *Am J Physiol Regul Integr Comp Physiol*, 300: R222–235.

Risinger FO, Boyce JM (2002) Conditioning tastant and the acquisition of conditioned taste avoidance to drugs of abuse in DBA/2J mice. *Psychopharmacology (Berl)*, 160: 225–232.

Roberts AJ, Cole M, Koob GF (1996) Intra-amygdala muscimol decreases operant ethanol self-administration in dependent rats. *Alcohol Clin Exp Res*, 20: 1289–1298.

Roberts AJ, Heyser CJ, Cole M, Griffin P, Koob GF (2000) Excessive ethanol drinking following a history of dependence: animal model of allostasis. *Neuropsychopharmacology*, 22: 581–594.

Robertson SD, Plummer NW, de Marchena J, Jensen P (2013) Developmental origins of central norepinephrine neuron diversity. *Nat Neurosci*, 16: 1016–1023.

Robinson DL, Lara JA, Brunner LJ, Gonzales RA (2000) Quantification of ethanol concentrations in the extracellular fluid of the rat brain: *in vivo* calibration of microdialysis probes. *J Neurochem*, 75: 1685–1693.

Robinson DL, Howard EC, McConnell S, Gonzales RA, Wightman RM (2009) Disparity between tonic and phasic ethanol-induced dopamine increases in the nucleus accumbens of rats. *Alcohol Clin Exp Res*, 33: 1187–1196.

Roberts AJ, Cole M, Koob GF (1996) Intra-amygdala muscimol decreases operant ethanol self-administration in dependent rats. *Alcohol Clin Exp Res*, 20: 1289–1298.

Rossetti ZL, Longu G, Mercuro G, Hmaidan Y, Gessa GL (1992) Biphasic effect of ethanol on noradrenaline release in the frontal cortex of awake rats. *Alcohol*, 27: 477–480.

Ryabinin AE, Criado JR, Henriksen SJ, Bloom FE, Wilson MC (1997) Differential sensitivity of c-Fos expression in hippocampus and other brain regions to moderate and low doses of alcohol. *Mol Psychiatry*, 2: 32–43.

Schier CJ, Dilly GA, Gonzales RA (2013) Intravenous ethanol increases extracellular dopamine in the medial prefrontal cortex of the Long-Evans rat. *Alcohol Clin Exp Res*, 37: 740–747.

Schwarz LA, Luo L (2015) Organization of the locus coeruleus-norepinephrine system. *Curr Biol*, 25: R1051–R1056.

Shnitko TA, Robinson DL (2015) Regional variation in phasic dopamine release during alcohol and sucrose self-administration in rats. *ACS Chem Neurosci*, 6: 147–154.

Simpson TL, Saxon AJ, Meredith CW, Malte CA, McBride B, Ferguson LC, Gross CA, Hart KL, Raskind M (2009) A pilot trial of the alpha-1 adrenergic antagonist, prazosin, for alcohol dependence. *Alcohol Clin Exp Res*, 33: 255–263.

Thiele TE, Cubero I, van Dijk G, Mediavilla C, Bernstein IL (2000) Ethanol-induced c-fos expression in catecholamine- and neuropeptide Y-producing neurons in rat brainstem. *Alcohol Clin Exp Res*, 24: 802–809.

Trantham-Davidson H, Centanni SW, Garr SC, New NN, Mulholland PJ, Gass JT, Glover EJ, Floresco SB, Crews FT, Krishnan HR, Pandey SC, Chandler LJ (2017) Binge-like alcohol exposure during adolescence disrupts dopaminergic neurotransmission in the adult prelimbic cortex. *Neuropsychopharmacology*, 42: 1024–1036.

Vena AA, Gonzales RA (2015) Temporal profiles dissociate regional extracellular ethanol *versus* dopamine concentrations. *ACS Chem Neurosci*, 6: 37–47.

Vena AA, Mangieri R, Gonzales RA (2016) Regional analysis of the pharmacological effects of acute ethanol on extracellular striatal dopamine activity. *Alcohol Clin Exp Res*, 40: 2528–2536.

Ventura R, De Carolis D, Alcaro A, Puglisi-Allegra S (2006) Ethanol consumption and reward depend on norepinephrine in the prefrontal cortex. *Neuroreport*, 17: 1813–1817.

Verplaetse TL, Rasmussen DD, Froehlich JC, Czachowski CL (2012) Effects of prazosin, an alpha1-adrenergic receptor antagonist, on the seeking and intake of alcohol and sucrose in alcohol-preferring (P) rats. *Alcohol Clin Exp Res*, 36: 881–886.

Vilpoux C, Warnault V, Pierrefiche O, Daoust M, Naassila M (2009) Ethanol-sensitive brain regions in rat and mouse: a cartographic review, using immediate early gene expression. *Alcohol Clin Exp Res*, 33: 945–969.

Vollstadt-Klein S, Wichert S, Rabinstein J, Buhler M, Klein O, Ende G, Hermann D, Mann K (2010) Initial, habitual and compulsive alcohol use is characterized by a shift of cue processing from ventral to dorsal striatum. *Addiction*, 105: 1741–1749.

Walker BM, Rasmussen DD, Raskind MA, Koob GF (2008) alpha1-noradrenergic receptor antagonism blocks dependence-induced increases in responding for ethanol. *Alcohol*, 42: 91–97.

Waller MB, Murphy JM, McBride WJ, Lumeng L, Li TK (1986) Effect of low dose ethanol on spontaneous motor activity in alcohol-preferring and -nonpreferring lines of rats. *Pharmacol Biochem Behav*, 24: 617–623.

Weinshenker D, Rust NC, Miller NS, Palmiter RD (2000) Ethanol-associated behaviors of mice lacking norepinephrine. *J Neurosci*, 20: 3157–3164.

Weiss F, Lorang MT, Bloom FE, Koob GF (1993) Oral alcohol self-administration stimulates dopamine release in the rat nucleus accumbens: genetic and motivational determinants. *J Pharmacol Exp Ther*, 267: 250–258.

Weiss F, Parsons LH, Schulteis G, Hyytia P, Lorang MT, Bloom FE, Koob GF (1996) Ethanol self-administration restores withdrawal-associated deficiencies in accumbal dopamine and 5-hydroxytryptamine release in dependent rats. *J Neurosci*, 16: 3474–3485.

Wise RA, You Z (2017) Microdialysis contributions to the understanding of reward and addiction. In *Compendium of In Vivo Monitoring in Real-time Molecular Neuroscience*. Volume 2: *Microdialysis and Sensing of Neural Tissues*. World Scientific Publishing, Singapore, pp. 217–248.

Yavich L, Forsberg MM, Karayiorgou M, Gogos JA, Mannisto PT (2007) Site-specific role of catechol-O-methyltransferase in dopamine overflow within prefrontal cortex and dorsal striatum. *J Neurosci*, 27: 10196–10209.

Yim HJ, Robinson DL, White ML, Jaworski JN, Randall PK, Lancaster FE, Gonzales RA (2000) Dissociation between the time course of ethanol and extracellular dopamine concentrations in the nucleus accumbens after a single intraperitoneal injection. *Alcohol Clin Exp Res*, 24: 781–788.

Yoshimoto K, Komura S (1993) Monitoring of ethanol levels in the rat nucleus accumbens by brain microdialysis. *Alcohol & Alcoholism*, 28: 171–174.

CHAPTER 14

THE CORTICOSTRIATAL COMMUNICATION PROBLEM IN HUNTINGTON'S DISEASE: DEFICITS IN ASCORBATE RELEASE AND GLUTAMATE UPTAKE

George V. Rebec

Program in Neuroscience, Department of Psychological and Brain Sciences, Indiana University, Bloomington, USA

rebec@indiana.edu

14.1 Introduction

Neurons are designed to communicate and communication between the cerebral cortex and the striatum plays a key role in behavior — from movement selection and habit formation to decision-making and emotional expression. The striatum, the main input structure of the basal ganglia, receives direct, monosynaptic input from virtually all areas of cerebral cortex via the release of glutamate, an excitatory amino acid transmitter. The striatum, in turn, communicates with cortex via multi-synaptic connections involving downstream basal ganglia structures that project to thalamus and then from thalamus back to cortex. Modulation of this network by complex clusters of interneurons, cortico-cortical connections, and ascending fibers from the brainstem, which release dopamine and other monoamine transmitters, ensures an efficient and well-regulated behavioral control system (Steiner and Tseng, 2010; Shepherd, 2013; Santos *et al.*, 2014; Hintiryan *et al.*, 2017).

Huntington's disease (HD), with its progressively worsening triad of motor, cognitive, and emotional symptoms, ravages this communication network. In most cases, symptoms emerge in the prime of life and end in death within 20 years after diagnosis. At autopsy, the end result is a massive loss of neurons — more than 90% in striatum and a large enough loss in cortex to produce profound thinning of cortical layers. But long before these neurons die, they become dysfunctional, creating a communication problem that eventually triggers the HD behavioral phenotype (Cepeda *et al.*, 2007; Raymond *et al.*, 2011; Estrada-Sanchez and Rebec, 2012; Andres *et al.*, 2017; Reiner and Deng, 2018). An understanding of the mechanisms underlying the onset and progression of this problem is needed to invigorate the search for an effective therapeutic strategy.

This chapter evaluates the communication problem based on evidence from *in vivo* monitoring of electrophysiological and neurochemical signals recorded from transgenic mice that model HD and their corresponding wild-type (WT) controls. The data from HD mice show dysfunctional cortical and striatal electrophysiological activity at the level of spiking patterns in single neurons and at the population level of local field potentials (LFPs). The underlying neurochemistry involves multiple transmitters but includes a dysfunctional glutamate transport system that operates in tandem with a failure to release ascorbate, an antioxidant vitamin found throughout the corticostriatal system. It is best to begin with an overview of HD, an autosomal dominant condition caused by a single-gene defect.

14.2 HD overview

George Huntington, a third-generation physician who grew up in East Hampton, Long Island, New York, was not the first person to describe the disease that would later bear his name, but he was the first to provide a detailed clinical account of both its symptoms and its pattern of inheritance (Huntington, 1872). The impact of the report, his only publication, was profound (Harper, 2002). His comprehensive 1 allowed HD to be identified as a unique condition separate from other disorders that might include some form of motor, cognitive, or emotional signs. The dominant pattern of inheriting HD suggested a genetic link. More than a hundred years later, the gene responsible for HD was identified (HD Collaborative Research Group, 1993).

14.2.1 HD genetics

The huntingtin gene (*HTT*) resides on the short arm of chromosome 4. The mutation involves an expansion of a CAG trinucleotide (cytosine, adenine, and guanine) repeat. In the human population, the number of CAGs in the repeat is typically less than 25, but as the number expands past 34, HD becomes increasingly likely. Full penetrance occurs at 39. HD typically begins in mid-life (late thirties to early fifties), but populations with an unusually large CAG repeat can develop symptoms as teenagers or even younger, a condition known as juvenile HD. Although an inverse correlation exists between the repeat number and the onset of symptoms, the relationship is not very strong owing to many moderating environmental and genetic factors (Bates *et al.*, 2015). Interestingly, the age of onset is closely related to the sex of the affected parent (Conneally, 1984). If the mother passes on the mutant gene (*mHTT*), symptom onset is more than five years later than if the father is responsible. In the case of the father, the repeat expansion increases in the offspring, suggesting an effect of HD on spermatogenesis. HD patients with the earliest onset have a number of males in the transmission line (e.g., an affected grandfather and an affected father). The overall prevalence of HD is roughly 5–10 per 100,000 with individuals of European ancestry clustered toward the higher end of this range (Harper, 2002).

There appears to be no relation between repeat size and the pattern of symptom expression. Motor symptoms typically prompt a visit to the neurologist and a subsequent HD diagnosis. In fact, Huntington's chorea, in recognition of the dance-like and often repetitive, involuntary movements, was an early name for HD. But cognitive and emotional (in the form of psychiatric) symptoms can also emerge early in the HD course. One symptom category may predominate if a relatively localized cortical region is most severely affected (Merhabi *et al.*, 2016). A focal deficit in prefrontal cortex (PFC), for example, might yield a symptom profile dominated by severe cognitive dysfunction.

The *HTT* gene contains 67 exons and encodes the very large (348 kDa) huntingtin protein, which is widely expressed throughout the body and brain (Bates *et al.*, 2015). The protein is found in both neurons and glia and plays important roles in multiple central nervous system functions, including axonal transport, vesicular release, and energy metabolism. The CAG repeat, which occurs in exon 1, is translated into a polyglutamine (poly-Q) stretch located just 17 amino acids from the N-terminus. When the mutant protein, with its expanded glutamine region, is cleaved, the misfolded N-terminal fragments are targeted for degradation. The fragments, however, persist and form aggregates found heterogeneously throughout the neuropil. Their role in HD neuropathology is still unclear.

No treatments are yet available to prevent HD or even slow its progression, but the isolation of the gene made it possible to study HD in animal models. This line of research has led to insights into the underlying neuropathology beginning with the observation that knockout of the mouse gene is embryonic lethal, suggesting a critical role for *HTT* in development (Bates *et al.*, 1997). Subsequent work has begun to unravel the corticostriatal dysfunction at the heart of the HD behavioral phenotype. Multiple mouse models of HD are now in use.

14.2.2 Mouse models of HD

Mice do not normally develop HD, but insertion of the entire human *mHTT* or its relevant fragment into the mouse genome can induce patterns of behavior and neuropathology that resemble the human condition. The onset and severity of neurological signs, however, vary widely depending on the transgenic model (for detailed reviews see Ernhoefer *et al.*, 2009; Menalled *et al.*, 2009; Crook and Houseman, 2011). Three general categories exist. In the first to be developed, truncated transgenics such as the R6/2 strain, which were created by introducing the N-terminal fragment of a juvenile HD patient's *mHTT* into the mouse genome, display an early and robust behavioral phenotype (Mangiarini *et al.*, 1996; Carter *et al.*, 1999). Although this phenotype is popular for neurobehavioral testing, the truncated model by definition lacks the full-length mutant protein, and the rapid progression of symptoms precludes assessment of many therapeutic strategies. Death can occur as soon as 12 or 13 weeks of age. In contrast, transgenic mice with random insertions of full-length human *mHTT*, such as the BACHD and YACHD strains (the large size of the HD gene requires the use of bacterial (BAC) or yeast artificial chromosome (YAC) DNA), show a wide range of symptom onset in adulthood and may live for well over a year (Van Raamsdonk *et al.*, 2005; Gray *et al.*, 2008). Thus, the window for neurobehavioral testing is much larger in BACHD and YACHD mice than with the truncated models, but symptom expression is not as robust. And although the full-length protein is present, it does not occupy its normal location in the mouse genome. A third transgenic category has the expanded CAG repeat inserted specifically into the endogenous mouse *HTT* gene (Menalled *et al.*, 2012). Unlike the other categories, this so-called knock-in model expresses the mutation in its appropriate genetic context. Although knock-ins have a large CAG expansion, as exemplified by the 140CAG and Q175 strains, the behavioral phenotype is surprisingly modest and may not emerge until the mice are well into adulthood. Lifespan is relatively normal at 18 months or longer.

With such genotypic and phenotypic variation among models, the choice of which one to use largely depends on the experimental question under investigation (Pouladi *et al.*, 2013; Rangel-Barajas and Rebec, 2018). Each model category has been used to assess the neuronal and glial abnormalities underlying the corticostriatal communication deficits induced by the mutant huntingtin gene. In subsequent sections, the focus is on model data obtained *in vivo*, an experimental approach that ensures corticostriatal circuitry is assessed in a relatively natural behavioral context.

14.3 Corticostriatal communication

14.3.1 Ascorbate

An interesting feature of ascorbate, the deprotonated form of vitamin C, is its relatively high concentration in the extracellular fluid of the striatum (Rebec and Pierce, 1994; Rice, 2000). This appears to be the case for all mammals even though some of them (e.g., humans) do not synthesize ascorbate in the liver but obtain it through diet. Mechanisms, including a sodium-dependent vitamin C transport system, are in place for transporting ascorbate from the bloodstream to the brain (Harrison and May, 2009). Basal levels at rest or under general anesthesia are in the 200–400 μM range (Basse-Tomusk and Rebec, 1991), which is above the extracellular level of glutamate and well above the low nanomolar range of dopamine. Like glutamate and dopamine, moreover, ascorbate levels change with behavioral state (O'Neill and Fillenz, 1985; Pierce and Rebec, 1990; Zetterstrom *et al.*, 1992). Extracellular ascorbate in striatum more than doubles during periods of intense behavioral activation. In fact, ascorbate in rat striatum increases after a tail pinch or an injection of amphetamine, a psychomotor stimulant. Conversely, a temporary reduction in extracellular ascorbate, induced by intra-striatal infusion of ascorbate oxidase, is accompanied by marked behavioral inactivity until striatal levels return to a normal range (Rebec and Wang, 2001).

The ability to measure ascorbate on a behaviorally relevant time scale comes from the application of voltammetric techniques to neuroscience in the 1970s (Adams, 1976). Originally intended as a way to measure dopamine and other monoamines on a rapid time scale, voltammetry is also well-suited for detecting ascorbate that, like the monoamines, acts as an electron donor (Rebec, 2007). Because ascorbate is easily oxidized, it is able to neutralize potentially damaging oxygen radicals (Padayatty *et al.*, 2003). In fact, a key physiological function of ascorbate is its role as an antioxidant, which, interestingly, has been linked to neuroprotection. The first *in vivo* assessment of striatal ascorbate in an HD mouse model revealed a decrease compared to WT controls but the decrease emerged only when the animals were behaviorally active (Rebec *et al.*, 2002). In the R6/2 model, for example, basal extracellular striatal ascorbate levels during anesthesia matched those of WT mice but failed to increase as the animals became behaviorally active. This failure of ascorbate release paralleled the decline in movement characteristic of strongly symptomatic R6/2s.

Importantly, the accumulation and storage of ascorbate in striatal tissue is not the problem. Tkac *et al.* (2007), using NMR spectroscopy, reported that cellular concentrations of ascorbate in R6/2 mice did not decrease relative to WT until 12 weeks of age when the behavioral phenotype is already severe. Moreover, restoring extracellular ascorbate to WT levels in striatum improves R6/2 motor patterns (Rebec *et al.*, 2003). Further support for a release deficit comes from evidence that cortical activation, which elicits a robust and prolonged increase in striatal ascorbate release in WT mice, produces only a weak and significantly shorter release response in R6/2s (Dorner *et al.*, 2009). Similarly, the amphetamine-induced increase in extracellular ascorbate in R6/2 striatum is blunted along with a blunted behavioral response relative to WT (Dorner *et al.*, 2009).

An interesting sex difference in striatal ascorbate release emerged in the 140CAG knock-in model, in which the mutant gene is knocked into or expressed in its exact genetic context. Males, tested at ages when multiple measures of motor activity begin to decline, show a strong decrease in striatal ascorbate release that does not appear in females tested at the same ages (Dorner *et al.*, 2007). In fact, the females show evidence of motor hyperactivity, especially in the dark phase of the diurnal cycle. In knock-in mice, a period of hyperactivity typically precedes the decline in movement (Menalled *et al.*, 2003), suggesting that female knock-in mice may have a later onset or slower progression of the HD phenotype than males (Dorner *et al.*, 2007).

Behavior-related changes in the concentration of striatal extracellular ascorbate indicate a role beyond that of simple antioxidant. At the level of individual neurons, ample evidence suggests that ascorbate directly modulates the electrophysiological response to sensorimotor processing (Cortright and Rebec, 2006) as well as to the application of glutamate or dopamine by iontophoresis (Kiyatkin and Rebec, 1998; Pierce and Rebec, 1995). A decline in striatal ascorbate release in HD mouse models, moreover, suggests a broader change in striatal function that may underlie the HD behavioral phenotype. To determine how HD alters striatal function, chronically implanted micro-wires have been used to record electrophysiological signals in HD and WT mice.

14.3.2 Electrophysiology

The most common neuronal type in the striatum is the medium spiny neuron (MSN), which accounts for >90% of the striatal neuronal population (Groves, 1983). The name is based on an abundance of dendritic spines and a relatively modest-sized cell body (10–15 μm diameter). Because MSNs are the main output neurons of the striatum, they are also known as "spiny projection neurons." MSNs are a critical part of basal ganglia circuitry. They project either directly to the substantia nigra reticulata (SNr) as part of the so-called direct pathway or indirectly via synapses in the external globus pallidus and subthalamic nucleus. The direct and indirect pathways to the SNr have been proposed to facilitate or inhibit movement, respectively (DeLong, 1990). Early assessments of basal ganglia dysfunction in HD, supported by evidence of initial loss of indirect-pathway MSNs (Reiner *et al.*, 1988), suggested that chorea arose from an overactive direct pathway unchecked by a failing or deteriorating indirect pathway (Albin *et al.*, 1989). Although the role of striatal MSNs in movement is still being refined, growing evidence suggests that neither pathway operates independently of the other. In fact, recordings from behaving animals indicate that motor output is shaped by co-activation of both pathways (Cui *et al.*, 2013). *In vitro* assessments of truncated and full-length HD mouse models are consistent with this view and indicate impairments in the way that both direct and indirect MSNs communicate with their downstream targets (Andre *et al.*, 2011; Cepeda *et al.*, 2013). Interestingly, synaptic communication between striatal MSNs is also altered in HD models, suggesting a change in axon collateral inhibition (Cepeda *et al.*, 2013). Thus, although striatal MSNs may project to different targets and degenerate at different rates in HD, it seems that functional alterations in both direct and indirect MSNs contribute to the HD behavioral phenotype.

In an initial assessment of presumed MSNs in behaving but symptomatic R6/2 mice, firing rate was significantly elevated relative to WT (Rebec *et al.*, 2006). This was the case even when the animals were anesthetized, arguing against a feedback effect of HD motor symptoms on MSN activity. Moreover, treatment with systemic injections of ascorbate at doses capable of increasing the extracellular level of ascorbate dramatically slowed firing in HD, but not WT, striatum. These results provided early support for dysfunctional striatal processing linked to a deficit in extracellular ascorbate as a mechanism underlying the HD behavioral phenotype.

HD-related deficits in striatal processing go well beyond a change in firing rate. In fact, an elevated rate seems unique to the R6/2 model, whereas changes in the pattern of MSN firing have been reported for both truncated and knock-in models (Miller *et al.*, 2011). For example, brief bursts of spike activity, which represent a form of plasticity and ensure reliable information transmission (Izhikevich *et al.*, 2003), are relatively rare in HD striatum compared to WT (Miller *et al.*, 2008b). Importantly, the structure of individual bursts, which includes their duration and the interval between individual spikes within a burst, is not changed in HD models, indicating that striatal neurons have the capacity to burst but are less likely to do so. Another aspect of firing pattern is coincident or correlated spiking between simultaneously recorded neuron pairs. In striatum, correlated firing is a dynamic process related to behavioral output and is also diminished in HD mouse models (Miller *et al.*, 2008b). That these effects have been identified in knock-in mice even before the onset of the HD behavioral phenotype suggests that striatal changes are an early contributor to dysfunctional behavioral coding.

Striatal neurons receive substantial glutamate input from cortical pyramidal neurons, which also show evidence of early changes in HD mouse models (Cepeda *et al.*, 2007; Estrada-Sanchez and Rebec, 2013). *In vitro* evidence indicates that pyramidal neurons undergo complex changes in ion channel function, membrane properties, and synaptic plasticity (Cummings *et al.*, 2009). In behaving HD mice, the corresponding electrophysiological changes appear to depend on the model and the cortical region being studied. In truncated, but not knock-in and full-length models, for example, elevated firing rates are common in primary motor cortex (M1) and in PFC, whereas M1 and PFC differ in the extent to which correlated firing is disrupted (Walker *et al.*, 2008; Miller *et al.*, 2011; Estrada-Sanchez *et al.*, 2015). Functional differences between these regions may account for differences in how HD neurological signs are manifest. In an assessment of conditioned fear, a behavioral response mediated by the prelimbic region of PFC (Quirk *et al.*, 2006), R6/2 mice acquired the task as rapidly as WT, but showed more rapid extinction, which correlated with a decline in prelimbic activity (Walker *et al.*, 2011). Impaired cognitive processing in HD — in this case altered fear expression in R6/2s — is directly linked to dysfunctional prelimbic processing of conditioned cues.

Given that HD impairs electrophysiological activity in both cortical pyramidal and striatal MSNs, albeit to different degrees in different models, an important issue is the role that cortical input exerts over striatal activity. In other words, is the presence of the mutant gene in MSNs sufficient to cause striatal dysfunction independent of cortical input (the so-called cell-autonomous mechanism) or does striatal dysfunction depend on the expression of the mutant gene in cortical pyramidal neurons (the so-called cell-interaction mechanism)? To address this question, MSN activity was recorded in behaving BACHD mice in which full-length *mHTT* was genetically reduced in glutamate projection neurons, including the cortical pyramidal neurons that project to striatum. In this case, the BACHD model was crossed to a cortex-specific Emx1-Cre/LoxP conditional mouse, in which the mutant gene can be switched off by Cre, a cyclization recombinase (Wang *et al.*, 2014). MSN spike activity and behavior were recorded from BACHD mice, BACHD/Emx1-Cre (BE mice), and corresponding WT controls (Estrada-Sanchez *et al.*, 2015). Mice were tested at regular intervals for several months to include periods before and after the development of the BACHD behavioral phenotype.

BE mice failed to show the decrease in MSN burst activity characteristic of BACHD and other HD models. In fact, BE matched WT on several parameters of burst firing analyzed across several months of recording (Estrada-Sanchez *et al.*, 2015). This finding is consistent with *in vitro* evidence that *mHTT* expression in cortical outputs disrupts the balance between excitatory and inhibitory control of MSN activity (Wang *et al.*, 2014). In addition, corresponding behavioral assessments showed that BE mice resembled WT more than BACHD across the same time span (Estrada-Sanchez *et al.*, 2015). Collectively, the data indicate that *mHTT* in cortical output neurons is critically involved in shaping the aberrant

MSN activity and behavioral symptoms characteristic of HD models. Although this outcome strongly supports a role for cell-interaction mechanisms in HD, not all aspects of BACHD electrophysiology and behavior were corrected in BE mice, suggesting that cell-autonomous mechanisms also contribute to the HD phenotype. The bottom line, however, is that *mHTT* in cortical output neurons plays a key role in striatal dysfunction and behavior.

Both cortex and striatum modulate behavior through the activity of large numbers of neurons, making it difficult to relate spike activity to spontaneously occurring movements. Recording LFP activity, which represents the ongoing oscillations of local neuronal networks including synaptic currents, is likely to provide more direct insight into the neuronal modulation of behavioral events than recordings of individual spikes (Buzsaki, 2006). Cortical and striatal LFPs, for example, have been implicated in both goal-directed and spontaneous movement (Berke *et al.*, 2004; Costa *et al.*, 2006). One aspect of this analysis has shown that as motor behavior becomes more predictable or inflexible, striatal LFP activity becomes less so. As R6/2 mice explore a four-arm plus-shaped maze, they become less likely than WT to turn left or right into a perpendicular arm upon entering the center or choice point of the maze and this behavioral event is reflected in increased unpredictability of striatal LFP activity (Hong *et al.*, 2012a). In follow-up work, simultaneous recording of LFPs in M1 and in its dorsal striatal target revealed that synchrony between these regions declined as R6/2 mice engaged in increasingly demanding motor behaviors from resting to grooming to exploring their open-field environment (Hong *et al.*, 2012b). WT mice showed the exact opposite pattern. Collectively, these data add to growing evidence of dysfunctional corticostriatal modulation of motor output in HD.

In light of evidence that striatal ascorbate release, which depends on cortical activation, is disrupted in HD (see above), it is useful to assess the neurochemical mechanisms by which cortex communicates with striatum. Although glutamate is the classical transmitter, its interaction with ascorbate cannot be ignored.

14.3.3 Glutamate uptake triggers ascorbate release

When glutamate is cleared from the synapse by an active uptake process, there is a corresponding increase in ascorbate release. Early evidence for such a relationship came from reports that in synaptosomes prepared from rat brain regions with high levels of ascorbate, including cortex and striatum, the addition of L- but not D-glutamate, evoked the release of ascorbate, which was blocked by the removal of sodium (Grunewald and Fillenz, 1984; Fillenz *et al.*, 1986). Because glutamate uptake is stereo-selective (favoring the transport of the naturally occurring L-isomer) and sodium dependent, the data suggested that glutamate uptake was the mechanism underlying ascorbate release. This view of a glutamate-ascorbate heteroexchange system was confirmed in subsequent experiments involving pharmacological and other manipulations (O'Neill *et al.*, 1983; Basse-Tomusk and Rebec, 1990).

The efficiency of glutamate clearance from the synapse is a key regulator of excitatory transmission and represents one form of protection against glutamate toxicity (Rice, 2000). Another form of protection arises when the ascorbate released by astrocytes is taken up by neurons via the sodium-dependent vitamin C transporter. Intra-neuronal ascorbate counteracts glutamate-induced free radical production by donating electrons to form dehydroascorbate (Wilson, 1997). In addition, the ascorbate that remains in extracellular fluid has neuromodulatory benefits by adjusting the strength of glutamate signaling (see above).

The primary glutamate transport protein in striatum and throughout the forebrain is termed GLT1 in rodents and EAAT2 in humans. It is found on astrocytes (Danbolt, 2001), although an isoform has been identified on the axon terminals of glutamate-releasing neurons (Berger *et al.*, 2005). Astrocytes, however, are the likely source of ascorbate release (Wilson *et al.*, 2000). They take up dehydroascorbate, the oxidized form of ascorbate that can be toxic to neurons, rapidly recycle it back to ascorbate (Lane and Lawen, 2009), and store ascorbate at concentrations of approximately 1 mM (Rice and Russo-Mena, 1998). Release is triggered by glutamate-induced cellular swelling. As astrocytes take up glutamate, volume sensitive anion channels allow ascorbate to diffuse into extracellular fluid (Lane and Lawen, 2013). Interestingly, a high concentration of extracellular ascorbate can also feedback to decrease glutamate uptake (Rebec *et al.*, 2005), suggesting a close interplay between neurons and glia mediated by glutamate and ascorbate. This interplay becomes largely dysfunctional in HD. It may even suggest a therapeutic strategy.

14.4 Correcting the glutamate-ascorbate problem

The decline in striatal ascorbate release in the striatum of truncated and knock-in mouse models suggests a problem with glutamate uptake. In line with this view, GLT1 is down-regulated in the striatum and glutamate uptake is impaired in R6/2 mice relative to WT (Behrens *et al.*, 2002). Impaired glutamate uptake also has been reported for brain tissue obtained from HD patients (Hassel *et al.*, 2008). Interestingly, ceftriaxone, a beta-lactam antibiotic, has been shown to increase brain GLT1 expression in rodents (Rothstein *et al.*, 2005). When R6/2 mice are treated with ceftriaxone, but not vehicle, both striatal GLT1 expression and glutamate uptake increase, while HD motor symptoms improve (Miller *et al.*, 2008a). And, in further confirmation of a link to ascorbate, ceftriaxone treatment also increases cortically evoked ascorbate release in R6/2 striatum (Miller *et al.*, 2012).

Astrocytes also express xCT, the catalytic subunit of the cystine-glutamate exchanger, which exchanges intracellular glutamate for extracellular cysteine (Baker *et al.*, 2002). This mechanism accounts for most extracellular glutamate not released from neurons. Astrocytes use cysteine in the production of glutathione, another antioxidant molecule. The metabotropic glutamate receptor, mGluR2/3, is located on axon terminals and responds to xCT-mediated glutamate release by decreasing depolarization-induced glutamate release from neurons (Xi *et al.*, 2002). Thus, GLT1 and xCT work in tandem to ensure tight control of glutamate transmission. Like GLT1, xCT is up-regulated by ceftriaxone and related beta-lactam antibiotics (Trantham-Davidson *et al.*, 2013; Althobaiti *et al.*, 2016), indicating that the improved HD behavioral phenotype induced by these drugs involves three closely interconnected mechanisms: an increase in glutamate uptake by astrocytes, a corresponding increase in ascorbate release from astrocytes, and a decrease in glutamate release from axon terminals.

The therapeutic potential of drug treatment, however, is limited by its relatively short-lived effect. In R6/2 mice, for example, behavioral improvements disappear within a week after cessation of treatment (Miller *et al.*, 2008a). Even more problematic would be reliance on a drug regimen for a nearly life-long condition. But if expression of the gene for the glutamate transporter could be increased, then a more permanent solution might be possible. In a ground-breaking report, Kasper and colleagues identified a viral vector that could deliver the GLT1/EAAT2 gene directly to the mouse central nervous system following intravascular administration (Faust *et al.*, 2009). Thus, corticostriatal delivery could be achieved without the complications of surgery. In fact, systemic injection of adeno-associated virus serotype 9 (AAV9) in Q175 mice has been shown to increase GLT1 immunofluorescence in both cortex and striatum (Vagner *et al.*, 2016). Transduction, moreover, occurs mainly in astrocytes.

Preliminary assessment of Q175 mice treated with intravascular AAV9-EAAT2 before HD neurological signs are expressed indicates that the animals maintain a WT-level of motor ability for at least a year (Rangel-Barajas *et al.*, 2016). For example, both rotarod performance and probability of turning in the plus-maze show little or no deterioration in heterozygous Q175 mice over a year of testing, while the behavior of untreated or vehicle-treated Q175s deteriorates significantly across the same time span. Additionally, striatal LFP activity in the Q175s shows a clear correlation between gamma band power and plus-maze turning. AAV9-EAAT2 treatment allowed Q175s to maintain the same level of gamma power at the choice point of the maze as WT mice, while also maintaining the same level of turning behavior. Without AAV9-EAAT2, Q175s turned less than WT and showed significantly less gamma band power. All mice were equally active in the maze, making it unlikely that the decline in turning probability reflected a general loss of movement.

Gamma band activity has been linked to attention and movement (Moisa *et al.*, 2014; Mably and Colgin, 2018), but perhaps more importantly, synchrony in the gamma band has been proposed to support ongoing communication between brain regions (Ahmed and Cash, 2013). Thus, it is tempting to speculate that upregulation of glutamate transport reinstates the WT level of corticostriatal communication disrupted by HD. The result, in the case of the plus maze, is improved motor flexibility. This could be due to a direct change in the extracellular dynamics of glutamate induced by an increase in uptake, a change in ascorbate release that in turn modulates how neurons respond to glutamate, or to other mechanisms sensitive to the changing levels of glutamate or ascorbate. Tellingly, an early assessment of AAV9-EAAT2 on protein expression in striatum indicates not only an increase in the glutamate transporter but also changes in xCT as well as in subunit expression of the N-methyl-D-aspartate (NMDA) glutamate receptor (Rangel-Barajas *et al.*, 2017). Similarly, ceftriaxone, the antibiotic that elevates GLT1 expression and increases glutamate uptake in R6/2 mice (Miller *et al.*, 2008a), also increases xCT expression (LaCrosse *et al.*, 2017). It is noteworthy, however, that an investigation of glutamate signaling after its synaptic release *in situ* showed no evidence of a decline in striatal glutamate uptake in the R6/2 model (Parsons *et al.*, 2016). Thus, further understanding of the molecular and neurochemical changes associated with glutamate transmission and its disruption by HD is required before a clear therapeutic strategy can emerge.

14.5 Conclusions

Identification of *mHTT* has not led to an immediate breakthrough in HD treatment. In fact, the drug prescribed for HD symptoms interferes with dopamine transmission and has only limited efficacy in reducing some early signs of loss of motor control (Scott, 2011). Dopamine, like glutamate, may be altered in HD models (e.g., Johnson *et al.*, 2006), but complexities in the extent and time course of the dopamine alteration are troublesome for treatment development (Rangel-Barajas and Rebec, 2017). Until gene suppression becomes available as a therapeutic approach — a possibility that seems increasingly likely with recent advances in medical genetic technology (Deverman *et al.*, 2018) — the stumbling block for any treatment search is the multitude of roles that *mHTT* plays in nervous system function. Its role as a participant in both cell-interaction and cell-autonomous functions is a further complication. The close relationship between glutamate uptake and ascorbate release is one of the functions disrupted by *mHTT* and it seems clear that this disruption is a major factor in the deterioration of corticostriatal communication in HD. Additional research is likely to reveal a therapeutic strategy and in the process reveal even more about the mechanisms of corticostriatal communication that have implications for a broad range of neurological and psychiatric diseases.

Acknowledgments

Generous support for research on ascorbate and glutamate, including *in vivo* voltammetry, *in vivo* electrophysiology, and corticostriatal protein expression in HD mouse models, was provided by the National Institutes of Health (NIH), the Hereditary Disease Foundation (HDF), and Cure Huntington's Disease Initiative (CHDI). This research would not be possible without the dedication of many talented technicians, students, and postdoctoral scholars.

References

Adams RN (1976) Probing brain chemistry with electroanalytical techniques. *Anal Chem*, 48: 1128A–1138A.

Ahmed OJ, Cash SS (2013) Finding synchrony in the desynchronized EEG: The history and interpretation of gamma rhythms. *Front Integr Neurosci*, 7: 58. DOI:10.3389/fnint.2013.00058.

Albin RL, Young AB, Penny JB (1989) The functional anatomy of basal ganglia disorders. *Trends Neurosci*, 12: 366–375.

Althobaiti YS, Alshehri FS, Almalki AH, Sari Y (2016) Effects of ceftriaxone on glial glutamate transporters in Wistar rats administered sequential ethanol and methamphetamine. *Front Neurosci*, 10.427. DOI: 10.3389/finins.2016.00427.

Andre VM, Cepeda C, Fisher YE, Huynh M, Bardakjian N, Singh S, Yang XW, Levine MS (2011) Differential electrophysiological changes in striatal output neurons in Huntington's disease. *J. Neurosci*, 31: 1170–1182.

Andres DS, Merello M, Darbin O (2017) Editorial: pathophysiology of the basal ganglia and movement disorders: gaining new insights from modeling and experimentation to influence the clinic. *Front Hum Neurosci*, 11:466. DOI: 10.3389/fnhum.2017.00466.

Baker DA, Xi ZX, Shen H, Swanson CJ, Kalivas PW (2002) The origin and neuronal function of *in vivo* nonsynaptic glutamate. *J Neurosci*, 22: 9134–9141.

Basse-Tomusk A, Rebec GV (1990) Corticostriatal and thalamic regulation of amphetamine-induced ascorbate release in the neostriatum. *Pharmacol Biochem Behav*, 35: 55–60.

Basse-Tomusk A, Rebec GV (1991) Regional distribution of ascorbate and 3,4-dihydroxyphenyacetic acid (DOPAC) in rat neostriatum. *Brain Res*, 538: 29–35.

Bates GP, Dorsey R, Gusella J, Haydem MR, Kay C, Leavitt BR, Lance M, Ross CA, Scahill RI, Wetzel R, Wild EJ, Tabrizi SJ (2015) Huntington disease. *Nat Rev Disease Primers*, 1: 1–21.

Bates GP, Mangiarini L, Mahal A, Davies SW (1997) Transgenic models of Huntington's disease. *Hum Molec Genet*, 6: 1633–1637.

Behrens PF, Franz P, Woodman B, Lindenberg KS, Landwehrmeyer GB (2002) Impaired glutamate transport and glutamate-glutamine cycling: Downstream effects of the Huntington mutation. *Brain*, 125: 1908–1922.

Berger UV, DeSilva TM, Chen W, Rosenberg PA (2005) Cellular and subcellular mRNA localization of glutamate transporter isoforms GLT1a and GLT1b in rat brain by in situ hybridization. *J Comp Neurol*, 492: 78–89.

Berke JD, Okatan M, Skurski J, Eichenbaum HB (2004) Oscillatory entrainment of striatal neurons in freely moving rats. *Neuron*, 43: 883–896.

Buzsaki G (2006) *Rhythms of the Brain*. Oxford University Press, New York.

Carter RJ, Lione LA, Humby T, Mangiarini L, Mahal A, Bates GP, Dunnett SB, Morton AJ (1999) Characterization of progressive motor deficits in mice transgenic for the human Huntington's disease mutation. *J Neurosci*, 19: 3248–3257.

Cepeda C, Galvan L, Holley SM, Rao SP, Andre VM, Botelho EP, Chen JY, Watson JB, Deisseroth K, Levine MS (2013) Multiple sources of striatal inhibition are differentially affected in Huntington's disease mouse models. *J. Neurosci*, 33: 7393–7406.

Cepeda C, Wu N, Andre VM, Cummings DM, Levine MS (2007) The corticostriatal pathway in Huntington's disease. *Prog Neurobiol*, 81: 253–271.

Conneally PM (1984) Huntington disease: Genetics and epidemiology. *Am J Hum Genet*, 36: 506–526.

Cortright JJ, Rebec GV (2006) Ascorbate modulation of sensorimotor processing in striatum of freely moving rats. *Brain Res*, 1092: 108–116.

Costa RM, Lin SC, Sotnikova TD, Cyr M, Gainetdinov RR, *et al.* (2006) Rapid alterations in corticostriatal ensemble coordination during acute dopamine-dependent motor dysfunction. *Neuron*, 52: 359–369.

Crook ZR, Houseman D (2011) Huntington's disease: Can mice lead the way to treatment? *Neuron*, 69: 923–935.

Cui G, Jun SB, Jin X, Pham MD, Vogel SS, Lovinger DM, Costa RM (2013) Concurrent activation of striatal direct and indirect pathways during action initiation. *Nature*, 494: 238–242.

Cummings DM, Andre VM, Uzgil BO, Gee SM, Fisher YE, Cepeda C, Levine MS (2009) Alterations in cortical excitation and inhibition in genetic mouse models of Huntington's disease. *J Neurosci*, 29: 10371–10386.

Danbolt NC (2001) Glutamate uptake. *Prog Neurobiol*, 65: 1–105.

DeLong MR (1990) Primate models of movement disorders of basal ganglia origin. *Trends Neurosci*, 13: 281–285.

Deverman BE, Ravina BM, Bankiewicz KS, Paul SM, Sah DWY (2018) Gene therapy for neurological disorders: Progress and prospects. *Nature Rev, Drug Disc* 7: 110. DOI: 10.1038/nrd.2018.110.

Dorner JL, Miller BR, Barton SJ, Brock TJ, Rebec GV (2007) Sex differences in behavior and striatal ascorbate release in the 140CAG knock-in mouse model of Huntington's disease. *Behav Brain Res*, 178: 90–97.

Dorner JL, Miller BR, Klein EL, Murphy-Nakhnikian A, Andrews RL, Barton SJ, Rebec GV (2009) Corticostriatal dysfunction underlies diminished striatal ascorbate release in the R6/2 mouse model of Huntington's disease. *Brain Res*, 1290: 111–120.

Ehrnhoefer DE, Butland SL, Pouladi MA, Hayden MR (2009) *Disease Mod Mechan*, 2: 123–129.

Estrada-Sanchez AM, Burroughs CL, Cavaliere S, Barton SJ, Chen S, Yang, XW, Rebec GV (2015) Cortical efferents lacking mutant huntingtin improve striatal neuronal activity and behavior in a conditional mouse model of Huntington's disease. *J Neurosci*, 35: 4440–4451.

Estrada-Sanchez AM, Rebec GV (2012) Corticostriatal dysfunction and glutamate transporter 1 (GLT1) in Huntington's disease: Interactions between neurons and astrocytes. *Basal Ganglia*, 2: 57–66.

Estrada-Sanchez AM, Rebec GV (2013) Role of cerebral cortex in the neuropathology of Huntington's disease. *Front Neural Circuits*, 7: 19.

Fillenz M, O'Neill RD, Grunewald RA (1986) Changes in brain extracellular ascorbate concentration as an index of excitatory amino acid release. In: *Monitoring Neurotransmitter Release during Behavior*, Joseph MH, Fillenz M, MacDonald IA, Marsden CA (eds.). Ellis Norwood, Chichester, pp. 144–163.

Foust KD, Nurre E, Montgomery CL, Hernandez A, Chan CM, Kasper BK (2009) Intravascular AAV9 preferentially targets neonatal neurons and adult astrocytes. *Nature Biotechnol*, 27: 59–65.

Gray M, Shirasaki I, Cepeda C, Andre VM, Wilburn B, Lu XH, Tao J, Yamazaki I, Li SH, Sun YE, Li XJ, Levine MS, Yang XW (2008) Full-length human mutant huntingtin with a stable polyglutamine repeat can elicit progressive and selective neuropathogenesis in BACHD mice. *J Neurosci*, 28: 6182–6195.

Groves PM (1983) A theory of the functional organization of the neostriatum and the neostriatal control of voluntary movement. *Brain Res*, 286: 109–132.

Grunewald RA, Fillenz M (1984) Release of ascorbate from synaptosomal fraction of rat brain. *Neurochem Int*, 6: 491–500.

Harper PS (2002) Huntington's disease: A historical background. In: *Huntington's Disease* (3rd ed.), Bates G, Harper P, Jones L (eds.). Oxford Univ. Press, New York, pp. 3–27.

Harrison FE, May JM (2009) Vitamin C function in the brain: Vital role of the ascorbate transporter SVCT2. *Free Radical Biol Med*, 46: 719–730.

Hassel B, Tessler S, Faull RL, Emson PC (2008) Glutamate uptake is reduced in prefrontal cortex in Huntington's disease. *Neurochem Res*, 33: 232–237.

Hintiryan H, Foster NN, Bowman I, Bay M, Song MY, Gou L, Yamashita S, Bienkowski MS, Zingg B, Zhu M, Yang XW, Shih JC, Toga AW, Dong HW (2016) The mouse cortico-striatal projectome. *Nature Neurosci*, 19: 1100–1117.

Hong SL, Barton, SJ, Rebec GV (2012a) Altered neural and behavioral dynamics in Huntington's disease: An entropy conservation approach. *PLoS ONE*, 7(1): e30879.

Hong SL, Cossyleon D, Hussain W, Walker LJ, Barton SJ, Rebec GV (2012b) Dysfunctional modulation of corticostriatal communication in the R6/2 mouse model of Huntington's disease. *PLoS ONE*, 7(10): e47026.

Huntington G (1872) On chorea. *Med. Surg. Reporter*, 26: 320–321.

Huntington's Disease Collaborative Research Group (1993) A novel gene containing a trinucleotide repeat that is expanded and unstable in Huntington's disease chromosomes. *Cell*, 72: 971–983.

Izhikevich EM, Desai NS, Walcott EC, Hoppenstaedt FC (2003) Bursts as a unit of neural information: Selective communication via resonance. *Trends Neurosci*, 26: 161–167.

Johnson MA, Rajan V, Miller CE, Wightman RM (2006) Dopamine release is severely compromised in the R6/2 mouse model of Huntington's disease. *J Neurochem*, 97: 737–746.

Kiyatkin EA, Rebec GV (1998) Ascorbate modulates glutamate-induced excitations of striatal neurons. *Brain Res*, 812: 88–106.

LaCrosse AL, O'Donovan SM, Sepulvada-Orengo MT, McCullumsmith RE, Reissner KJ, Schwendt M, Knackstedt LA (2017) Contrasting the role of xCT and GLT-1 upregulation in the ability of ceftriaxone to attenuate the cue-induced reinstatement of cocaine seeking and normalize AMPA receptor subunit expression. *J Neurosci*, 37: 5809–5821.

Lane DJR, Lawen A (2009) Ascorbate and plasma membrane electron transport — enzymes *vs* efflux. *Free Radical Biol Med*, 47: 485–495.

Lane DJR Lawen A (2013) The glutamate aspartate transporter (GLAST) mediates L-glutamate-stimulated ascorbate release via swelling-activated anion channels in cultured neonatal rodent astrocytes. *Cell Biochem Biophys*, 65: 107–119.

Mably AJ, Colgin LL (2018) Gamma oscillations in cognitive disorders. *Curr Opin Neurobiol*, 16: 9.

Mangiarini L, Sathasivam K, Seller M, Cozens B, Harper A, Hetherington C., Lawton M, Trottier Y, Lehrach H, Davies SW, Bates GP (1996) Exon 1 of the HD gene with an expanded CAG repeat is sufficient to cause a progressive neurological phenotype in transgenic mice. *Cell*, 87: 493–506.

Menalled L, El-Khodor BF, Patry M, Suarez-Farinas M, Orenstein SJ, Zahasky B, Leahy C, Wheeler V, Yang XW, MacDonald M, Morton AJ, Bates G, Leeds, J, Park L, Howland D, Signer E, Tobin A, Brunner D (2009) Systemtic behavioral evaluation of Huntington's disease transgenic and knock-in mouse models. *Neurobiol Dis*, 35: 319–336.

Menalled L, Kudwa AE, Miller S, Fitzpatrick J, Watson-Johnson J, Keating N, Ruiz M, Mushlin R, Alosio W, McConnell K, Connor D, Murphy C, Oakeshott S, Kwan M, Beltran J, Gavami A, Brunner D, Park LC, Ramboz S, Howland D (2012) Comprehensive behavioral and molecular characterization of a new knock-in mouse model of Huntington's disease: zQ175. *PLoS ONE*, 7(12).

Menalled L, Sison JD, Dragatsis I, Zeitlin S, Chesselet MF (2003) Time course of early motor and neuropathological anomalies in a knock-in mouse model of Huntington's disease with 140 CAG repeats. *J Comp Neurol*, 465: 11–126.

Mehrabi NF, Waldvogel HJ, Tippett LJ, Hogg VM, Sinek BJ, Faull RLM (2016) Symptom heterogeneity in Huntington's disease correlates with neuronal degeneration in the cerebral cortex. *Neurobiol Dis*, 96: 67–74.

Miller BR, Dorner JL, Bunner KD, Gaither TW, Klein EL, Barton SJ, Rebec GV (2012) Up-regulation of GLT1 reverses the deficit in cortically evoked striatal ascorbate efflux in the R6/2 mouse model of Huntington's disease. *J Neurochem*, 121: 629–638.

Miller BR, Dorner JL, Shou M, Sari Y, Barton SJ, Sengelaub DR, Kennedy RT, Rebec GV (2008a) Up-regulation of GLT1 expression increases glutamate uptake and attenuates the Huntington's disease phenotype in the R6/2 mouse. *Neuroscience*, 153: 329–337.

Miller BR, Walker AG, Barton SJ, Rebec GV (2011) Dysregulated neuronal activity patterns implicate corticostriatal circuit dysfunction in multiple rodent models of Huntington's disease. *Front Syst Neurosci*, 5: 26.

Miller BR, Walker AG, Shah AS, Barton SJ, Rebec GV (2008b) Dysregulated information processing by medium spiny neurons in striatum of freely behaving mouse models of Huntington's disease. *J Neurophysiol*, 100: 2205–2216.

Moisa M, Polania R, Grueschow M, Ruff CC (2014) Brain network mechanisms underlying motor enhancement by transcranial entrainment of gamma oscillations. *J Neurosci*, 36: 12053–12064.

O'Neill RD, Fillenz M (1985) Circadian changes in extracellular ascorbate in rat cortex, accumbens, striatum, and hippocampus: Correlations with motor activity. *Neurosci Lett*, 60: 331–336.

Padayatty S, Katz A, Wang Y, Eck P, Kwon O, Lee J, Chen S, Corpe C, Dutta A, Dutta S, Levine M (2003) Vitamin C as an antioxidant: Evaluation of its role in disease prevention. *J Am Coll Nutr*, 22: 18–35.

Parsons MP, Vanni MP, Woodard CL, Kang R, Murphy TH, Raymond LA (2016) Real-time imaging of glutamate clearance reveals normal striatal uptake in Huntington's disease mouse models. *Nat Commun*, 7: 11251.

Pierce CP, Rebec GV (1990) Stimulation of both D1 and D2 dopamine receptors increases behavioral activation and ascorbate release in neostriatum of freely moving rats. *Eur J Pharmacol*, 191: 295–302.

Pierce CP, Rebec GV (1995) Iontophoresis in the neostriatum of awake, unrestrained rats: Differential effects of dopamine, glutamate, and ascorbate on motor and non-motor-related neurons. *Neuroscience*, 67: 313–324.

Pouladi M, Morton AJ, Hayden MR (2013) Choosing an animal model for the study of Huntington's disease. *Nature Rev Neurosci*, 14: 708-721.

Quirk GJ, Garcia R, Gonzalez-Lima F (2006) Prefrontal mechanisms in extinction of conditioned fear. *Biol Psychiat*, 60: 337–343.

Rangel-Barajas C, Bunner KD, Barton SJ, Rebec GV (2016) Effect of GLT1 up-regulation on striatal local field potential activity and motor performance. *Soc Neurosci Abstr*, 418: 10.

Rangel-Barajas C, Rebec GV (2016) Dysregulation of corticostriatal connectivity in Huntington's disease: A role for dopamine modulation. *J Huntington Dis*, 5: 303–331.

Rangel-Barajas C, Rebec, GV (2018) Overview of Huntington's disease models: Neuropathological, molecular and behavioral differences. *Curr Protoc Neurosci*, 83: e47.

Raymond LA, Andre VM, Cepeda C, Gladding CM, Milnerwood AJ, Levine MS (2011) Pathophysiology of Huntington's disease: Time-dependent alterations in synaptic and receptor function. *Neuroscience*, 198: 252–273.

Rebec GV (2018) Corticostriatal network dysfunction in Huntington's disease: Deficits in neural processing, glutamate transport, and ascorbate release. CNS *Neurosci Ther*, 14: 1–11.

Rebec GV (2007) From interferant anion to neuromodulator: Ascorbate oxidizes its way to respectability. In: *Electrochemical Methods for Neuroscience*, Michael AC, Borland LM (eds). CRC Press, Boca Raton, FL, pp. 149–165.

Rebec GV, Barton SJ, Ennis MD (2002) Dysregulation of ascorbate release in the striatum of behaving mice expressing the Huntington's disease gene. *J Neurosci*, 22: RC202.

Rebec GV, Barton SJ, Marseilles AM, Collins K (2003) Ascorbate treatment attenuates the Huntington behavioral phenotype in mice. *NeuroReport*, 14: 1263–1265.

Rebec GV, Conroy SK, Barton SJ. (2006) Hyperactive striatal neurons in symptomatic Huntington R6/2 mice: Variations with behavioral state and repeated ascorbate treatment. *Neuroscience*, 137: 327–336.

Rebec GV, Pierce CP (1994) A vitamin as neuromodulator: Ascorbate release into the extracellular fluid of the brain regulates dopaminergic and glutaminergic transmission. *Prog Neurobiol*, 43: 537–565.

Rebec GV, Wang Z (2001) Behavioral activation in rats requires endogenous ascorbate release in striatum. *J Neurosci*, 21: 668–675.

Rebec GV, Witowski SR, Sandstrom MI, Rostand RD, Kennedy RT (2005) Extracellular ascorbate modulates cortically evoked glutamate dynamics in rat striatum. *Neurosci Lett*, 378: 166–170.

Reiner, A., Albin, R.L., Anderson, K.D., D'Amato, C.J., Penney, J.B., and Young, A.B. (1988). Differential loss of striatal projection neurons in Huntington disease. *Proc Natl Acad Sci USA*, 85: 5733–5737.

Reiner A, Deng YP (2018) Disrupted striatal neuron inputs and outputs in Huntington's disease. *CNS Neurosci Ther*, 24: 250–280.

Rice ME (2000) Ascorbate and its neuroprotective role in the brain. *Trends Neurosci*, 23: 209–216.

Rice ME, Russo-Menna I (1998) Differential compartmentalization of brain ascorbate and glutathione between neurons and glia. *Neuroscience*, 82: 1213–1223.

Rothstein JD, Patl S, Regan MR, Haenggeli C, Huang YH, Bergles DE, Jin L, Dykes HM, Vidensky S, Chung DS, Toan SV, Bruijn LI, Su ZZ, Gupta P, Fisher PB (2005) Beta-lactam antibiotics offer neuroprotection by increasing glutamate transporter expression. *Nature*, 433: 73–77.

Santos L, Opris I, Hampson R, Godwin DW, Gerhardt G, Deadwyler S (2014) Functional dynamics of cortco-striatal networks during volitional movements. *Front Syst Neurosci*, 8: 27.

Scott LJ (2011) Tetrabenazine: For chorea associated with Huntington's disease. *CNS Drugs*, 25: 1073–1085.

Shepherd GMG (2013) Corticostriatal connectivity and its role in disease. *Nature Rev Neurosci*, 14: 278–291.

Steiner H, Tseng KY (2010) *Handbook of Basal Ganglia Structure and Function*. Academic Press, San Diego, CA.

Tkac I, Dubinsky JM, Keene CD, Grutter R, Low WC (2007) Neurochemical changes in R6/2 mouse striatum detected by *in vivo* ^{1}H-NMR spectroscopy. *J Neurochem*, 100: 1397–1406

Trantham-Davidson H, Lalumiere RT, Reissner KJ, Kalivas PW, Knackstedt LA (2013) Ceftriaxone normalizes nucleus accumbens synaptic transmission, glutamate transport and export following cocaine self-administration and extinction training. *J Neurosci*, 32: 12406–12410.

Vagner T, Dvorzhak A, Wojtowicz AM, Harms C, Grantyn R (2016) Systemic application of AAV vectors targeting GFAP-expressing astrocytes in z-Q175-KI Huntington's disease mice. *Mol Cell Neurosci*, 77: 76–86.

Van Raamsdonk JM, Murphy Z, Slow EJ, Leavitt BR, Hayden MR (2005). Selective degeneration and nuclear localization of mutant huntingtin in the YAC128 mouse model of Huntington disease. *Hum Mol Genet*, 14: 3823–3835.

Walker AG, Miller BR, Fritsch JN, Barton SJ, Rebec GV (2008) Altered information processing in the prefrontal cortex of Huntington's disease mouse models. *J Neurosci*, 28: 8973–8982.

Walker AG, Ummel JR, Rebec GV (2011) Reduced expression of conditioned fear in the R6/2 mouse model of Huntington's disease is related to abnormal activity in prelimbic cortex. *Neurobiol Dis*, 43: 379–387.

Wang N, Gray M, Lu XH, Cantle JP, Holley SM, Greiner E, Gu X, Shirasaki D, Cepeda C, Li Y, Dong H, Levine MS, Yang XW (2014) Neuronal targets for reducing mutant huntingtin expression to ameliorate disease in a mouse model of Huntington's disease. *Nat Med*, 20: 536–541.

Wilson JX, Peters CE, Sitar SM, Danust P, Gelb AW (2000) Glutamate stimulates ascorbate transport by astrocytes. *Brain Res*, 858: 61–66.

Wilson JX (1997) Antioxidant defense of the brain: A role for astrocytes. *Canad J Physiol Pharmacol*, 75: 1149–1163.

Xi ZX, Baker DA, Shen H, Carson DS, Kalivas PW (2002) Group II metabotropic glutamate receptors modulate extracellular glutamate in the nucleus accumbens. *J Pharmacol Exp Ther*, 300: 162–171.

Zetterstrom T, Wheeler DB, Boutelle MG, Fillenz M (1992) Striatal ascorbate and its relationship to dopamine receptor stimulation and motor activity. *Eur J Neurosci*, 3: 940–946.

CHAPTER 15

NEUROPHYSIOLOGICAL MONITORING OF POTASSIUM

Michael Thompson

Department of Chemistry and Institute for Biomaterials and Biomedical Engineering, University of Toronto, Canada
m.thompson@utoronto.ca

15.1 Introduction

Both potassium and sodium cations are crucial for the control of cellular homeostasis in terms of health, where abnormalities can lead to serious disorders and disease (Pohl *et al.*, 2013; Clausen *et al.*, 2013). The maintenance of pH, regulation of osmotic pressure and water distribution, regulation of the function of the heart and other muscles, regulation of brain excitability, participation in oxidation-reduction reactions, and their general role as cofactors for enzymes are among the physiological functions of these ions. Normal levels of cellular concentration for sodium and potassium in humans follow, where intracellular potassium comprises some 90% of the total distribution of the ion in the body:

K^+ 150 mM (intracellular)
K^+ 3.5–5 mM (extracellular)
Na^+ 10 mM (intracellular)
Na^+ 135–145 mM (extracellular)

The potassium ion is the major cation of the intracellular fluid, where it is present in the form of potassium bicarbonate buffer. In nerve cells, sodium-potassium flux and the famous sodium-potassium pump are responsible for generating cellular electrical potential and are involved in the generation of nerve impulses. Cellular uptake of potassium is regulated by the pump, while exodus of the ion is controlled by membrane permeability and electrical and chemical gradients. In addition to its role in neurochemistry, potassium is critically important in the processes of protein synthesis, glucose metabolism, and muscle building.

The control of sodium and potassium homeostasis results from a highly complex system of physiological processes and if an imbalance or abnormality is evident it can cause a variety of medical issues. Low potassium levels, termed *hypokalaemia*, can cause, for example, muscle weakness, slow reflexes, skin disease, nerve disorders, cardiac arrhythmia, decreased heart rate and, in severe situations, death. High levels of potassium, termed *hyperkalaemia*, are just as dangerous in that the issue can actually cause cardiac arrest.

In this chapter, following further detail on the biological function of the potassium cation, we discuss selected neurophysiological issues where real time monitoring of the ion would be highly desirable. This is followed by a review of efforts that have been conducted to fabricate devices capable of this task. Included is an evaluation of our research, which involves an attempt to produce micro-sensors that can monitor nerve cell behavior, and which include crucial anti-fouling surface chemistry.

It is explicitly understood that when the ion is referred to in the common short form of K^+, it is actually present biologically as either a hydrate, $K^+(H_2O)_n$ (intra- or extra-cellular fluid) or a non-hydrate (channel). The term *extracellular fluid* refers to both the aqueous matrix that is a major component of blood plasma, and fluid that is in closer proximity or bathes cells such as nerves. The cerebrospinal fluid (CSF) is an example of the latter. The concentration of potassium in this type of liquid is often denoted, $[K^+]_0$.

The scientist from the medicine, biology or physiology communities will wish to skip Sections 15.2 and 15.3. However, the biosensor or analytical chemistry specialist may find it instructive to read these sections before moving on to the discussion of the *in vivo* monitoring of potassium by device technology, where the emphasis is on neurophysiology.

15.2 A summary of potassium physiology

In order to appreciate the necessity of effecting *in vivo* measurement of potassium it is appropriate to briefly review the bio-electrochemistry and neuro-chemistry of the cation. This section and the subsequent one on pathological aspects are presented to achieve this aim.

15.2.1 Potassium membrane permeability and channel

We begin with a necessary summary of the origins of the membrane electrical potential in all cells, with particular emphasis on the role played by K^+. The

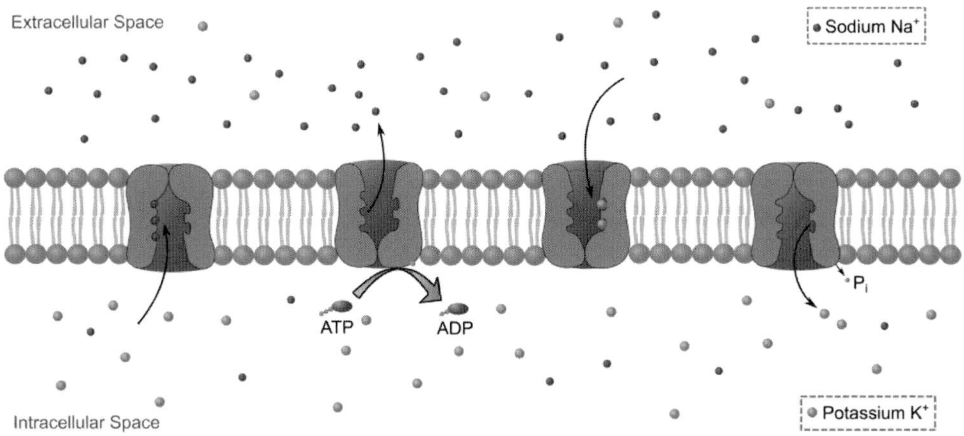

Figure 15.1 Schematic of the sodium-potassium pump.

resting potential for most cells varies from –9 to –100 mV depending on the type of tissue, with the value for most neurons being –70 mV. The basic (non-channel) membrane permeability for K^+ (2×10^6 cm sec^{-1}) is approximately 50 to 100 times higher than the value for sodium. For K^+ the concentration gradient is outward and the electrical gradient inward. The biological mechanism responsible for the active transport of sodium from the cell and potassium into the cell is the sodium-potassium pump (Figure 15.1). This system, which requires energy, is fuelled by enzyme hydrolysed ATP and has been the subject of research for many years.

In both excitable and non-excitable cells, passive transport crossing of the cellular lipid-bilayer membrane by potassium occurs via gated channels, whether of the voltage, ligand, or tandem-pore variety (Kuang *et al.*, 2015). Indeed, the nature and structure of channels for K^+ movement remains an active field of research, reflected by the extensive literature in the field. (Research has associated channel malfunction with disorders such as epilepsy (see later), cardiac arrhythmia, and other medical issues.) Great progress has been made over recent years with respect to the structure and ionic specificity of K channels, especially involving research on bacteria. Notably, the Nobel Prize was awarded to McKinnon and his group for their pioneering work on the K channel. Briefly, from X-ray crystallographic (Doyle *et al.*, 1998; Zhou *et al.*, 2001) and other studies it has been shown that the structure of the potassium channel (Figure 15.2) is based upon a homotetramer of protein residues that surround a pore designed for ion passage. Amazingly, a 5-residue sequence, TVGYG, is responsible for providing the necessary selectivity for the potassium cation; this sequence is thus often referred to as the *selectivity filter*. The sequence is within a loop and involves removal of the water of hydration from the ion via the provision of a structure reflective of the hydration shell (Sauer *et al.*, 2011; Hellgren *et al.*, 2006). Ion passage through the 10-angstrom pore is

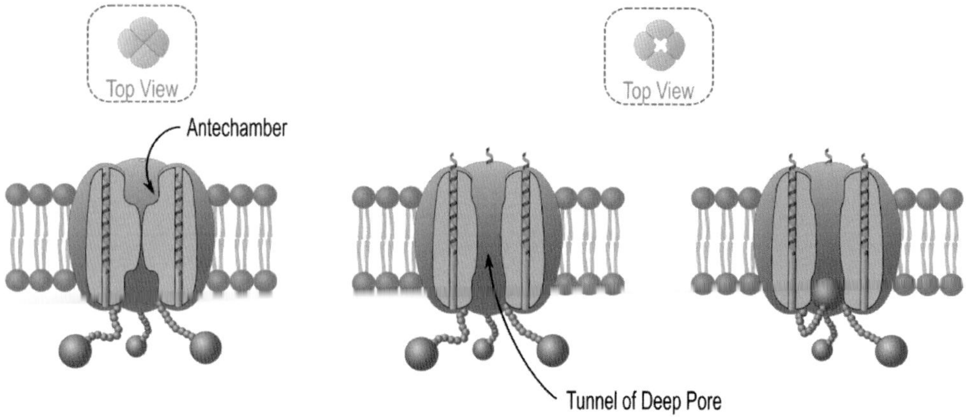

Figure 15.2 Schematic of a typical voltage-gated K^+ channel with one subunit of a tetramer missing to show inside structure. Voltage detection is by the helix structure.

facilitated by the water present there, which generates a high dielectric zone. Finally, as mentioned above, the system is gated (channel opening and closing) by residues that operate largely via electrostatic or ligand mechanisms.

15.2.2 Excitable tissue

Neurons (of which there are some 100 billion in a human body), muscle, and heart cells are capable of supporting action potentials, i.e. electrical excitation. Concisely, action potentials in neurons are associated with a depolarization of the membrane potential that results in the opening of Na^+ channels, which are voltage-gated. This allows the entry of sodium ions into the cell. This is followed by the opening of afore-mentioned K^+ channels that permit the exit of these ions from the cell. The inward flow of sodium ions increases the concentration of positively charged cations in the cell and causes depolarization, where the potential of the cell is higher than the resting potential of the cell. The sodium channels close at the peak of the action potential, while potassium continues to leave the cell. The efflux of potassium ions increases the membrane potential or hyperpolarizes the cell. The sodium current is always transient because the Na-channel has both activation and inactivation gates. These gates are voltage dependent but, crucially, operate at different speeds, i.e. time constants. Thus, for any steady state voltage, sodium channels are closed; however, when voltage changes, they can open briefly because the activation gate is quicker to open (time constant, tau, ~1 ms at –70 mV) before the inactivation closes the channel (tau ~8 ms at –70 mV). The conductivity of the potassium ion channel proceeds in proportion with membrane voltage and (for some types) ligand concentration. The overall process is depicted schematically in

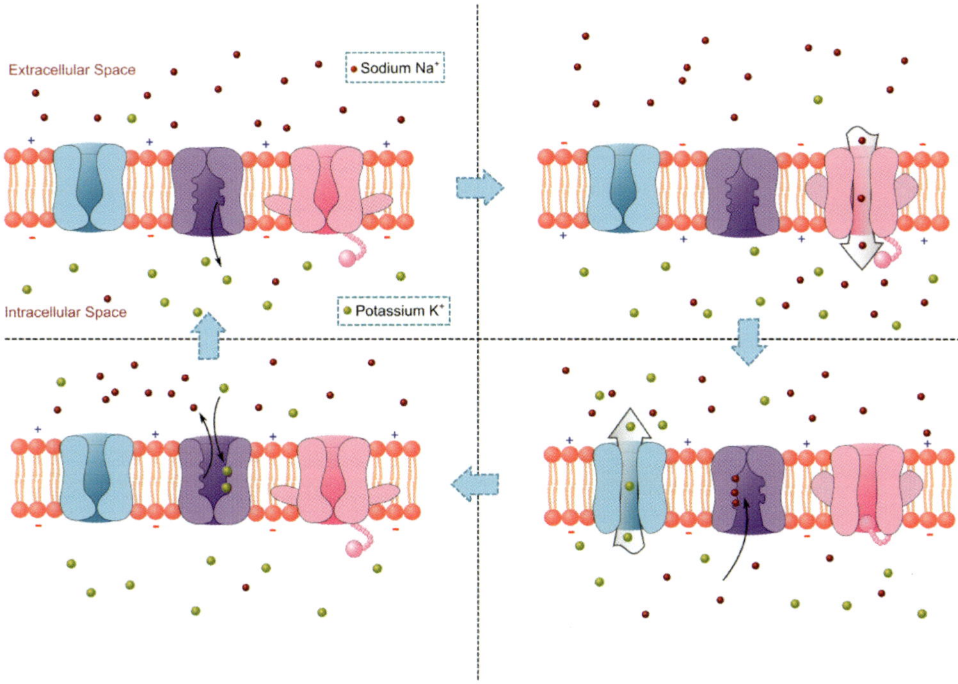

Figure 15.3 Potassium and sodium ion movement during the action potential at a typical cell membrane.

Figure 15.3. The firing rate of the neuron represents the frequency at which the cell produces action potentials.

The action potential characteristics of cardiac cells are somewhat different from those of neurons outlined above (a schematic of the heart is depicted in Figure 15.4). The cardiac action potential is initiated by cells present in the right atrium, or *sinoatrial node*. These cells fire automatically in causing a contraction. The cells, which are often called *pacemaker cells*, are linked via gap junctions. The atrial cells contract together, which then causes an analogous contraction of the ventricular cells. It is the latter that are responsible for conduction of the cardiac action potential. The behavior of these ventricular myocyte cells has been studied extensively in animals. There are thought to be five quite complex phases of operation associated with these cells, which are described in considerable detail by Grunnet in 2010. In the *diastole phase* (4), the cell is in a resting state (unlike pacemaker cells) where the cell potential is mainly determined by K^+ membrane passage in and out of the cell via rectifying potassium channels. In the next phase (0), depolarization occurs, resulting in movement in 2 msec to a positive value for the cell membrane. This is caused by the opening of sodium channels, with influx of the cation into the cell. This

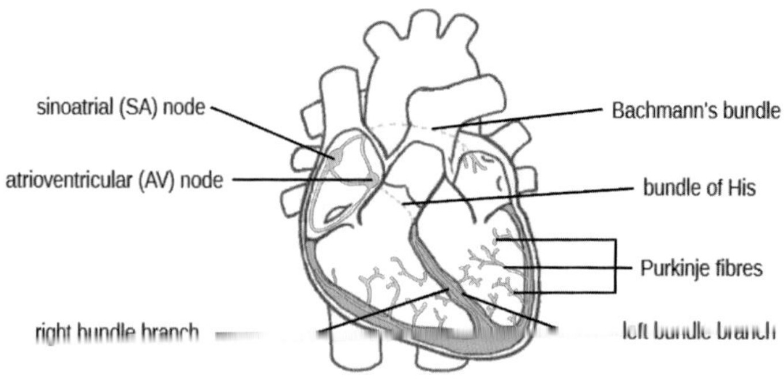

Figure 15.4 Electrical configuration in the heart.

Table 15.1 Example currents during the cardiac ventricular action potential (adapted from Sherwood 2012).*

Ion	Current	α Subunit protein	Phase/role
Na^+	I_{Na}	$Na_v1.5$	0
Ca^{2+}	$I_{Ca(L)}$	$Ca_v1.2$	0–2
K^+	I_{to1}	$K_v4.2/4.3$	1, notch
K^+	I_{Ks}	$K_v7.1$	2, 3
K^+	I_{Kr}	$K_v11.1$	3
K^+	I_{K1}	$K_{ir}2.1/2.2/2.3$	3, 4

*Note: The subscript "v" refers to a "voltage-gated channel," "ir" refers to the term "inward rectified" potassium channel, and "s" refers to a "single" channel. The numbers are associated with the gene that codes for the — α subunit protein.

process is instigated by activation from a neighboring cell via the aforementioned gap junction. Phase 1 involves rapid reduction of sodium movement into the cell and is coupled with the fast opening and closing of potassium channels. This is referred to as a "notch" on the action potential waveform (Santana *et al.*, 2010). The next phase (2) involves slow membrane repolarization. In this phase, in addition to channel-based passage of potassium from the cell, calcium channels are activated; this activation actually causes the contraction of the heart. Furthermore, Cl^- channels are also activated which in turn allows these anions into the cell. Accordingly, voltage change caused by K^+ is opposite to that induced by Ca^{2+} and Cl^-. In the last phase (3), the action potential is repolarized via the closing of Ca^{2+}channels, with the potassium channels remaining open. Thus, the outward current results in a final negative potential for the membrane. In Table 15.1, I present certain characteristics of the various channels involved in the ventricular action potential (Sherwood, 2012).

Finally, in this section I mention skeletal muscle cells, which are also excitable and depolarized by the action of acetylcholine, which is released at the neuromuscular junction. Skeletal muscle is composed of fibers, which are long, multinucleated, and cylindrical in shape. Each fiber is made up of fibrils, which are divisible into filaments of the protein molecules myosin, actin, troponin, and tropomyosin. The cell membrane is called the *sarcolemma*, and its cytoplasm is accordingly termed the *sarcoplasm*. When observed under light, skeletal muscle exhibits striations, which are associated with repeating units of actin and myosin (termed the *sarcomere*). With regard to the excitation of nerve and muscle cells at the neuromuscular junction, especially the role played by channels, Goodman (2008) has published an excellent didactic review. In summary, the action potential of the nerve associated with muscular excitation opens nerve terminal calcium channels that are voltage-gated. This process in turn causes release of acetylcholine into the synaptic cleft. This molecule binds to receptors on the end plate of the skeletal muscle cell, thus influencing voltage-gated sodium channels and the attendant generation of the muscle cell action potential.

15.3 A summary of potassium neuropathology

In this section, I outline *selected* neuropathological conditions that cause or are associated with abnormalities in potassium concentration, whether from the extra- or intra-cellular perspective. There are, of course, numerous neurological disorders, including Parkinson's, multiple sclerosis, cerebral palsy, and dementia. Here, our main focus is to indicate cases where *in vivo* measurement of potassium concentration could prove to be of significant value in terms of understanding the role played by the ion in neurochemical processes that have gone awry.

15.3.1 Epilepsy

Epileptic seizures are the result of excessive and abnormal synchronized neuronal activity in the cortex of the brain (Alarcon *et al.*, 2009). Epileptic events, termed termed *ictal states*, can range from mild symptoms to violent shaking over quite long periods of time. The condition affects literally millions of individuals worldwide. A myriad of factors such as brain trauma, stroke, and brain tumors are thought to be associated with seizures, and genetics also play a role (Berkovic *et al.*, 2006). However, the reality is that little is known, despite many years of research, about the specific factor(s) that actually instigate, sustain, and terminate an epileptic event. There has been interest for some time in the possible implication that potassium concentration is connected to disorders of the cerebrospinal

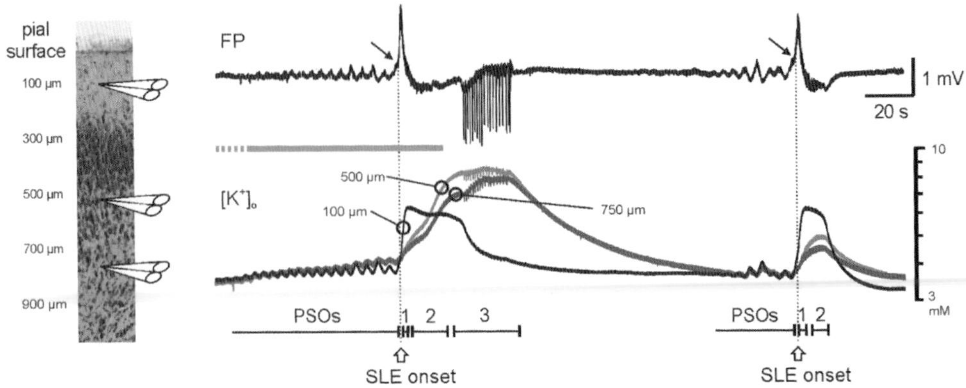

Figure 15.5 Correlation between seizure-like events (SLEs) and extracellular potassium concentration ($[K^+]_o$). Left: Schematic drawing of the positions of the three K^+-sensitive double barrel electrodes is depicted on the left. The distance from the pial surface of each electrode is indicated on a coronal microphotograph of the PC stained with thionine. Right: Simultaneous field potential (FP) recording at 500 μm depth and $[K^+]_o$ shifts recorded at 100, 500, and 750 μm from the pial surface (black, light gray, and dark gray traces, respectively) during SLEs induced by arterial application of 4-aminopyridine (50 μm; horizontal gray bar). SLE phases are reported on the bottom. The second SLE (right) is shorter and does not present phase 3. The arrow of FP trace indicates the deflection that marks the initiation of SLEs after periodic slow oscillations. (Reprinted by kind permission of the Society for Neuroscience, Washington, DC, USA.)

fluid (CSF) that bathes the brain, which is intimately connected to neuronal potassium channel behavior (Smart *et al.*, 1998). (In addition to K^+, the fluid is composed of water, other ions, glucose, and proteins and is maintained in a homeostatic fashion (Bradbury *et al.*, 1970; Iliff *et al.*, 2013)). An example of the link between seizures and extra-cellular potassium concentration is depicted in Figure 15.5 (Uva *et al.*, 2017).

In recent years, considerable interest has been directed at the effect that alterations by mutation and malfunctions of a variety of K^+ channels have on epileptic seizures (Rudiger *et al.*, 2016; Villa *et al.*, 2016). Indeed, the definition "K^+ channelepsies" has been specified in an attempt to describe the connection between channel characteristics and the epileptic condition (D'Adamo *et al.*, 2013). An important recent review evaluates the possible connection between K^+ channel abnormalities and extra-cellular potassium concentration (de Curtis *et al.*, 2018). In this work the authors detail carefully the characteristics of potassium dynamics during epileptic events. They conclude that "it is now well-recognized that enhanced (concentration) of extra-cellular potassium is both a consequence of seizure activity *and* the cause of brain tissue hyperexcitability"; italics and bracket term from this author. It is specified that much remains to be learned with respect to how potassium actually regulates brain hyperexcitability. Finally, an interesting and highly relevant opinion with regard to the instigation of seizures has been expressed to this author recently by an expert in the field (Carlen, 2018): "Epilepsy is not an electrographic disease, but rather a complex manifestation of underlying biochemical and biophysical abnormalities associated with electrographic abnormalities which reflect as well as possibly contribute to the underlying brain dysfunction." This implies that abnormalities in potassium homeostasis in the cerebral blood flow are strongly implicated in inducing the ictal state.

15.3.2 Neuromuscular disorder and ataxia

Gene mutations of ion channels and the presence of autoantibodies to channel proteins have been strongly connected with disorders evident at the neuromuscular junction. Such abnormalities in channel biology are often termed *channelopathies*." Several diseases have been associated with potassium channel abnormalities, including episodic ataxia type 1 (EA1), Andersen-Tawil syndrome, and acquired neuromyotonia (Figure 15.6). The first of these conditions involves imbalance and uncoordinated movements, often accompanied by nausea and headaches (Riant *et al.*, 2011). It is induced by emotion or stress and often involves myokymia (involuntary quivering of muscles) both during and between attacks, which last from a few seconds to minutes. Several mutations in the gene that encodes for a voltage-gated potassium channel ($K_V1.1$) are responsible for

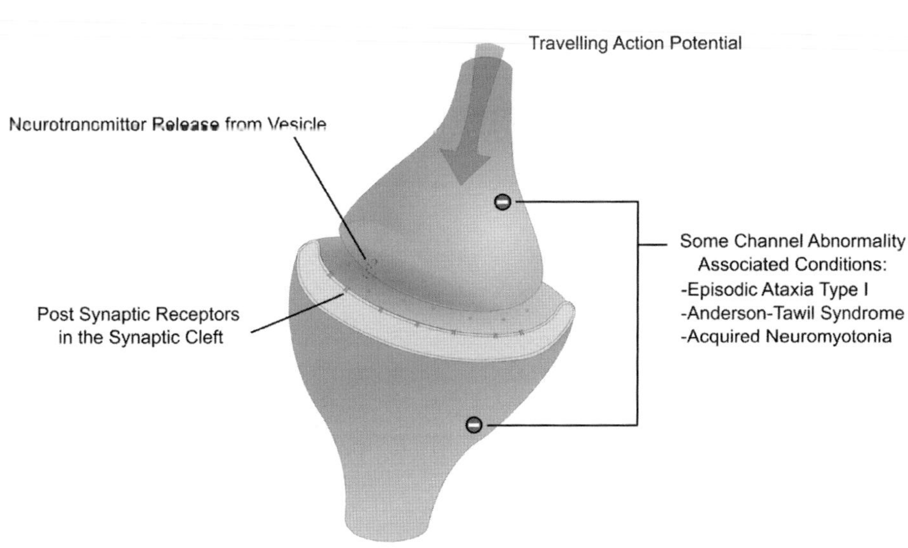

Figure 15.6 Neuromuscular diseases.

EA1 (Brown *et al.*, 1994). (These mutations have been identified following several years of gene research work on the potassium channel.) The $K_V1.1$ channel is present in cells that form GABAergic synapses on Purkinje cells, which are among the largest in the brain. These cells forward inhibitory projections to the deep cerebellar nuclei, and, therefore, constitute the output of motor coordination in the cerebellar cortex.

Andersen-Tawil syndrome, which is a very rare condition, again involves gene mutated alteration of a protein in potassium channels in muscle cells (Nguyen *et al.*, 2013). This is especially the case for skeletal and cardiac muscles; characteristics of the syndrome include periodic paralysis and irregular heart rhythm. It is fair to say that much remains to be researched with regard to this particular syndrome at the current time. Finally in this section, we note that certain aspects of the sodium-potassium pump referred to above have also been connected to forms of ataxia (Forrest *et al.*, 2012).

15.3.3 Cardiac arrhythmia

Cardiac arrhythmia (CA), which manifests itself in irregular, too fast, or too slow heartbeats, involves a very large number of different conditions. It affects millions of individuals; indeed, on a global basis, deaths from cardiovascular disease, which is associated with CA, constitute 15% of all mortalities (Mehra, 2007). There are four main types of CA: premature or extra beats, supraventricular tachycardias, ventricular arrhythmias, and bradyarrhythmias (NHLBI, 2012). The terms *tachycardias* and *bradycardias* refer to fast and slow heart rates, respectively. The conditions, which can be detected in many cases by ECG, are generally treated by surgery, pacemaker installment, or drugs where appropriate.

Given that much of the CA condition is associated with heart muscle electrical effects, it is not surprising that potassium has figured prominently in studies of CA diseases. The cation plays a crucial role, for example, in maintaining myocardial electrical stability. As is the case with many of the conditions mentioned above, the mechanism of cardiac cell repolarization via potassium channels is thought to be critically important. The currents involved in both ventricular and atrial action potentials are shown in Figure 15.7 (Ravens *et al.*, 2008). Channel malfunctions in these tissues in terms of both gain *and* loss of K^+ channel function can apparently lead to arrhythmia. With regard to ventricular arrhythmia, gene-based mutation resulting in modified channel protein species can delay the repolarization process. In contrast, the majority of mutations for atrial tissue lead to a gain of channel function, which in turn generates an abbreviation of action potential duration.

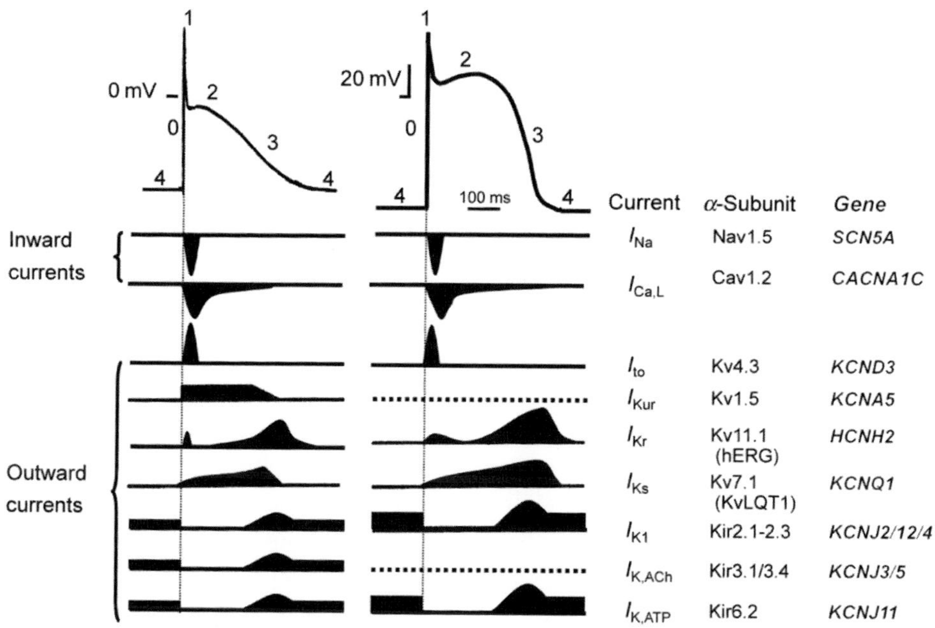

Figure 15.7 Inward, depolarizing and outward, repolarizing currents that underlie the atrial and ventricular action potential. Inward currents: INa sodium current; ICa,L L-type calcium current; Ito transient outward current; IKur ultra rapidly activating delayed rectifier current; IKr and IKs rapidly and slowly activating delayed rectifier current; IK1 inward rectifier current; IK,ACh acetylcholine-activated potassium current. Note that IKur is present in atria only. Phase 0, rapid depolarization; phase 1, rapid early repolarization phase; phase 2, slow repolarization phase (plateau phase); phase 3, rapid late repolarization phase; phase 4, resting membrane potential. (Reprinted by kind permission of the European Society of Cardiology, Sophia Antipolis, Biot, France.)

Recently, the connection between macroscopic serum potassium levels and ventricular arrhythmia through K^+ measurement correlated ECGs was studied (Patel *et al.*, 2017). Interestingly, it was found that potassium levels < 3.5 mEq/L and > 5 mEq/L both conferred an increased risk of cardiovascular-caused death (via tachyarrhythmias and bradyarrhythmias — see above).

15.3.4 Alzheimer's disease

There is a vast literature on this disease, with publications detailing many aspects of the condition. However, the role of the potassium ion, if any, has not received quite as much attention. Alzheimer's is a poorly understood, chronic neurodegenerative disease that appears have its origin in a number of factors such as genetic issues, head injuries, obesity, and hypertension, although its precise etiology is unknown (Burns *et al.*, 2009). Annually, there are over 30 million new cases of the disease worldwide; the death toll is approximately 2 million individuals. Some years ago, the disease was associated with plaques and tangles that form in the brain and disturb normal neural physiology (Hardy *et al.*, 1990). The cause of the genesis of such structures has received much attention, resulting in the appearance of a number of hypotheses. Among these, the amyloid and tau mechanisms have figured prominently, especially the former. In this case, it has been thought for some time that extracellular amyloid beta (A_β) deposits are responsible for the disease (Holmes *et al.*, 2008; Lacor *et al.*, 2007).

As is the situation with a number of neural conditions, cellular K^+ channels have been linked with the disease (Etcheberrygaray *et al.*, 1993; de Silva *et al.*, 1998). In the early 2000s, Williams specifically drew attention to the fact that dysfunction of potassium channels associated with altered proteins could be connected with a number of neurodegenerative diseases such Huntington's, spinocerebellar ataxia, and Alzheimer's (Williams, 2006). Subsequently, a number of studies on the channel and its link to the disease have appeared. For example, the Kv3.4 potassium channel and its ancillary subunit, MiRP2, are involved in neuronal death and are up-regulated by Aβ peptide in a nuclear factor κB-dependent fashion at the transcriptional level (Choi *et al.*, 2007). In a comprehensive study, it was shown that expression of Kv1.3 (a voltage-gated potassium channel) is increased in the frontal cortex of humans with the disease as compared to non-Alzheimer's controls. Kv1.3 channels are expressed by microglia in the vicinity of Aβ plaques (Rangaraju, 2015). Here, the authors indicate that the system represents a possible therapeutic target, which appears to be a significant aim of much of the work on the K^+ channels.

15.4 History of real-time measurement of K^+ concentration *in vivo*

Here we provide a concise summary of examples of selected past attempts to measure the concentration of K^+ in biological fluid *in vivo* and *in real time*. (A number of methods, mostly involving ion-selective electrodes, have been available for some time that can assay the cation concentration in biological samples on a static and *in vitro* basis.) This section is not designed to constitute a detailed compendium, but is intended to present a look at the general progression in the field with respect *specifically* to K^+ detection. This chapter does not, for the most part, deal with imaging techniques, which are often not performed in real time.

Measurements *in vivo* date back to the 1970s; they usually involve standard glass electrode technology (e.g. Suga *et al.*, 1970). Endolymph mentioned in this work is a fluid containing a relatively high concentration of potassium. It is present in the membranous labyrinth of the inner ear. One of the first works to study real time effects was that on the assay of both Ca^{2+} and K^+, using ion selective electrodes with 2 micron tips, during seizures instigated in the cortex of an animal (Heinemann *et al.*, 1977). The concentration of the calcium ion decreased in contrast to potassium which increased; these effects were discussed in terms of their relationship to epilepsy.

The famous group of Ernő Pungor in Hungary developed a new crown ether for selective binding of K^+ in a 10-μm microelectrode for *in vivo* measurements (Tarcali *et al.*, 1985). The potassium electrode and a closely attached micro-reference electrode were introduced into the cortex of anaesthetized rats. Introduction of the electrodes produced a relatively high potential, reflecting a high potassium activity. A small crystal of veratrine was then placed on the cortical surface very close to the electrode. Administration of the drug resulted in regular oscillations: periodic increases of potassium activity followed by its return to near the original value. In the cerebellum of a fish, an anaesthetized skate, the concentration of potassium was studied during excitation by an electrode (Rice *et al.*, 1988). In this case K^+ measurements in the cortex were achieved using a standard valinomycin-based ion-selective electrode. With regard to stimulation, a bipolar stimulating electrode was positioned on the cerebellar surface to activate the parallel fiber-Purkinje cell system and evoked fields were monitored on a digital oscilloscope. It was observed that during parallel fiber activation, $[K^+]_0$ increased by 3.6 mM to an upper limit of 12–14 mM with an approximately linear dependence on stimulation frequency (1–20 Hz).

Again using micro-pipette based electrodes, both extracellular calcium and potassium concentrations were measured in great detail in the cerebral cortex of cats under ketamine and xylazine anesthesia during patterns of slow sleep oscillations and spike-wave seizures (Amzica *et al.*, 2002). In summary, measurements obtained during the slow oscillation indicated that glial cells phasically take up part of the extracellular K^+ extruded by neurons during the depolarizing phase of the slow oscillation. During this condition, the redistribution of potassium appeared to be local. Large steady increases in K^+ and phasic potassium accumulations were measured during spike-wave seizures. The same group, in a particularly ground-breaking set of experiments, assayed both intra- and extra-cellular potassium concentrations in anaesthetized rats using a micro-optode (Dufour, 2011). (Determinations of intra-cellular ionic concentrations *in vivo* are quite rare.) The biosensor system consisted of a potassium microelectrode for extracellular assays paired with both a second microelectrode for field potential determinations and a tapered fiber optic device for intracellular assay of the cation. The overall configuration had a diameter of 15 μm. The experimental set-up, which is based in part on conventional fluorescence microscopy, is depicted in Figure 15.8. The recorded luminescence was obtained from potassium-binding benzofuran isophthalate, a

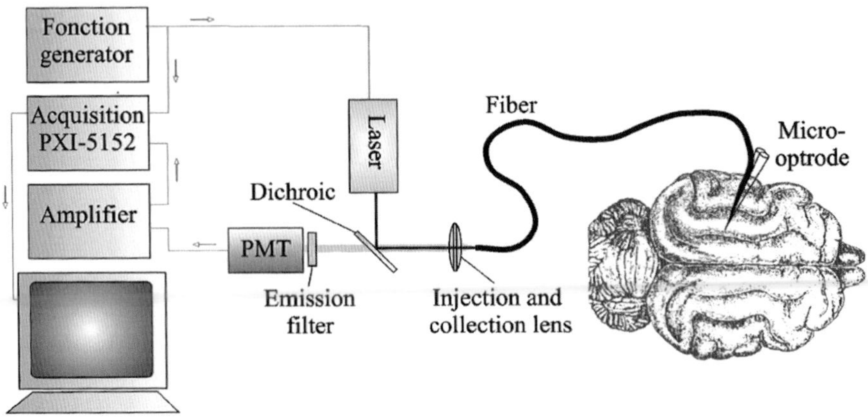

Figure 15.8 Excitation light generated by the laser is reflected by a dichroic filter and directed through an aspheric injection lens toward the multimode delay fiber and further to the micro-optode. After excitation of the tissue, fluorescence is collected with the same optical fiber, re-collimated by the aspheric lens, directed through the dichroic mirror, filtered with a 40 nm large band pass filter cantered on 500 nm, to finally reach a photomultiplier tube (PMT) detector. The PMT signal is amplified and digitized. (Reprinted by kind permission of Elsevier.)

K^+-sensitive dye. The performance of the optode configuration was assessed via experiments involving animal sleep-like patterns, induced seizures (with penicillin), and anaesthesia comatose states. An example of the results obtained is given in Figure 15.9. In summary, the authors found that in the majority of cases paroxysmal discharges were associated with positive variations of the intracellular fluorescence corresponding to extracellular K^+ increases. In the remaining cases, intracellular K^+ was reduced for an extracellular K^+ increase. Comatose states were accompanied by an extracellular drop of K^+ of 1.31 ± 0.13 mM, which was reflected, in all cases, by an intracellular K^+ increase of $39 \pm 4\%$.

Emphasis was placed on the design of a neural probe for experiments within an animal cortex in 2013 (Vander Wouden *et al.*). The device was used to examine the instigation of cortical spreading depression, which is thought to be a component of the migraine condition. As in many cases involving stimulation of the cortex, an increase of potassium concentration in a transient fashion was observed. Recently, research has appeared describing the development of genetically encoded fluorescent probes for K^+ that allow the real time detection of the cation *in vitro* and *in vivo* (Bischof *et al.*, 2017). The basic concept, which has its core detection by Förster resonance energy transfer (FRET), is indicated in Figure 15.10. Starting with a bacterial potassium binding protein, the authors were able to produce K^+ chimeras, which they call genetically encoded potassium ion indicators (GEPIIs). The response of the FRET ratio to K^+ covered a 3.2-fold range, which

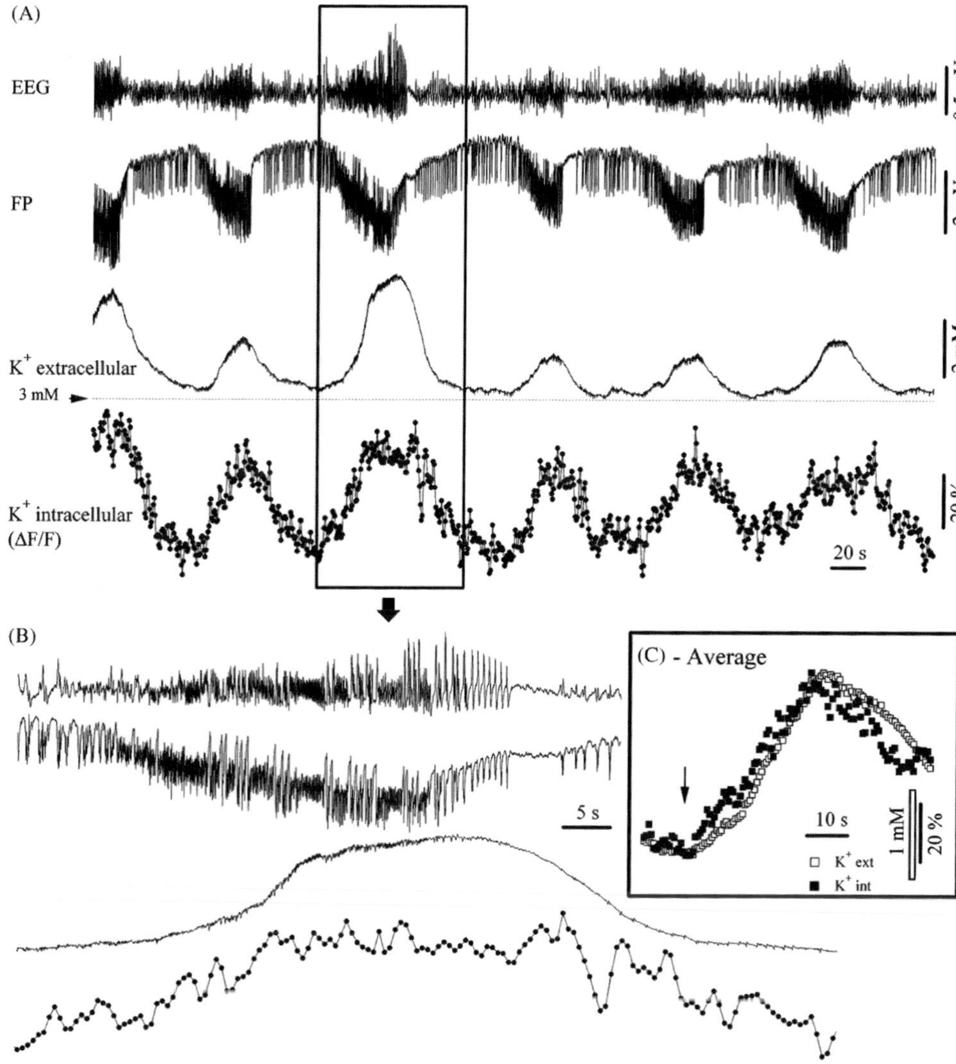

Figure 15.9 Cortical intracellular and extracellular K^+ variations during epileptic seizures induced by topical application of penicillin on the cortex. (A) Recurrent seizures, occurring every 1–1.5 min, are characterized by increased amplitude and accelerated waves in the EEG and additional negative DC shift potentials in the field potentials (FP). Our intracellular optical recordings displayed in this case simultaneous increases of the intracellular K^+. The period within the square is expanded in (B) to emphasize the details of the seizures. Note presence of fast runs and SW complexes during the seizures. (C) Inset with grand averages of seizures (n = 7) from intra- (black squares) and extracellular (white squares) K^+ signals. The vertical arrow marks the triggering point as detected from the onset of the seizure in the FP trace. Note a rise in the intracellular K^+ than in the extracellular K^+ (Reprinted by kind permission of Elsevier.)

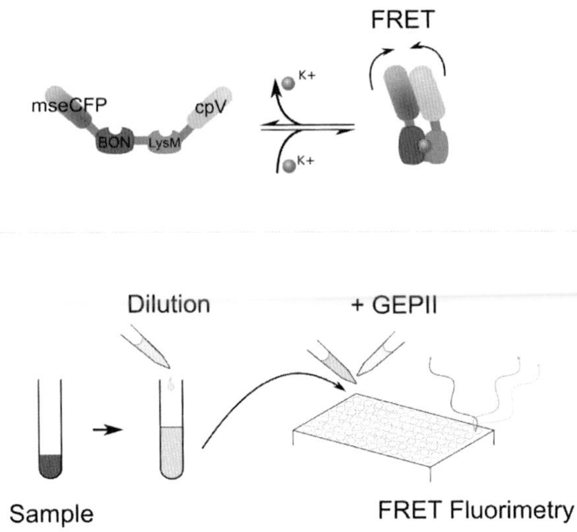

Figure 15.10 Mechanism of K^+ detection by FRET and genetically-encoded potassium ion indicator and experimental system. (Adapted from Bischof *et al.*, 2013.)

was ascribed to a conformational rearrangement in the protein from an elongated to a spherical structure upon K^+ binding. GEPIIs expressed in cells enabled the time-lapse and real-time recordings of global and local intracellular K^+ signals. Extracellular fluctuations of the cation *in vivo* could also be visualized by 2-photon microscopy.

15.5 Spatio-temporal measurement of extra-cellular potassium in the brain

In this section, I review our own research on the real-time detection of potassium in biological fluid, with particular attention to extracellular fluid. Our efforts emphasize applications in both research on epilepsy and therapy by nerve stimulation. As outlined above, there is considerable interest in the possible link between potassium and the occurrence of epileptic seizures. Accordingly, a device that can assay the cation extra-cellularly in a spatio-temporal manner in the brain's extra-cellular cerebrospinal fluid would be of significant value.

15.5.1 Anti-fouling and potassium probe surface chemistry

The important criteria for respect to the fabrication of our micro-sensor for operation in CSF are: (1) the construction of a *selective* probe for the cation that is capable of surface attachment on the gold electrode of the device (see below), (2) the combination of the probe with a surface agent that can reduce fouling by protein and other entities present in CSF, and (3) the design and fabrication of a real-time signaling sensor with electrodes small enough to implant in an animal for research purposes.

The strategy for criteria 1 and 2 was to produce a mixed monolayer of thiolated probe and antifouling agent for attachment to the device's gold electrode. With regard to the latter component of the surface chemistry, we have established that an ultra-thin (6 angstroms) covalently-bound monoethyleneglycol species (MEG-OH) on silica dramatically reduces adsorption of moieties from serum (Sheikh *et al.*, 2012). The introduction of the ethlyleneglycol functionality is based on the long-known use of macroscopic layers of polyethylene glycol as an antifouling compound. The highly significant reduction in fouling by proteins is associated with a hydration structure that penetrates into the monolayer, as evidenced by neutron spectrometry and molecular dynamics studies (Pawlowska *et al.*, 2014; Sheikh *et al.*, 2015). The hydration mechanism has been expressed in terms of a water barrier to protein adsorption but it still remains to be established if this is actually the role of the water layer. Clearly, the interaction of protein functional groups with those on the substrate surface becomes energetically unfavorable.

In an analogous fashion to MEG-OH mentioned above, a thiol-containing molecule with the ethylene glycol group (MEG-SH) for attachment to gold also reduces protein adsorption although not quite as efficiently as the –OH version (Avci *et al.*, 2013). With regard to a probe for potassium, a crown ether (Gokel *et al.*, 2004) was modified by us synthetically to allow binding to a gold surface. The complete configuration of surface chemistry is depicted in Figure 15.11. The tandem successful attachment of MEG-SH to the probe was confirmed by X-ray photoelectron spectroscopy and surface free energy analysis (Machado *et al.*, 2016). The surface chemistry was imposed on the gold electrodes of a fabricated microarray as described below.

15.5.2 Micro-sensor array

This device (Figure 15.12) was designed and constructed by the group of Genov in the Department of Electrical Engineering at the University of Toronto (Machado *et al.*, 2018). Microelectrodes (6) were fabricated on a thick polyimide substrate with dimensions of 3 mm long by 200 μm wide. Gold layers of 800 nm thickness were deposited using a direct current sputtering technique and were patterned using photolithography and wet etching. The configuration included both reference (2) and working (4) electrodes. The electrodes were wire-bonded to the

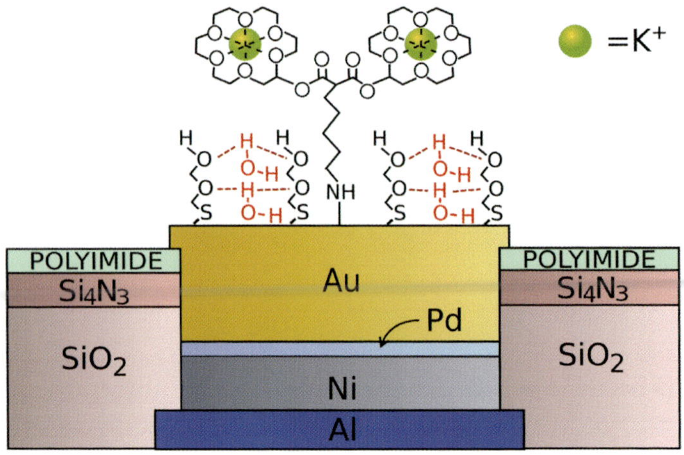

Figure 15.11 Surface chemistry for attachment of K^+-sensitive crown ether to impedimetric gold electrodes in tandem with anti-fouling capability.

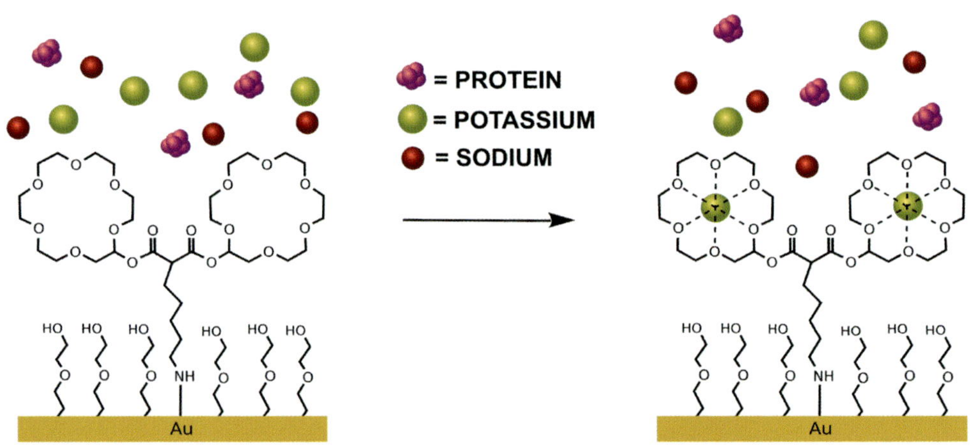

Figure 15.12 Multichannel gold microelectrode device and structure with imposed surface chemistry.

soldering pads of the microelectrode array. The area of the solder pads that was not covered by epoxy was soldered to wires that connect the electrode signals to the bench top instruments. The connecting wires, which run over the probe tip and holder, interface the electrode with the function generator, the power supply, and the data acquisition terminals. Impedance spectroscopy via a sweep from 100 Hz to 1 kHz was performed in real-time. Notably, the peak input amplitude of 4 mV was chosen to avoid affecting animal neuronal behaviour. The four array outputs, three working and one reference, were averaged by 10 nF capacitors immediately and then digitized by a data acquisition unit. Real-time $[K^+]_0$ variations from the outputs of the four channels, as well as voltage and current sweeps, were processed by a system identification algorithm. The algorithm extracted the resistance (R) and capacitance (C) parameters of the series RC model of the electrode–tissue interface. The continuous time-impedance model employed in these experiments is approximated in the discrete-time domain. It is described in detail in Machado *et al.*, 2018.

15.5.3 Spatio-temporal monitoring in mouse brain

These experiments were conducted by the group of Carlen at the Krembil Neuroscience Institute at the Toronto Western Hospital (Machado *et al.*, 2016). Two types of *in vivo* research were conducted on the cerebral cortex of anaesthetized mice. Firstly, transient recording was performed during optogenetic stimulation to evoke extracellular potassium rises in mice, where the neurons of these genetically modified mice expressed a light-sensitive cation channel, which can be activated by blue light (Ritter *et al.*, 2014). Secondly, transient potassium accumulation was also recorded after accompanying seizure-like events were initiated by the application of a 4-aminopyridine (4-AP) solution onto the exposed cortical surface. For comparison purposes, extracellular $[K^+]_0$ was concurrently recorded with a glass capillary ion-sensitive recording electrode in all experiments (see above). In terms of surgery on the mouse, an incision was performed over the parietal bone to expose the somatosensory cortex of the right hemisphere via 3 × 3 mm craniotomy (the craniotomy was centred 2 mm caudal to bregma and 2 mm lateromedial). A K^+-sensitive microelectrode made of a pulled glass capillary and a glass local field potential electrode (filled with saline) were inserted into cortical layer II/III. The gold microelectrode system coated with the anti-fouling potassium probe was carefully deposited onto the cortical surface with a micromanipulator in proximity to the glass electrodes.

Optogenetic stimulation of the mouse cortex revealed concentrations of potassium that initially decreased slightly at the onset of stimulation by approximately 1–2 mM, followed by a sharp increase of 2 mM for a period of less than 100 ms (Figure 15.13). Return to equilibrium at 4 mM was immediately observed for each subsequent stimulated event. Relative to the glass capillary electrode, the coated gold microelectrode exhibited reduced signal width and more consistent K^+ peak spikes between seizure events. The difference in measuring speed between the capillary and gold microelectrode appeared to be approximately 50%, on the order of hundreds of milliseconds.

The microarray-based electrode offered an interesting modest spatial resolution in addition to real-time measurement of the cation. The time-averaged $[K^+]_0$ signals recorded from three coated gold microelectrode channels showed large

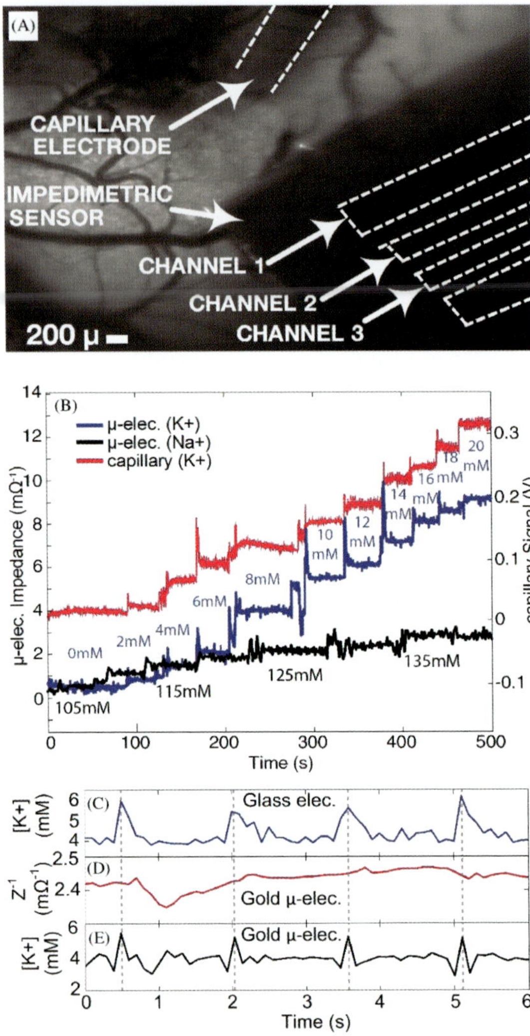

Figure 15.13 (A) Raw fluorescence image of a mouse cranial window, with 10X magnification: the glass capillary electrode (top), and monolayer gold coated electrode (right); (B) calibration curve of both the glass capillary electrode (red) and coated gold microelectrode (blue) for varying concentrations of KCl and NaCl. Sodium concentrations were also evaluated for the gold microelectrode using NaCl in simulated CSF to determine alternative cation sensitivity. While a slight increase in signal was achieved, it was not as significant as for potassium. Simulated CSF solutions containing 2.5–22.5 mM KCl were changed in 2 mM increments. Each solution also contained: NaCl 120 mM, glucose 10 mM, $NaHCO_3$ 22 mM, and 2.6 mM $NaH_2PO_4{\cdot}H_2O$, $MgSO_4$, and $CaCl_2$. The pH was adjusted to 7.4. The potassium-sensitive glass capillary electrode was connected to the same power supply set as the gold microelectrode, which had a built-in voltage gain to amplify the signal, resulting in heightened sensitivity to the cation concentration; (C–E) *In vivo* recordings of extracellular potassium during optogenetically elicited responses for both glass capillary (blue) and coated gold microelectrodes (black). Raw impedance recordings for the coated gold microelectrode were also represented for comparison (red). A blue light driver was pulsed every 1.5 s to initiate a potassium spike. Gold microelectrode measurements were recorded at 300 Hz. (Reprinted by kind permission of MDPI, Basel, Switzerland.)

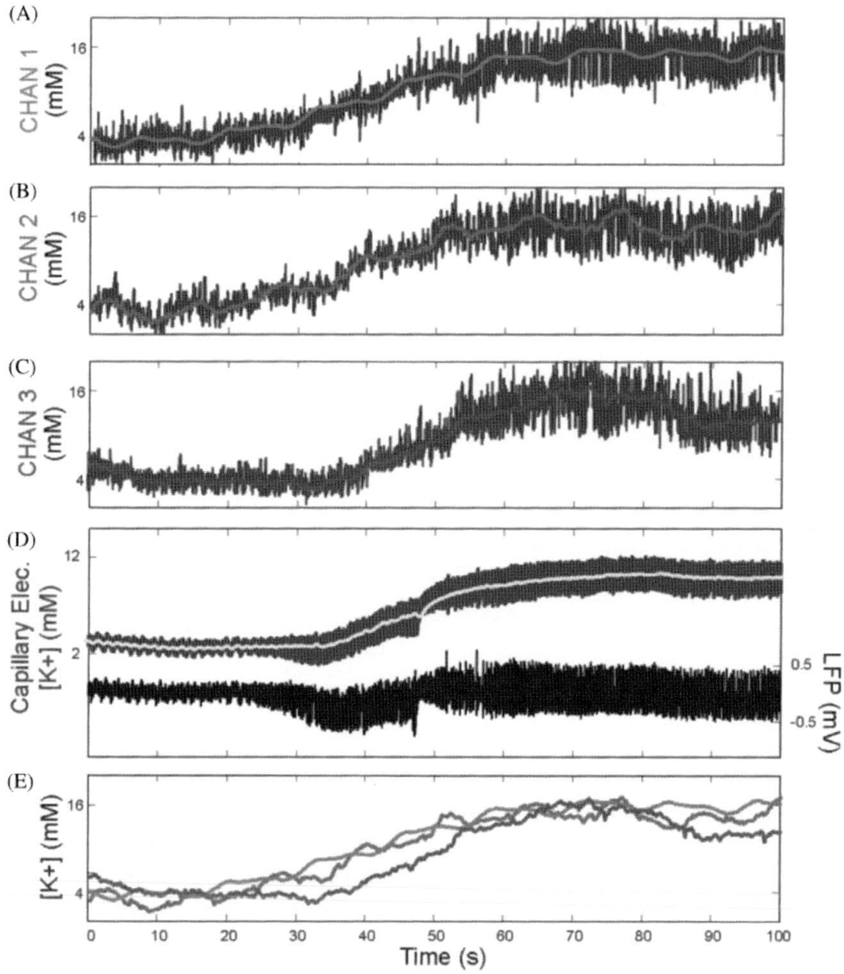

Figure 15.14 (A–C) At the onset of a 4-AP–induced focal neocortical seizure, $[K^+]_0$ was measured with the coated gold microelectrodes for three individual channels, as per Figures 15.13(A)–15.13(C) with smoothed signals overlying the raw signal; (D) $[K^+]_0$ measurements for capillary electrode (left axis) and electroencephalogram signal from the reference (right axis); and (E) overlay of all three coated gold microelectrode channels with $[K^+]_0$ smoothed measurements, showing a delayed response in channel 3, which was apparently farthest from the seizure-onset zone, here demonstrating spatiotemporal specificity. Due to the plateau of the signal at each electrode, only the first 100 s of the 5 min measurement are displayed. (Reprinted by kind permission of MDPI, Basel, Switzerland.)

overlap (Figure 15.14). For chemically induced epileptic seizures in mice, a rise in concentration to approximately 10 mM for each channel was observed, although the rise of $[K^+]_0$ in channel 3 lagged behind those of the other two channels. This spatio-temporal specificity may indicate that epileptic seizures occur in an entire region of the brain, rather than at a very specific starting point. Finally, the $[K^+]_0$

of all three channels was maintained at ~10 mM throughout the entire seizure. Accordingly, it is possible that the rise in potassium concentration caused a signalling cascade throughout the neighbouring regions of the brain. In summary, it is clear that much remains to be understood regarding what transpires before, during and after epileptic seizures, especially as they involve the potassium cation.

15.6 Towards measurement in nerves

The direct, real-time detection of ions in peripheral nerve tissue offers the potential to not only monitor nerve activity, but also, importantly, to provide physiological feedback. Chemical signals, in general, offer better signal-to-noise ratio than electrical signaling although on a slower time-scale. This type of neural activity recording is interesting for closed-loop neuromodulation systems, with intelligent timing and nerve stimulation dosage (see below). An example of this sort of research is the measurement of hydronium ion concentration in the vagus nerve (Cork *et al.*, 2018). Briefly, the vagus ("wandering") nerve is the longest cranial nerve, running all the way from the brain stem to part of the colon. The nerve supplies motor parasympathetic fibers to many organs and is involved in the control of a number of muscles including those in the heart. Stimulation of the nerve has been employed in the treatment of both epilepsy and obesity (the termination of feeding involves the nerve) (Yuan *et al.*, 2016; Browning *et al.*, 2017). A 125-μm iridium wire was used for the successful real-time assay of pH in the exposed subdiaphragmatic vagus nerve of an anaesthetized rat. The oxidized metal is sensitive to pH changes and is thought to be biocompatible enough for operation in biological media.

As is the case for the measurement of pH in the nerve, there is significant interest in the detection of changes in potassium concentration in a real-time fashion. Alterations in the concentration of the cation during nerve signaling are considered to be more dramatic than the alterations observed in hydronium ion changes. There is scant literature on ion selective (other than pH) and biosensor devices involving iridium. We have produced the necessary chemistry on the metal for the selective binding of K^+ (De La Franier *et al.*, 2017). This is a modified chemistry from that shown above for the gold electrode microarray (Figure 15.15). Physiology experiments are planned to examine the sensitivity to potassium in the vagus nerve.

15.7 Future perspective

15.7.1 Potassium probe

Since the days of valinomycin as a binder for K^+, electrochemists in particular have directed a significant amount of research at improving the stability and selectivity

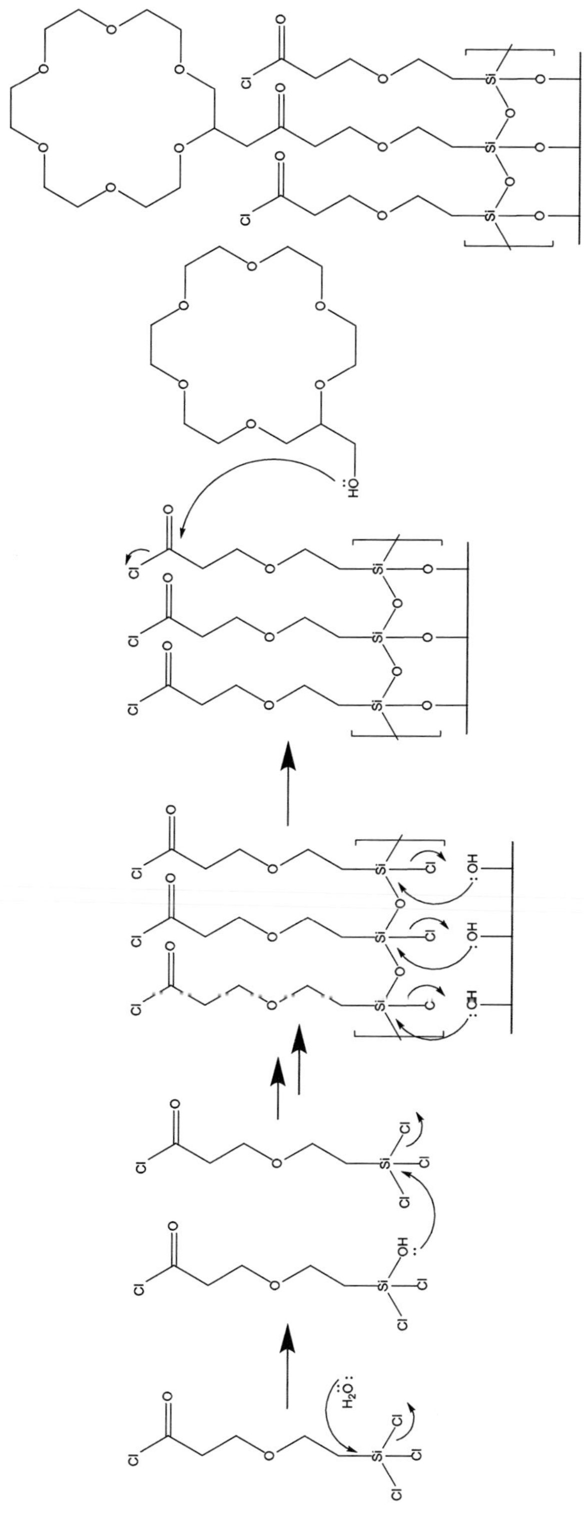

Figure 15.15 Immobilization of a K^+-sensitive crown ether to iridium-based biosensor surface with anti-fouling capability.

of probes for the cation, especially against sodium. Hence, the aforementioned work on crown ethers where some of the research is conducted on solution-based experiments. However, the employment of such species on electrochemical, acoustic wave, and optical transducers implies that their attachment to substrate surfaces will see the cation binding selectivity retained. In our experience this is generally far from the case. Evolution has given us the selectivity filter in biological potassium channels. Accordingly, given the modern emphasis on mimicking biological systems in biotechnology, one wonders whether the K^+ channel proteins can yield a better chemistry for biosensor development. A possible hand in this research might be found in the microarray approach, which is performed on surfaces and has been employed successfully in the design of molecular imprinted polymers for sensor applications (Kutner *et al.*, 2018). Finally, the recent work on genetically-encoded K^+ sensitive fluorescence probes mentioned above may provide a unique methodology for assaying the concentration of the cation in biological samples (Bischof *et al.*, 2017). It remains to be seen how competitive this approach will be compared to electrochemical and other technologies, especially in terms of facility of application.

15.7.2 Anti-fouling sensor surface chemistry

The fouling of surfaces by unwanted interfering moieties in biological fluids has definitely constituted something of an Achilles' Heel with respect to bioanalytical applications of biosensor technology. In our research, described above, we employed an anti-fouling agent in a 90% format in tandem with the cation probe. However, such a strategy does not avoid the attachment of undesired entities to the probe molecule itself. One solution might be detection via the selective instigation of structural and conformation changes in a probe by the analyte, rather than detection via a simple binding event. The physics of such a mechanism may allow signaling via a critical attachment of the analyte, rather than an attachment of the interferent. We used this approach in the sensing of cocaine in serum by acoustic wave sensor (Neves *et al.*, 2015). In this case structural changes in nucleic probes were responsible for the selective signal.

From a more fundamental standpoint, much needs to be achieved with regard to anti-fouling surface chemistry; attention must be paid especially to the role played by interfacial hydration (this has been called *kosmotrophic* layer). There is a rapidly burgeoning literature in this area, focusing on both new coatings and physico-chemical studies. Research in anti-fouling surface chemistry is also highly applicable to the production of biocompatible materials. This is not surprising, considering that there is much common ground between these fields (Blaszykowski *et al.*, 2012).

15.7.3 Extra-cellular v. intra-cellular measurement

It is certainly the case that, by far, the majority of K^+ measurements *in vivo* have been made in an extra-cellular manner, as summarized above. Given the fact there is significant research interest in internal cell biophysical chemistry, and the fact that most of the body's K^+ is in this location, measurement of the concentration of the cation intra-cellularly may prove to be extremely useful. Such experiments present a challenge because most human cells have a diameter in the range of 5–150 μm, which implies that a micro device is required *if* the cell wall is to be penetrated. There are other single cell techniques which interrogate the interior of the cell without the need to disturb the membrane, but these are most often applied to isolated single cells rather than in an *in vivo* fashion. One example among numerous of this type is the use of combined surface-enhanced Raman scattering and linear sweep voltametry in the study of PC12 cells. In this study, the cells are immobilized on gold nanodots/indium tin oxide substrate surfaces (El-Said, 2015). There is no doubt that such experiments will lead to new important information regarding the behavior of the biophysical chemistry of cells. However, the approach obviously does not deal with the massive concerted activity of cells that occur *in vivo*.

15.7.4 Spatiotemporal multianalyte measurements

In our experiments, outlined above, we demonstrated the significant potential for performing spatiotemporal studies of mouse brain via a modest three electrode measurement. Clearly, implementing much larger arrays for deployment, for example, in brain tissue would constitute a very strong advance even if all the electrodes were responding to a single analyte. This is especially the case if the system can be interrogated wirelessly. A configuration of this type may well assist in enhanced characterization of medical conditions such as epilepsy, especially if there is an appropriate time resolution involved. Such a multi-electrode array, which has been produced by the group of Roman Genov of the University of Toronto, provides a large number of sensing pads. The combination of this sort of device with the capability to assay a plethora of analytes such as pH, Na^+, glucose, GABA, and other species would obviously offer considerable insights into biochemical events *in vivo*. However, such a system will require the installation of differential surface chemical probes on the device, which represents a challenge.

Looking ahead, a rather futuristic notion regarding a wireless, wearable system for interrogating the brain is depicted in Figure 15.16 (Sawan, 2013). Here, the concept was the achievement of simultaneous near-infrared spectrometry (NIRS) and scalp electro-encephalography (EEG) for noninvasive monitoring and

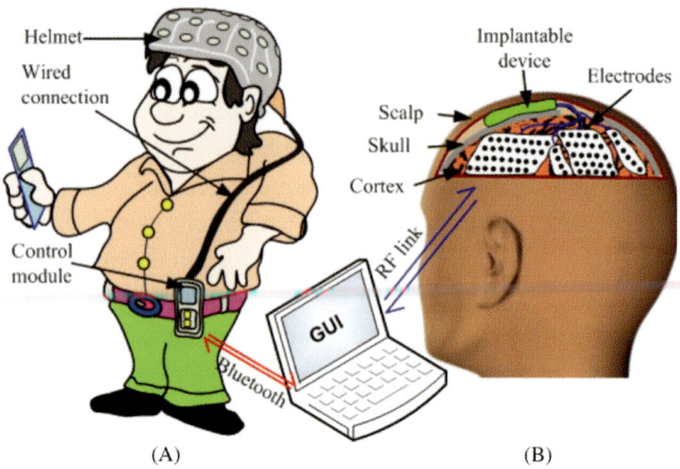

Figure 15.16 Wireless recording systems: From noninvasive EEG-NIRS to invasive EEG devices. (Reprinted by kind permission of IEEE, Piscataway, NJ, USA.)

intracerebral EEG (icEEG) for invasive monitoring. Although the system is not configured for analyte detection, such an approach may well be feasible in the future.

15.7.5 Closed-loop engineering

A distinct possible application of sensing arrays is the direct use of signals to interface with physiological treatment of patients. The legendary interest in the development of the so-called "artificial pancreas" system is one example of the potential medical applications of this sort of approach. In this case, the long term measurement of blood glucose concentration would be achieved via an implanted biosensor which would then "give feedback" to an insulin delivery system (reservoir, pump and calculation chip, etc.) for the diabetic patient. Medtronic began marketing the claimed first closed-loop system available for diabetic patients in 2017 (MiniMed 670G). Despite this significant advance, not all are convinced of the efficacy of the technology. For example, consider the following from Hussain (2018): "it seems like a fantastic pump with a new feature, rather than the 'The World's First Hybrid Closed Loop System.'"

As mentioned above, the concept of physiological feedback through pH assay has in fact been established with respect to vagus nerve stimulation (Cork *et al.*, 2018). Changes in pH are known to occur during stimulation of the nerve (Endres *et al.*, 1986). In these experiments, it was demonstrated that pH detection in the nerve could be used to control simulation of such in a feedback mechanism in order to provide satiety signals. Such an approach to medical treatment involving the array-based multianalyte configuration described above may provide a powerful tool in the future.

Acknowledgments

Our research group and colleagues gratefully acknowledge the Natural Sciences and Engineering Council of Canada and Canadian Institutes of Health Research for financial support. The author is very much indebted to Ruben Machado of The Hospital for Sick Children, Alex Romaschin of St. Michael's Hospital and Peter Carlen of the Krembil Neuroscience Center, all of Toronto, and Konstantin Nikolic (Institute of Biomedical Engineering) and Simon Cork (Hammersmith Hospital) of Imperial College, London, UK and for their kind support and very helpful remarks. The author also wishes to thank Roman Genov of the University of Toronto for graciously allowing the inclusion of the array device picture, and Thomas Vincent for his work in preparing graphics.

References

Alarcon G, Nashef L, Cross H, Nightingale J, Richardson S (2009) *Epilepsy, Oxford Specialist Handbooks on Neurology*, Oxford University Press, Oxford UK.

Amzica F, Massimini M, Manfridi A (2002) Spatial buffering during slow and paroxysmal sleep oscillations in cortical networks of glial cells *in vivo*. *J Neurosci*, 22: 1042–1053.

Avci C, Sheikh S, Blaszykowski C, Thompson M (2013) Critical role of surface hydration on the dynamics of serum adsorption studied with monoethylene glycol adlayers on gold. *Chem Comm*, 49: 466–467.

Berkovic SF, Mulley JC, Scheffer IE, Petrou S (2006). Human epilepsies: Interaction of genetic and acquired factors. *Trends Neurosci*, 29(7): 391–398.

Bischof H, Rehberg M, Stryeck S, Artinger K, Eroglu E, Waldeck-Weiermair M, Gottschalk B, Rost RT, Deak AT, Niedrist T, Vujic N, Lindermuth H, Prassl R, Pelzmann B, Groschner K, Kratky D, Eller K, Rosenkranz AR, Madl T, Plesnila N, Graier WF, Mallil R (2017) Novel genetically encoded fluorescent probes enable real-time detection of potassium *in vitro* and *in vivo*. *Nature Comms*, DOI: 10.1038/s41467-017-01615-z.

Blaszykowski C, Sheikh S, Thompson M (2012) Surface chemistry to minimize fouling from blood-based fluids. *Chem Soc Rev*, 41: 5599–5612.

Bradbury M, Barbora S (1970) Efflux mechanism contributing to the stability of the potassium concentration in cerebrospinal fluid. *J Physiol*, 208: 415–430.

Browne D, Gancher S, Nutt J, Brunt E, Smith E, Kramer P, Litt M (1994) Episodic ataxia/myokymia syndrome is associated with point mutations in the human potassium channel gene, KCNA1. *Nat Genet*, 8(2): 136–140.

Browning KN, Verheijden S, Boeckxstaens GE (2017) The vagus nerve in appetite regulation, mood, and intestinal inflammation. *Gastroenterology*, 152: 730–744.

Burns A, Iliffe S (2009) Alzheimer's disease. *BMJ*, 338: b158.

Carlen P (2018) Krembil Research Institute, Toronto Western Hospital, Toronto, Ontario, personal communication.

Choi E, Abbott GW (2007) *Mol Pharmacol*, 72: 499–501.

Clausen M, Jakob V, Poulsen H (2013) Chapter 3, Sodium/Potassium homeostasis in the cell. In Banci, L (Ed.). *Metallomics and the Cell. Metal Ions in Life Sciences*. Volume 12, Springer.

Cork SC, Eftekhar A, Mirza KB, Zuliani C, Nikolic K, Gardiner JV, Bloom SR, Toumazou C (2018) Real time pH monitoring for use in closed-loop vagus nerve stimulation. *J Neural Eng*, 15: 016001.

D'Adamo MC, Catacuzzeno L, Di Giovanni G, Franciolini F, Pessia M (2013) K($^+$) channelepsy: Progress in the neurobiology of potassium channels and epilepsy. *Front Cell Neurosci*, 7: 134.

De Curtis M, Uva L, Gnatkovsky V, Labrizza L (2018) Potasiium dynamics and seizures: Why is potassium ictogenic? *Epilepsy Res*, 143: 50–59.

De La Franier B, Cork SC, Nikolic K, Thompson M, to be published.

De La Franier B, Jankowski A, Thompson M (2017) Functionalizable self-assembled trichlorosilyl-based monolayer for application in biosensor technology. *Appl Surf Sci*, 414: 435.

de Silva HA, Aronson JK, Grahame-Smith DG, Jobst KA, Smith AD (1998) Abnormal function of potassium channels in platelets of patients with Alzheimer's disease. *Lancet*, 352: 1590–1593.

Doyle DA, Morais Cabral J, Pfuetzner RA, Kuo A, Gulbis JM, Cohen SL, Chait BT, MacKinnon R (1998). The structure of the potassium channel: Molecular basis of K^+ conduction and selectivity. *Science*, 280(5360): 69–77.

Dufour S, Dufour P, Chever O, Vallée R, Amzica F (2011) *In vivo* simultaneous intra- and extracellular potassium recordings using a micro-optrode. *J Neurosci Methods*, 194: 206–217.

Eller K, Rosenkranz AR, Madl T, Plesnila N, Graier WF, Mallil R (2017) Novel genetically encoded fluorescent probes enable real-time detection of potassium *in vitro* and *in vivo*. *Nature Comms*, DOI: 10.1038/s41467-017-01615-z.

El-Said WA, Kim TH, Chung YH, Woo J (2015) Fabrication of new single cell chip to monitor intracellular and extracellular redox state based on spectroelectrochemical method. *Biomaterials*, 40: 80–87.

Endres W, Grafe P, Bostock H, Ten Bruggencate G (1986) Changes in extracellular pH during electrical stimulation of isolated rat vagus nerve. *Neurosci Lett*, 64: 201–205.

Etcheberrigaray R, Ito E, Oka K, Tofel-Grehl B, Gibson GE, Alkon DL (1993) Potassium channel dysfunction in fibroblasts identifies patients with Alzheimer disease. *Proc Natl Acad Sci USA*, 90(17): 8209–8213.

Forrest MD, Wall MJ, Press DA, Feng J (2012) The sodium-potassium pump controls the intrinsic firing of the cerebellar Purkinje neuron. *PLoS ONE*, 7(12): e51169.

Gokel G, Leevy W, Weber L (2004) Crown ethers: Sensors for ions and molecular scaffolds for materials and biological models. *Chem Rev*, 104: 2733–2750.

Goodman BE (2008) Channels active in the excitability of nerves and skeletal muscles across the neuromuscular junction: Basic function and pathophysiolog. *Physiol Educ*, 32: 127–135.

Grunnet M (2010) Repolarization of the cardiac action potential. Does an increase in repolarization capacity constitute a new anti-arrhythmic principle? *Acta Physiol*, 198 Suppl 676: 1–48.

Hardy J, Allsop D (1991) Amyloid deposition as the central event in the aetiology of Alzheimer's disease. *Trends Pharmacol Sci*, 12(10): 383–388.

Heinemann U, H.D. Lux HD, Gutnick MJ (1977) Extracellular free calcium and potassium during paroxysmal activity in the cerebral cortex of the cat. *Exp. Brain Res*, 27: 237–243.

Hellgren M, Sandberg L, Edholm, O (2006). A comparison between two prokaryotic potassium channels (KirBac1.1 and KcsA) in a molecular dynamics (MD) simulation study. *Biophys Chem*, 120(1): 1–9.

Holmes C, Boche D, Wilkinson D, Yadegarfar G, Hopkins V, Bayer A, Jones RW, Bullock R, Love S, Neal JW, Zotova E, Nicoll JA (2008) Long-term Effects of Abeta42 Immunisation in Alzheimer's disease: Follow-up of a randomised, placebo-controlled phase I trial. *Lancet*, 372(9634): 216–223.

Hussain TS (2018) *Medcity News*, February 2018.

Iliff J, Wang M, Liao Y, Plogg B, Peng W, Gunderson G, Benveniste H, Vates G, Deane R, Goldman S (2013) Homeostatic control of brain Function. *Sci Transl Med*, 4: 146.

Kuang O, Purhonen P, Hebert H (2015) Structure of potassium channels. *Cell Mol Life Sci*, 72: 3677–369.

Kutner W, Sharma PS (Eds.) (2018) *Molecularly Imprinted Polymers for Analytical Chemistry Applications*. RSC, Cambridge, UK.

Machado R, Soltani N, Dufour S, Carlen PL, Genov R, ElAnsary M, Thompson M (2018) 50nW 5KHz BW Opamp-less impedance analyser for brain neurochemistry monitoring. *IEEE ISSCC*, 2018.

Machado R, Soltani N, Dufour S, Salam MT, Carlen PL, Genov R, Thompson M (2016). Antifouling impedimetric sensor for the array high-resolution extracellular potassium monitoring in the brain. *Biosensors*, 6: 53–64.

Mehra, R (2007) Global public health problem of sudden cardiac death. *J Electrocardiology*, 40: (6 Suppl): S118–122.

National Heart, Lung, and Blood Institute US (2012) Types of arrhythmia. *https://www.nhlbi.nih.gov/health-topics/arrhythmia*.

Neves MAD, Blaszykowski C, Bokhari S, Thompson M (2015) Ultra-high frequency piezoelectric aptasensor for the label-free detection of cocaine. *Biosens Bioelec*, 72: 383–392.

Nguyen HL, Pieper GH, Wilders R (2013) Andersen-Tawil syndrome: Clinical and molecular aspects. *Int J Cardiol*, 170: 1–16.

Pascale N. Lacor PN, Buniel MC, Furlow PW, Sanz A, Clemente PT, Velasco PT, Wood M, Kirsten KL, Klein WL (2007) Aß Oligomer-induced aberrations in synapse composition, shape, and density provide a molecular basis for loss of connectivity in Alzheimer's disease. *J Neurosci*, 27 (4): 796–807.

Patel RB, Tannenbaum S, Viana-Tejedor A, Guo1 J, KyungAh I, Morrow DA, Benjamin M Scirica BM (2017) Serum potassium levels, cardiac arrhythmias, and mortality following non-ST-elevation myocardial infarction or unstable angina: Insights from MERLIN-TIMI 36. *Eur Heart J Acute Cardiovasc Care*, 6: 18–25.

Pawlowska NM, Fritzsche H, Vezvaie M, Blaszykowski C, Sheikh S, Thompson M (2014) Probing the hydration of ultra-thin antifouling silane adlayers using neutron reflectometry. *Langmuir*, 30: 1199–1204.

Pohl HR, Wheeler JS, Murray HE (2013) Sodium and potassium in health and disease. *Met Ions Life Sci*, 13: 29–47.

Rangaraju S, Gearing M, Jin L-W, Levey A (2015) Potassium channel Kv1.3 is highly expressed by microglia in human Alzheimer's disease. *J Alzheimer's Dis*, 44(3): 797–808.

Ravens U, Cerbai E (2008) Role of potassium currents in cardiac arrhythmias. *Europace*, 10: 1133–1137.

Riant F, Vahedi K, Tournier-Lasserve E (2011) Hereditary episodic ataxia. *Rev Neurol (Paris)*, 167(5): 401–407.

Rice ME, Nicholson C (1988) Behavior of extracellular K^+ and pH in skate *(Raja erinacea)* cerebellum. *Brain Res*, 461: 328–334.

Ritter L, Golshani K, Takahashi K, Dufour S, Valiante T, Kokaia M (2014) WONOEP Appraisal: Optogenetic tools to suppress seizures and explore the mechanisms of epileptogenesis. *Epilepsia*, 55: 1693–1702.

Rudiger R, Wolfart J (2006) Potassium channels in epilepsy. *Cold Spring Harbor Perspectives in Medicine*, Cold Spring Harbor Press. NY, USA.

Santana LF, Cheng EP, Lederer WJ (2010) How does the shape of the cardiac action potential control calcium signaling and contraction in the heart? *J Mol Cell Cardiol*, 49(6): 901–903.

Sauer, DB, Zeng, W, Raghunathan, S, Jiang, Y (2011). Protein interactions central to stabilizing the K^+ channel selectivity filter in a four-sited configuration for selective K^+ permeation. *Proc Nat Acad Sci USA*, 108(40): 16634–16639.

Sawan M, Salam MT, Le Lan J, Kassab A, Gélinas S, Vannasing P, Lesage F, Lassonde M, Nguyen DK (2013) Wireless recording systems: From noninvasive EEG-NIRS to invasive EEG devices. *IEEE Trans Biomed Circuits Systems*, 7: 186–195.

Sheikh S, Blaszykowski C, Thompson D, Thompson M (2015) On the hydration of subnanometric antifouling organosilane adlayers: a molecular dynamics simulation. *J Coll Interface Sci*, 437: 197–204.

Sheikh S, Yang DY, Blaszykowski C, Thompson M (2012) Single ether group in a glycol-based ultra-thin layer monolayer prevents surface fouling from undiluted serum. *Chem Comm*, 48: 1305–1307.

Sherwood L. (2012). *Human Physiology, From Cells to Systems* (8th ed.). Cengage Learning, Boston, USA.

Silverthorn DU (2007) *Human Physiology: An Integrated Approach*. Pearson Education, San Francisco, CA, USA.

Smart SL, Valeri Lopantsev, Zhang CL, Robbins CA, Wang H, Chiu SY, Schwartzkroin PA (1998) Deletion of the KV1.1 potassium channel causes epilepsy in mice. *Neuron*, 20: 809–819.

Suga F, Nakashima T, Snow JB Jr (1970) Sodium and potassium ions in endolymph. *In vivo* measurements with glass microelectrodes. *Arch Otolaryngol*, 91(1): 37–43.

Tarcali J, Nagy G, Tóth K, Pungor E, Juhász G, Kukorelli T (1985) *In vivo* measurements of potassium with an ion-selective microelectrode based on a new bis (Crown ether). *Analyt Chim Acta*, 178: 231–237.

Uva L, Saccucci S, Chikhladzi L, Tassi L, Gnatkovsky V, Milesi G, Morbin M, de Curtis M (2017) A novel focal seizure pattern generated in superficial layers of the olfactory cortex. *J Neurosci*, 37: 3544–3554.

Van der Wouden EJ, Tolner EA, M. Odijk1 M, Olthuis, van den Maagdenburg AMJM, M.D. Ferrari MD, van den Berg A (2013) Development of a neural probe to measure potassium during a cortic spreading depression *in vivo*. *IEEE Transducers*, 2145–2148.

Villa C, Combi R (2016) Potassium channels and human epileptic phenotypes: An updated overview. *Front Cell Neurosci*, 10: 81.

Williams R (2006) Switch to the potassium channel. *Nature Rev Neurosci*, 7: 330.

Yuan H, Silberstein SD (2016) Vagus nerve and vagus nerve stimulation, a comprehensive review: part II. *Headache*, 56: 259–266.

Zhou Y, Morais-Cabral JH, Kaufman A, MacKinnon R (2001) Chemistry of ion coordination and hydration revealed by a K^+ channel-Fab complex at 2.0 Angstrom resolution. *Nature*, 414 (6859): 43–48.

CHAPTER 16

MONITORING BRAIN INJURY WITH MICROELECTRODE BIOSENSORS

Clélia Allioux*,†, Anne Meiller*,‡, Baptiste Balança*,‡,§ and Stéphane Marinesco*,†,‡,¶

*INSERM U1028, CNRS UMR5292,
Lyon Neuroscience Research Center — CRNL, Team TIGER, and
†AniRA-Neurochem Technological Platform, Lyon, France
‡Université Claude Bernard Lyon 1, Lyon, France
§Hospices Civils de Lyon, Lyon, France
¶Stephane.marinesco@univ-lyon1.fr

16.1 Introduction

Brain injuries caused by stroke or trauma can have devastating consequences and sometimes result in patients remaining comatose or with impaired consciousness for extended periods of time. Such patients represent a difficult challenge for physicians because they cannot respond to even simple clinical examination. Consequently, medical staff must rely on monitoring techniques to gain information about the state of the patient's brain. This is especially important during the first weeks following the initial injury, when unnoticed neurological complications can produce long-term disability or death.

16.1.1 Primary injury

Brain injury typically develops in two phases. Primary lesions occur at the time of the initial insult whereas secondary lesions start shortly afterward and can continue to develop for several weeks. Primary injuries include ischemic or haemorrhagic insults occurring after a stroke, as a consequence of vessel occlusion or rupture. Early brain injuries often involve a global decrease in cerebral blood flow (CBF) and hydrocephaly if the presence of blood in the ventricles impairs the clearance of the cerebrospinal fluid; in subarachnoid haemorrhage (SAH), bleeding occurs in the subarachnoid space often in response to aneurysm rupture (Tholance, 2014; Grasso *et al.*, 2017). In traumatic brain injury (TBI), resulting from an external mechanical force applied to the skull, primary lesions are much more complex and heterogeneous. Diffuse TBI, characterized by global brain oedema immediately following the impact, is usually distinguished from focal TBI, which is characterized by extra and/or intraparenchymal bleeding. In focal TBI, part of the initial injury resembles that induced by haemorrhagic stroke (Zoerle *et al.*, 2017). The severity of all these early injuries is evaluated during the first physician contact using the Glasgow coma score. Patients who survived a severe initial injury and display persistent consciousness impairment are then transferred to a neurological critical care unit where clinicians will face two major challenges: (1) optimize brain energy supply to prevent secondary injury, and (2) detect the onset of new lesions in order to initiate rapid treatment.

16.1.2 Secondary injury

The primary phase of injury is often followed by a secondary phase characterized by an increased risk for neurological deterioration. Such neurological complications can occur between 1 and 15 days after admission to critical care. The likelihood of secondary lesions increases with the severity of the initial injury, as does the probability of lifelong disability (Bilgin *et al.*, 2012). Possible complications include progression of the initial insult into surrounding brain areas, epileptic seizures, or the occurrence of new lesions such as rebleeding in the case of hemorrhagic stroke. Delayed cerebral ischemia (DCI) can occur in 30% of SAH patients (Roos *et al.*, 2000; Tholance, 2014) and often leads to irreversible cerebral infarcts, resulting in death or severe disability in about half of the affected patients (Vergouwen *et al.*, 2010). DCI is probably caused by multiple factors acting in synergy, such as cerebral vasospasm and microvascular thrombosis (Vergouwen *et al.*, 2008), spreading depolarizations and spreading ischemia (Dreier *et al.*, 2009), impaired autoregulation of cerebral vascular tone (Otite *et al.*, 2014), and activation of inflammatory pathways in the brain (McMahon *et al.*, 2013).

Because secondary injuries can have devastating consequences, their prevention is of major importance. To this end, physicians need reliable biomarkers of brain function in comatose patients who are unable to communicate with clinical staff. Understanding the state of the patient's brain and its evolution after injury relies heavily on monitoring techniques that provide continuous physiological measurements through a set of sensors permanently fitted on the brain. To improve the accuracy of diagnosis, several physical/chemical parameters are recorded in these patients, providing multimodal monitoring over several days.

16.2 Multimodal neurological monitoring in severely brain-injured patients

Upon admission to a neurological care unit, the patient undergoes standard monitoring of vital signs and peripheral markers, such as heart rate, mean arterial pressure, respiratory rate, body temperature, blood gases, and metabolites. In severely brain injured patients, who are often comatose and cannot communicate with caregivers, the state of the brain is monitored through multiple parameters including intracranial pressure, electrophysiological activity, brain hemodynamics, oxygenation, and metabolism (Figure 16.1). Secondary complications can develop

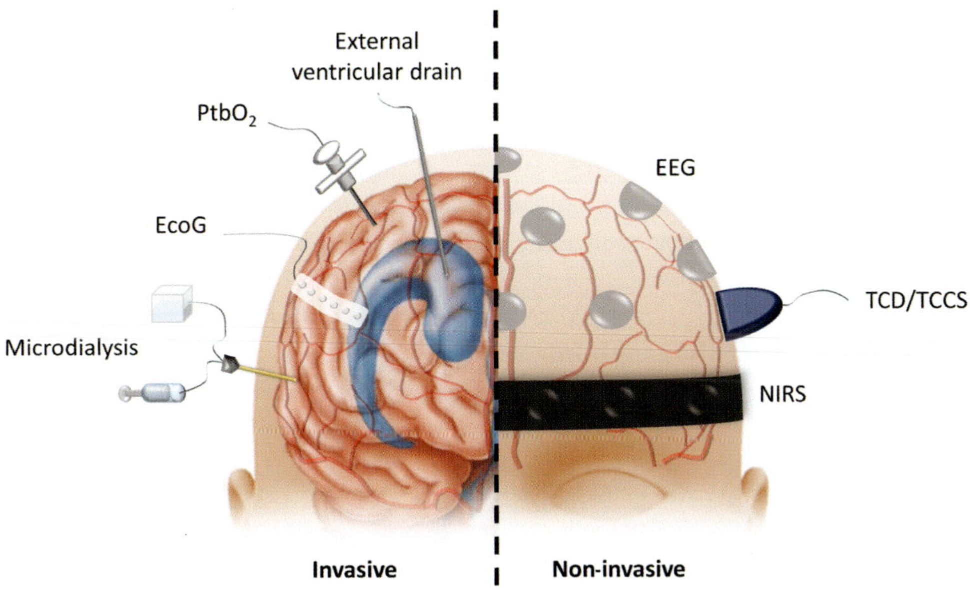

Figure 16.1 Schematic representation of multimodal monitoring techniques in neurointensive care. Different invasive (left) and non-invasive (right) tools can measure physiological parameters in the brain simultaneously. An external ventricular drain consists of a catheter inserted into the ventricles (blue) to extract excess cerebrospinal fluid in case of hydrocephaly and is often equipped with an intracranial pressure (ICP) sensor. Brain tissue oxygen partial pressure ($PbtO_2$) probes are implanted into the white matter. Near-infrared spectroscopy (NIRS) is a non-invasive tool used to estimate oxygen saturation in blood vessels near the skull. Cerebral blood flow (CBF) can be measured non-invasively using transcranial Doppler flowmetry coupled to color-coded duplex sonography (TCD/TCCS). Brain electrical activity can be detected via electroencephalography (EEG) and/or strip electrodes for electrocorticography (ECoG). Microdialysis probes are implanted in the parenchyma to monitor extracellular fluid chemistry (e.g. glucose, lactate, pyruvate concentrations).

rapidly and unexpectedly, and multimodal monitoring guarantees that the patient's brain is under constant surveillance so that appropriate measures can be taken in a timely manner (Roh and Park, 2016). Moreover, the state of the specific brain area in which the probes are placed can be a good indicator of the whole brain, especially after diffuse injury (Bouzat *et al.*, 2015).

16.2.1 Intracranial pressure

Intracranial pressure is of paramount importance since it determines cerebral perfusion pressure (CPP), defined as the mean arterial pressure minus intracranial pressure. As intracranial pressure increases, most often due to brain swelling and increases in blood or cerebrospinal fluid volume, CPP decreases and can lead to cerebral infarction when CBF falls below 15–20 ml/100 g/min (Botteri *et al.*, 2008). Typical physiological intracranial pressure values range between 3 and 15 mmHg with a pathological threshold usually set at 20 mmHg (Bratton *et al.*, 2007). Moreover, clinical outcome correlates well with the total duration of high intracranial pressure episodes (Guiza *et al.*, 2015). The gold standard technique for monitoring intracranial pressure is the external ventricular drain, a small pressure sensor mounted on a catheter implanted in the lateral ventricle. This system allows the physician to drain excess interstitial fluid and relieve intracranial pressure whenever it reaches a pathological level. In patients who do not need ventricular drainage, intraparenchymal microsensors (fiber-optic devices, strain gauge devices, or pneumatic sensors) are implanted. Depending on the mechanism of intracranial hypertension, several therapeutic strategies allow physicians to improve CPP, such as controlled hypertension, sedation, mild hypothermia, or decompressive craniotomy (Ragland and Lee, 2016). Intracranial pressure monitoring is part of the guidelines for the management of severe TBI and remains the subject of clinical research throughout the world (Chesnut *et al.*, 2012; Sarrafzadeh *et al.*, 2014).

16.2.2 Brain tissue oxygen

The brain is critically dependent on a continuous supply of oxygen from the blood. Oxygen is transported either dissolved in the serum or bound to haemoglobin in erythrocytes. The amount of oxygen available for brain cells is directly proportional to its partial pressure in the tissue (brain tissue oxygen partial pressure or $PbtO_2$). The value of $PbtO_2$ at any given point results from highly variable local factors such as distance from nearest vessels, intravascular PO_2, local O_2 diffusion, and consumption rates. Hence, at a micrometric scale, physiological oxygen tension varies within a broad range of 6 to 30 mmHg as determined by two-photon microscopy in mice (Sakadzic *et al.*, 2010). In the human brain, $PbtO_2$ monitoring is achieved by implantable optical or electrochemical probes. Optical methods usually consist of an optical fiber terminated by a ruthenium dye that changes its fluorescence intensity in response to changes in $PbtO_2$ (Griffiths and Robinson, 1999). Alternatively, electrochemical methods rely on oxygen reduction by a platinum or carbon electrode, similar to the "Clark electrode" developed in the 1960s (Clark and Lyons, 1962; Bolger and Lowry, 2005). Given the large size of these probes (200 to 800 μm in diameter), $PbtO_2$ measurements in the ICU are a mix of capillary, intra- and extracellular oxygen concentrations. $PbtO_2$ probes are usually implanted in white matter where the physiological $PbtO_2$ ranges between 20 and 35 mmHg (De Georgia, 2015; Okonkwo *et al.*, 2017), and pathological thresholds indicative of brain hypoxia range between 10 and 20 mmHg. $PbtO_2$ below 15 mmHg has been associated with brain ischemia in patients with SAH (Hinzman *et al.*, 2010; Chesnut *et al.*, 2012) and $PbtO_2$ monitoring is recommended in TBI patients at risk for ischemic complications (Le Roux *et al.*, 2014).

A more recent non-invasive technique called near-infrared spectroscopy (NIRS) presents an elegant alternative to invasive probes. NIRS measures regional cerebral oxygen saturation (rSO_2) through the attenuation of a near-infrared light source emitted from the scalp. Because NIRS measures the saturation of blood haemoglobin with oxygen, it provides an indirect estimate of $PbtO_2$ in the cortical grey matter close to the scalp. Typically, values of rSO_2 <60% are considered pathological and probably result in a hypoxic brain (Davies *et al.*, 2015).

A decrease in $PbtO_2$ reflects a mismatch between brain oxygen supply and consumption. The first strategy usually adopted when $PbtO_2$ is below 20 mmHg includes CPP optimisation, a decrease in oxygen consumption (i.e. treatment of seizure or fever) and, eventually, an increase in oxygen delivery via an increased fraction of inspired oxygen and/or red blood cell transfusion (De Georgia, 2015; Okonkwo *et al.*, 2017). Overall, the combination of intracranial pressure and $PbtO_2$ monitoring represents a promising approach to improve long-term outcomes in acutely brain-injured patients (Stiefel *et al.*, 2005).

16.2.3 Brain hemodynamics

After acute brain injury, CBF often displays a heterogeneous distribution because of ischemic stroke, oedema, contusions, or other lesions. It is common practice to distinguish the ischemic core, in which brain perfusion is impaired and brain tissue is injured beyond recovery, from well-perfused remote areas. The penumbra corresponds to ischemic areas surrounding the ischemic core, in which brain tissue is still viable for rescue (Cunningham *et al.*, 2005). In TBI, there is increasing evidence for the existence of a "traumatic penumbra" of tissue most at risk of secondary ischemic neuronal injury surrounding primary lesions (Botteri *et al.*, 2008). Therefore, assessment and monitoring of CBF is critical for both brain energy supply optimization and detection of new ischemic insults such as DCI after SAH (Stein *et al.*, 2006; Pluta *et al.*, 2009) or small vessel occlusion after traumatic injury (Vergouwen *et al.*, 2008).

Cerebral blood flow can be measured by a variety of imaging techniques. For example, computed tomography (CT) perfusion imaging reveals the penetration of inhaled Xenon into the brain parenchyma and provides useful information on regional CBF. Alternatively, angiography can visualize major blood vessels after administration of a contrast agent visible under X-ray radiation. Magnetic resonance imaging (MRI) can also measure cerebral perfusion by detecting brain penetration of Gadolinium injected intravenously. All these brain imaging techniques provide essential information about CBF but, unfortunately, they can only offer a snapshot of brain perfusion at a single time and cannot be repeated at short intervals (Wintermark *et al.*, 2005). These imaging techniques are therefore performed when continuous multimodal monitoring suggests a decrease in CBF.

Large vessels such as the middle cerebral artery or the internal carotid artery can be monitored at the bedside non-invasively using transcranial Doppler flowmetry coupled with color-coded duplex sonography (Proust *et al.*, 1999). This method measures the velocity of erythrocytes in large blood vessels; cerebral vasospasm presents as an increase in mean velocity. More specifically, an increased velocity in intracranial arteries coupled with normal velocity in the internal carotid is indicative of cerebral vasospasm. However, this technique cannot assess small vessels such as penetrating arterioles or capillaries, which would be more relevant to secondary injuries or DCI. Alternatively, parenchymal thermal diffusion probes can estimate local CBF from the dissipation of heat between two thermistors embedded in the brain parenchyma (Vajkoczy *et al.*, 2000).

Monitoring cerebral perfusion can be instrumental in the detection of new ischemic injuries in brain injured patients. However, consensus has not yet been reached regarding the pathological threshold values that could trigger medical intervention, and confirmation by brain imaging techniques is often considered mandatory in order to initiate therapy (Le Roux *et al.*, 2014).

16.2.4 Brain electrophysiological activity

The electrical signals generated by neurons during synaptic transmission and action potential propagation can be detected from the scalp by electroencephalography (EEG). EEG is mostly used for diagnostic and prognostic purposes (Korbakis and Vespa, 2017), but in some cases may also play a preventative role.

Frequency analysis of EEG signals can reveal a decrease in CBF before neuronal injury occurs. In particular, brain hypoperfusion is associated with an enrichment in low EEG frequencies that can be detected through Fourier analysis (Jordan, 2004; Balanca *et al.*, 2018). Moreover, EEG can detect epileptic seizures, a common neurological complication in brain injured patients, and trigger the timely administration of anti-epileptic drugs (Kurtz *et al.*, 2009). In addition to the detection of cerebral vasospasm and epileptic seizures, EEG can also reveal a disruption of sleep architecture that has been correlated to poor outcome in high-grade SAH patients (Claassen *et al.*, 2006). Specific patterns in EEG signals monitored in TBI patients can also reveal thalamic injury associated with poor outcomes (Hebb *et al.*, 2007).

An interesting development in electrophysiological brain monitoring has been the introduction of subdural strip electrodes, which monitor brain electrical activity from under the dura mater (a method called electrocorticography or ECoG), coupled to direct current (DC) recording. Monitoring ultra-low frequency electrophysiological events using DC ECoG has allowed the detection of cortical spreading depolarizations, waves of massive neuronal and glial depolarization that travel through the cortex at a speed of 2 to 4 mm/min (Dreier *et al.*, 2012). Spreading depolarizations are frequently detected in severely brain injured patients and are associated with poor outcome (Hartings *et al.*, 2011b). They are one of the mechanisms underlying cytotoxic oedema and occur whenever the cortical matter is injured (Dreier *et al.*, 2018). Therefore, their detection in comatose, brain-injured patients can help direct therapeutic intervention and detect the onset of new secondary lesion.

16.2.5 Brain energy metabolism

Glucose is by far the most important energy substrate of the brain. Glucose consumption by the brain accounts for almost 20% of the total glucose metabolism in the human body (Mergenthaler *et al.*, 2013). Once taken up by neurons, glucose is quickly metabolized into glucose-6-phosphate and pyruvate. Then, pyruvate enters the tricarboxylic acid cycle (also called the Krebs cycle) in the presence of oxygen which produces 30–34 molecules of adenosine-triphosphate (ATP) or undergoes non-oxidative glycolysis to produce lactate and only two molecules of ATP.

Analysing the chemical composition of the brain interstitial fluid and monitoring the concentrations of the major energy metabolites glucose, lactate and pyruvate over time can reveal critical information on the physiological/pathological state of the brain. For example, a healthy, well-oxygenated brain typically displays glucose extracellular concentrations in the 1.5–2.5 mM range, lactate in the 2.5–3.5 mM range, and pyruvate around 0.15 mM (Reinstrup *et al.*, 2000; Schulz *et al.*, 2000). During periods of high neuronal activity, such as neuronal stimulation or cortical spreading depolarization, glucose demand increases and is usually partly compensated by a hyperemic response to increase glucose supply to the brain. Consequently, glucose extracellular concentration decreases, while pyruvate and lactate are produced. This situation of high cellular metabolism is called hypermetabolism. It is characterized by low glucose extracellular concentrations, and high lactate and pyruvate concentrations, and is usually associated with favourable outcome. In contrast, during hypoxia or ischemia, oxidative metabolism is impaired, and pyruvate is quickly degraded into lactate through non-oxidative glycolysis. Because this pathway produces less ATP than oxidative glycolysis, glucose consumption increases to meet the energetic demand of the brain tissue. This situation also occurs in cases of mitochondrial dysfunction, is referred to as metabolic crisis, and is associated with poor outcome. To help distinguish between these different brain metabolic states, the ratio of lactate over pyruvate concentrations (L/P) has been introduced (Vespa *et al.*, 2005) to reflect the level of oxidative glycolysis, which is itself a good indicator of the health of the brain tissue. L/P values above 40 are usually considered pathological and may indicate secondary neurological complications in brain injured patients (Stahl *et al.*, 2001).

The global metabolic state of the whole brain can be estimated by analysing the glucose, lactate, and oxygen concentrations in jugular blood and comparing them to arterial concentrations (Barcelos *et al.*, 2013). However, the most precise method is probably cerebral microdialysis. This method consists of implanting an intracerebral probe terminated by a semi-permeable dialysis catheter allowing the diffusion of small molecules through the probe. Artificial cerebrospinal fluid is circulated through the system and equilibrates with the endogenous interstitial medium before being collected and analysed by chromatography and/or mass spectrometry (Chefer *et al.*, 2009; Rogers and Boutelle, 2013; Hutchinson *et al.*, 2014). This technique is widely implemented in pre-clinical animal research, but also frequently used for the bedside monitoring of brain injured patients in neurological intensive care units (Carpenter *et al.*, 2017). Intracerebral microdialysis allows hourly estimates of glucose, lactate, and pyruvate at the bedside, and can also provide continuous measurements with the help of microfluidic devices and online chemical sensors (Rogers and Boutelle, 2013; Booth *et al.*, 2018). The major drawback of local metabolic monitoring is that brain metabolism can vary widely from one region to another. For example, microdialysis probes located away from areas subjected to DCI are unable to detect the metabolic changes characteristic of ischemia. This problem can be circumvented in SAH patients in whom the location of infarcted brain areas can be predicted statistically based on the location of the initial bleeding (Tholance *et al.*, 2017). In addition, intracerebral microdialysis suffers from the large size of the probes implanted into the brain, with external diameters ranging from 210 to 500 μm (Venton *et al.*, 2006; Lietsche *et al.*, 2015). Microdialysis sampling is a rather invasive technique that can damage the blood-brain barrier and tissue surrounding the probe (Clapp-Lilly *et al.*, 1999; Bungay *et al.*, 2003; Borland *et al.*, 2005). Moreover, microdialysis probes likely induce inflammatory reactions in their vicinity, as revealed by the presence of increased interleukin-1 and -6 concentrations in the dialysates (Woodroofe *et al.*, 1991). Overall, the presence of compromised tissue and inflammation around the probe could significantly alter the local concentrations of energy metabolites measured by microdialysis.

Microelectrode biosensors could present a promising avenue of research to provide miniaturized probes, minimize tissue damage, and perform more accurate measurements. This technology is not yet approved for clinical use in patients but has already been implemented in pre-clinical research involving animal models of brain injury.

16.3 Microelectrode biosensors

According to the International Union of Pure and Applied Chemistry (IUPAC) an electrochemical biosensor is a self-contained integrated device, capable of providing specific quantitative or semi-quantitative analytical information using a biological recognition element (biochemical receptor) which is retained in direct spatial contact with an electrochemical transduction element (Thévenot *et al.*, 2001). For brain interstitial fluid analysis, biosensors are implanted within the brain parenchyma and their size cannot exceed a few hundreds of micrometers, justifying the term "microelectrode biosensor." The range of biological elements potentially available for biosensing is theoretically very large, and includes enzymes, antibodies, nucleic acid sequences and aptamers, molecularly imprinted polymers, and even whole cells or biological tissues (Thévenot *et al.*, 2001; Hamdan and Mohd Zain, 2014; Perumal and Hashim, 2014). However, brain monitoring is a specific case of analysis that requires continuous surveillance of vital signs and biochemical parameters over time. Single use electrochemical biosensors using antibodies, nucleic acid sequences, molecularly imprinted polymers, or protein receptors to achieve one-time recognition of a molecule are therefore poorly suited for monitoring applications. The only type of electrochemical biosensors currently used for continuous brain monitoring are enzyme-based microelectrode biosensors.

16.3.1 Principles of enzyme-based electrochemical biosensors

Enzyme-based microelectrode biosensors rely on the immobilization of an oxidoreductase enzyme on a microelectrode. These enzymes catalyze redox reactions of the type:

$$A^- + B \rightarrow A + B^- \quad (1)$$

A is the electron donor (oxidizing agent) and B is the electron acceptor (reducing agent). The basic principle of enzyme-based electrochemical biosensors is therefore to shuttle some of the electrons involved in the redox reaction catalyzed by the enzyme to the electrode, thereby translating the concentration of the enzyme substrate into a measurable electrical current. Potentially hundreds of natural oxidoreductases are available for biosensor construction. However, only oxidases have been successfully immobilized on microelectrodes for brain monitoring so far. There are 143 natural enzymes which have been classified as oxidases (*http://www.brenda-enzymes.org*). They use oxygen as an electron acceptor and catalyze redox reactions of the type:

$$\text{Substrate} + O_2 \rightarrow \text{Oxidized product} + H_2O_2 \text{ or } H_2O \quad (2)$$

Reactions catalyzed by oxidases usually involve a cofactor, embedded within the protein backbone, which introduces an intermediate step. The most frequent cofactor is flavin adenine dinucleotide (FAD). Glucose oxidase, glutamate oxidase, choline oxidase, and D-amino acid oxidase, all widely used for microelectrode biosensor preparation, use FAD as their cofactor (Keilin and Hartree, 1952; Sato *et al.*, 1979; Pollegioni *et al.*, 1992; Arima *et al.*, 2003). A few enzymes use other cofactors, such as l-lactate oxidase, which contains flavin mononucleotide (FMN) (Yorita *et al.*, 1997) and xanthine oxidase, which binds two FAD molecules, two molybdenum atoms, and eight iron atoms (Enroth *et al.*, 2000). Consequently, the redox reaction catalyzed by oxidases is better described by a two-step reaction:

$$\text{Substrate} + \text{Enzyme-FAD} \rightarrow \text{Oxidized product} + \text{Enzyme-FADH}_2 \quad (3)$$

$$\text{Enzyme-FADH}_2 + O_2 \rightarrow \text{Enzyme-FAD} + H_2O_2 \quad (4)$$

There are three strategies to transfer electrons to an electrode, corresponding to the so-called first, second, and third generation biosensors. In first generation biosensors, H_2O_2 is reoxidized into H_2O and O_2 at the electrode surface, with or without electron mediators. Second generation biosensors transfer electrons from the reduced cofactor (usually $FADH_2$) to the electrode by means of electron-conductive polymers that make them independent of ambient O_2. Finally, third generation biosensors provide direct electron transfer from the enzyme cofactor to the electrode by means of metal nanoparticles or carbon nanotubes (Borgmann *et al.*, 2012; Chatard *et al.*, 2018).

16.3.2 First generation biosensors

Hydrogen peroxide reoxidation into molecular oxygen follows this scheme:

$$H_2O_2 \rightarrow 2\,H^+ + O_2 + 2\,e^- \quad (5)$$

Platinum (Pt) is the most widely used electrode material to catalyze H_2O_2 reoxidation. The mechanism of this reaction involves a first step of H_2O_2 adsorption on the Pt surface through Pt(OH) or $Pt(OH)_2$ species that are quickly oxidized into O_2 and H^+ (Hall *et al.*, 1999; Katsounaros *et al.*, 2012). The electrons released in the process generate an oxidation current that can be detected and measured to provide an estimate of the concentration of H_2O_2 produced by the enzyme and, hence, of the enzyme substrate in the ambient medium (Figure 16.2A).

H_2O_2 oxidation on Pt occurs at holding potentials above 200–300 mV *versus* Ag/AgCl (Vasylieva and Marinesco, 2013). Other noble metals like ruthenium or rhodium can also be employed. However, carbon fibers, which are widely used to make microelectrodes for brain dopamine detection, require higher holding potentials, of about 800 mV *versus* Ag/AgCl or more, to oxidize H_2O_2. Carbon fiber microelectrodes can nonetheless be used for fast-scan cyclic voltammetric detection of H_2O_2 in the brain (Sanford *et al.*, 2010; Meunier *et al.*, 2018).

A major limitation of first generation biosensors is their oxygen dependence. In the brain, PO_2 is on average between 10 and 40 mmHg, which represents about 50–150 μM of dissolved O_2 in the interstitial fluid. This relatively low concentration means that if the concentration of enzyme substrate is in the 10^{-4}–10^{-3} M range, as is the case for brain glucose or lactate, the biosensor response will be limited by ambient oxygen. A common solution to this problem is to cover the enzymatic membrane with an outer layer of polyurethane that reduces the diffusion of enzyme substrate to the biosensor. Consequently, the enzyme is exposed to lower substrate concentrations than those present in the ambient medium, so that O_2 is no longer limiting. Using this strategy, one can extend the linear range of the biosensor (the concentration range in which the biosensor's response is linear with concentration) by a factor of three or four, to reach 5–10 mM and allow accurate glucose or lactate detection (Yu *et al.*, 2006; Vasylieva *et al.*, 2015). Another interesting consequence is that the rate of the reaction at the biosensor is limited by mass transfer through the polyurethane layer, rather than by the enzyme kinetics. Consequently, the biosensor becomes less dependent on temperature, because the

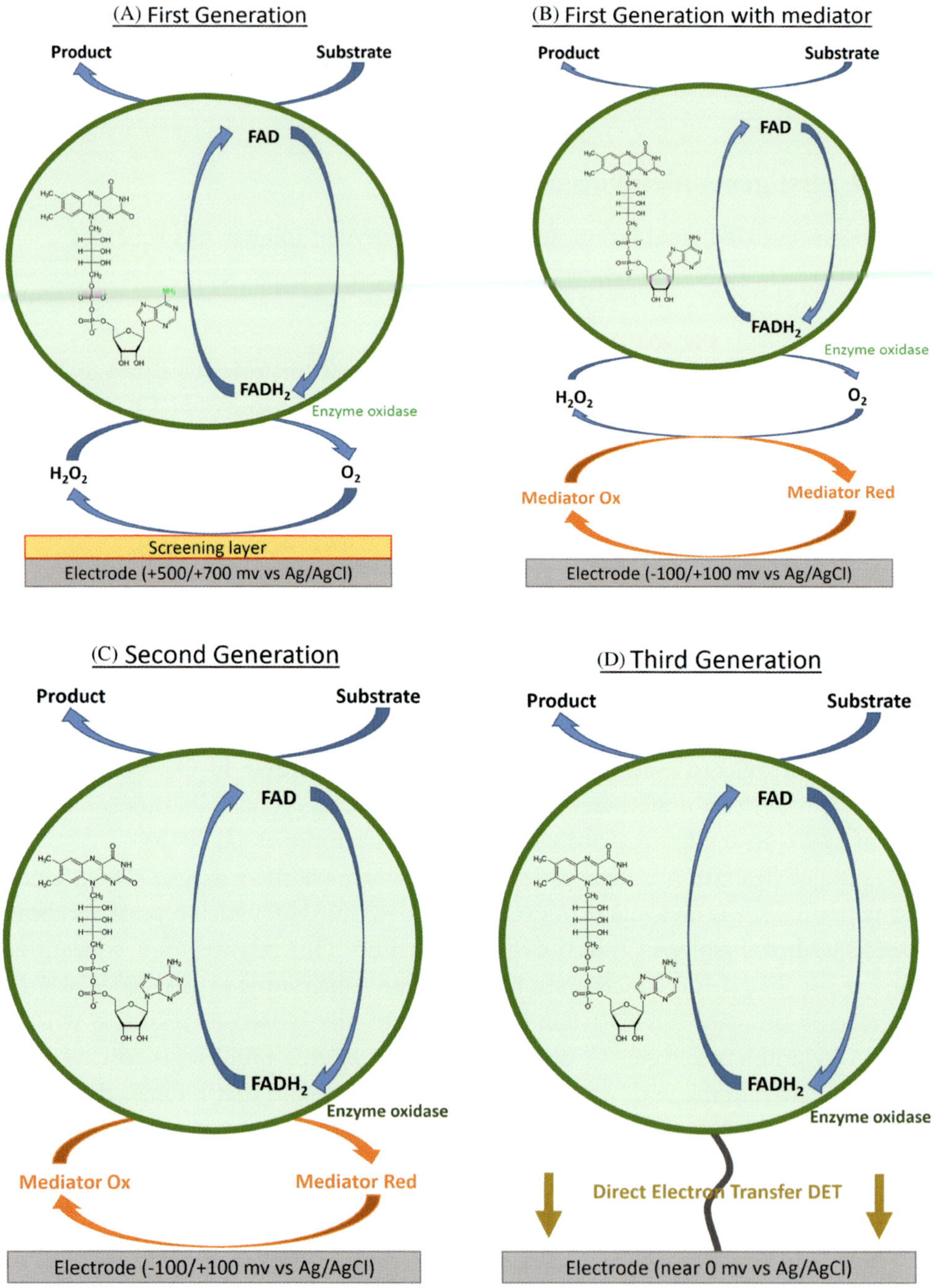

Figure 16.2 (*Continued*)

mass transfer process is less temperature dependent (about 2–3%/°C), than enzyme kinetics (about 8–10%/°C) (Bindra *et al.*, 1991; Wilson and Zhang, 2009). However, first generation microelectrode biosensors remain poorly suited to monitor the brain in situations of low ambient O_2 such as ischemia or hypoxia.

Another limitation of first generation biosensors is selectivity. To oxidize H_2O_2 optimally, microelectrode biosensors are polarized at 500 to 700 mV *versus* Ag/AgCl. However, many endogenous molecules present in the brain can also oxidize in this potential range. Antioxidant molecules like ascorbic acid or uric acid oxidize at 100–200 mV and 300 mV *versus* Ag/AgCl, respectively, and are present at high concentrations (up to 500 μM for ascorbic acid) (Miele and Fillenz, 1996; Euser *et al.*, 2009; Chatard *et al.*, 2018). In addition, neurotransmitters such as dopamine and serotonin and their metabolites (3,4-dihydroxyphenylacetic acid [DOPAC] and 5-hydroxyindoleacetic acid [5-HIAA], respectively) can also oxidize at carbon or Pt microelectrodes below 500 mV *versus* Ag/AgCl, and create non-specific signals related to their rapidly changing extracellular concentrations. There are two major strategies to overcome this problem: (1) a polymeric screening layer that allows H_2O_2 diffusion and blocks interfering molecules, or (2) chemical mediators that facilitate H_2O_2 oxidation and lowers its oxidation below that of interfering molecules.

16.3.2.1 Screening layers

Because H_2O_2 is a very small molecule compared to most endogenous oxidizable metabolites and neurotransmitters present *in vivo*, it is possible to use selective membranes that allow the diffusion of low molecular weight species while blocking larger molecules. Poly-phenylenediamine or poly-pyrrole can be readily electropolymerized on a microelectrode to create a steric barrier blocking the diffusion of molecules larger than 200 Da while allowing H_2O_2 to reach the electrode

Figure 16.2 (Figure on facing page) Schematic representation of electron transfer mechanisms in three generations of biosensors. Biosensors are based on the oxidation of a molecule of interest (substrate) by an oxidase enzyme that produces H_2O_2. (A) In first generation biosensors, H_2O_2 diffuses to the electrode surface and oxidation occurs with a holding potential ranging from +500 mV to +700 mV *versus* Ag/AgCl. A screening layer deposited on the electrode surface blocks interfering endogenous redox molecules such as ascorbic acid, uric acid, dopamine, or serotonin. (B) In some first generation biosensors an electron mediator lowers the oxidation potential of H_2O_2 to −100 mV to +100 mV *versus* Ag/AgCl in order to avoid electrochemical interference from endogenous redox molecules. (C) Second generation biosensors incorporate an electron mediator that exhibits higher affinity for the enzyme cofactor compared with O_2. Oxygen is therefore bypassed and the biosensor is independent of O_2. (D) In third generation biosensors, direct electron transfer (DET) is possible between the enzyme cofactor and the electrode surface. We should note that in second and third generation biosensors, dehydrogenases can also be used in addition to oxidases.

surface (Malitesta *et al.*, 1990; Quinto *et al.*, 2000; O'Neill *et al.*, 2004; Schuvailo *et al.*, 2006). Nafion, a perfluoro-sulfonated polymer, is also widely used as a screening layer on microelectrode biosensors. It is a permselective membrane that allows the diffusion of cations but not anions. It is efficient at blocking the non-specific detection of negatively charged molecules such as ascorbate and urate anions, as well as DOPAC and 5-HIAA. However, it will not prevent detection of cations such as dopamine and serotonin and might even increase the sensitivity of the biosensor for these molecules (Gerhardt *et al.*, 1984).

16.3.2.2 Electrochemical mediators

Another way of improving the selectivity of microelectrode biosensors is to lower their holding potential below that of oxidizable molecules present in the brain. One way to achieve this is to incorporate an electrochemical mediator to the enzyme layer. The oxidized form of the mediator can catalyze H_2O_2 oxidation while its reduced form can be reoxidized on the electrode surface at a lower potential than H_2O_2 (Figure 16.2B). For example, when horseradish peroxidase is added to the enzyme layer it can catalyze H_2O_2 oxidation into H_2O, and be reoxidized at the electrode surface at 0 to 100 mV *versus* Ag/AgCl (Belay *et al.*, 1999; Lindgren *et al.*, 2000; Zhu *et al.*, 2007). Hexaferrocyanates such as prussian blue or ruthenium purple can also mediate H_2O_2 oxidation and lower the biosensor holding potential to –50 mV (Karyakin *et al.*, 1995, 2000; Tian and Dale, 2013). An adenosine-triphosphate microelectrode biosensor incorporating ruthenium purple in the enzyme layer has been used in brain slices (Tian *et al.*, 2007).

16.3.3 Second generation biosensors

Second generation biosensors are those in which the electron mediator achieves a higher rate of electron transfer with the enzyme cofactor than with O_2 (Figure 16.2C). Thus, reoxidation of the enzyme cofactor by ambient O_2 is bypassed and the biosensor is independent of brain PO_2 (Wang, 2008). This is especially important for monitoring the injured brain in situations of hypoxia or ischemia. Mediators such as ferrocene derivatives or poly(vinylimidazole), and poly (vinylpyridine) can achieve this goal (Frew and Hill, 1988), but such freely-diffusing mediators are at risk of leaking out of the enzyme layer. It is possible, however, to covalently tether the mediator to the enzyme layer by means of a polymeric mesh immobilized on the microelectrode. One of the early second generation biosensor designs involves poly(1-vinylimidazole), $[Os(bpy)2Cl]^+$ ions, and glucose oxidase cross-linked with poly(ethylene glycol) diglycidyl ether (PEGDE)

(Ohara *et al.*, 1993, 1994). Similar devices have been developed for the monitoring of brain glutamate (Kulagina *et al.*, 1999; Oldenziel and Westerink, 2005).

It is possible to completely eliminate the need for O_2 or an electron mediator by directly connecting the redox active center of the enzyme to the electrode, a process called direct electron transfer (DET, Figure 16.2D). Such devices have already been proposed, although never for *in vivo* implantation (Willner *et al.*, 1996; Bollella *et al.*, 2018). To achieve DET in a biosensor, a thorough understanding of the mechanisms of electron transfer within the enzyme is necessary. For example, Lee *et al.* (2018) achieved DET between a mutated FAD-dependent glucose dehydrogenase and an electrode through a self-assembled monolayer, and this was only possible through specific mutations in the enzyme, creating an electron transfer subunit, and tight control on the thickness of the monolayer. These advances in molecular design and bioengineering may lead someday to the development of implantable third generation microelectrode biosensors.

16.4 Microelectrode designs

During microelectrode biosensor fabrication, the material of the underlying electrode must be carefully chosen. First generation biosensors that rely on H_2O_2 reoxidation on the electrode surface must use Pt (or possibly ruthenium or rhodium) to catalyze this reaction; Pt is preferred as it is inert and non-toxic. When an electron mediator is incorporated into the enzyme layer, a wider variety of materials can be considered. For example, ruthenium purple biosensors for the detection of ATP or hypoxanthine in the brain were fabricated from gold microelectrodes (Tian *et al.*, 2007). Carbon is poorly suited for H_2O_2 detection because of the large overpotential required for its oxidation (usually between 0.8 and 1.2 *versus* Ag/AgCl), but carbon fibers are commercially available as extremely small fibers (5–10 μm external diameter) which are easy to manipulate by hand. Carbon fiber microelectrodes have been proposed for first generation biosensors using fast-scan cyclic voltammetry (Sanford *et al.*, 2010; Lugo-Morales *et al.*, 2013; Smith *et al.*, 2018), as well as in redox hydrogel-coated glutamate biosensors (Kulagina *et al.*, 1999; Oldenziel and Westerink, 2005; Oldenziel *et al.*, 2006b, a).

Notwithstanding these differences in electrode material, an important distinction in microelectrode biosensor design lies in the number of analytes that can be detected simultaneously by the device.

16.4.1 Single channel devices

The first microelectrodes designed for brain implantation in the 1950s consisted of a metallic wire in a glass capillary (Figures 16.3A and 16.3B). This design still

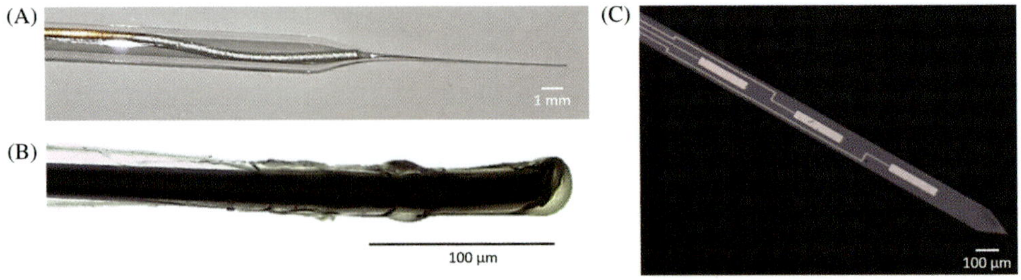

Figure 16.3 Examples of a single- and multi-channel biosensors. (A) and (B) Single enzymatic electrochemical biosensor is composed of a 25 μm diameter platinum-iridium wire extending 100 μm outside a pulled glass capillary. The biosensor tip is biofunctionalized with polyphenylenediamine and enzyme layers. Scale bars: 1 mm and 100 μm, respectively. (C) Electron microscopy image of a silicon/SU8 multichannel electrode as developed by Vasylieva *et al.* (2015). Scale bar: 100 μm. (Reproduced with permission from Vasylieva *et al.*, 2015.)

provides the thinnest microelectrodes to date, with external diameters ranging from 7 to 100 μm. It allows probe implantation in deep brain structures with minimal tissue damage (Gonon *et al.*, 1980; Netchiporouk *et al.*, 1995; Pernot *et al.*, 2008). Early glucose and lactate biosensors developed by Hu and Wilson (1997) included a 178 μm diameter Pt/Iridium (Ir) wire sealed into a glass micropipette and covered with glucose or lactate oxidase (Hu and Wilson, 1997). This microelectrode design is still in use today (Naylor *et al.*, 2012). More recent microelectrode designs have used small Pt/Ir or gold wires as well as carbon fibers to obtain even smaller devices in the 7–25 μm range (Kawagoe *et al.*, 1991; Tian *et al.*, 2007; Pernot *et al.*, 2008; Lugo-Morales *et al.*, 2013). Alternatively, commercially available teflon-coated Pt/Ir wires have also been proposed for use in enzyme biosensors (Lowry and O'Neill, 1994; Ryan *et al.*, 1997; McMahon and O'Neill, 2005). Including enzyme and polymer layers the outer diameter of such single-channel devices is in the 40–50 μm range. Unfortunately, such devices often suffer from poor reproducibility and high production costs due to their being hand-made. In addition, they can only detect a single analyte at a time. Many neuroscience applications demanding the detection of several analytes therefore require the implantation of several devices side by side, which can be a significant limitation in small laboratory animals.

16.4.2 Multichannel electrodes

Multichannel electrodes were developed with the advent of microfabrication and photolithographic techniques in the 1970s (Wise *et al.*, 1970; Pochay *et al.*, 1979). It is now possible to obtain multielectrode probes that minimize tissue damage through the use of flexible biocompatible materials coupled with small dimensions

(Weltin *et al.*, 2016). The microfabrication procedure usually involves (1) an initial step of metal deposition on a flat supporting material such as silicon, glass or ceramic, followed by (2) patterning with photolithography and lift-off techniques to define the number of electrodes, as well as their shape and connecting lines, and (3) etching of the substrate to obtain a needle-shaped device that is suitable for brain implantation (Figure 16.3C) (Frey *et al.*, 2010; Vasylieva *et al.*, 2015; Weltin *et al.*, 2016).

The first *in vivo* multielectrode biosensors, based on a ceramic substrate, were used for multisite dopamine (Sreenivas *et al.*, 1996) and glutamate (Burmeister *et al.*, 2000) detection in the brain. Silicon wafers have also been widely used for glutamate, lactate, and glucose detection (Wassum *et al.*, 2008; Vasylieva *et al.*, 2013). Such devices are larger than single channel microelectrodes embedded in glass micropipettes. Recent years have seen flexible substrates introduced to the field of *in vivo* brain biosensors. For example, glucose biosensors based on polyimide (Weltin *et al.*, 2014) or parylene substrates (Welch *et al.*, 2015) provide high flexibility and accommodate small brain movements in animals. This presents a promising avenue of research to improve the biocompatibility of brain implantable microelectrode biosensors.

Typical size of ceramic or silicon needles is at least $50 \times 100\ \mu m^2$, much larger than the 40 μm diameter achievable using glass micropipettes. The tissue lesions created upon brain implantation are therefore potentially larger with multichannel microelectrodes. However, such devices are necessary for monitoring several molecules simultaneously, for example glucose and lactate (Vasylieva *et al.*, 2015; Lourenco *et al.*, 2017), choline and glutamate (Frey *et al.*, 2007, 2010), or choline and acetylcholine (Burmeister *et al.*, 2008). In addition, for neurotransmitters and metabolites present at low concentrations in the interstitial fluid, oxidation currents resulting from enzymatic activity are very low and other non-specific currents resulting from pH changes, local field potentials, or endogenous oxidizable molecules could interfere with selective detection of the molecule of interest. This is the case, for example, for glutamate and D-serine, two amino acids typically detected at concentrations between 1 and 3 μM (Pernot *et al.*, 2012; Maucler *et al.*, 2013; Vasylieva *et al.*, 2013). To perform these measurements as accurately as possible it is necessary to use a second control electrode covered with bovine serum albumin (BSA) instead of an enzyme, in order to measure these non-specific currents. Non-specific currents are then subtracted from the signal detected by the enzyme electrode to measure precisely the enzymatic currents and compute the concentration of the enzyme substrate around the electrode. This procedure is referred to as self-referencing (Burmeister and Gerhardt, 2001). This is often accomplished by implanting two microelectrodes side by side, but multielectrode devices offer the opportunity to perform this complex task with only one implantation.

One important issue arising with multielectrode devices is the possibility of overlap of the diffusion layers of each electrode. Consequently, one electrode can decrease the concentration of oxygen (or enzyme substrate) available for the next electrode, or instead, release H_2O_2 outside the enzyme layer that could reach neighboring recording sites. However, such electrode cross-talk can be avoided by creating a recess within the ceramic or silicon needle, and by spacing adjacent microelectrodes by 200 μm (Vasylieva *et al.*, 2015).

16.5 Enzyme immobilization

A second important aspect of enzyme biosensor technology is immobilization of the enzyme on the microelectrode. Some biosensor systems incorporate more than one enzyme to allow the recognition of a molecule for which no specific oxidase is available. In this case, the reaction taking place on the microelectrode is:

$$\text{Substrate} \xrightarrow{\textit{Enzyme 1}} \text{Product 1} \xrightarrow{\textit{Oxidase}} \text{Product 2} + H_2O_2 \qquad (6)$$

For example, there is no specific oxidase for acetylcholine (Ach) recognition, but it is possible to mix acetylcholine esterase with choline oxidase to obtain an acetylcholine biosensor through the following reaction:

$$\text{Ach} \xrightarrow{\textit{Ach esterase}} \text{Choline} \xrightarrow{\textit{Choline oxidase}} \text{Betaine aldehyde} + H_2O_2 \qquad (7)$$

Similarly, ATP can be detected using a two-enzyme mix:

$$\text{ATP} + \text{glycerol} \xrightarrow{\textit{Glycerol kinase}} \text{glycerol-3-phosphate} \qquad (8)$$

$$O_2 + \text{glycerol-3-phosphate} \xrightarrow{\textit{Glycerol-3-phosphate oxidase}} \text{glycerone-phosphate} + H_2O_2 \qquad (9)$$

and adenosine, with a mixture of three enzymes:

$$\text{Adenosine} \xrightarrow{\textit{Adenosine deaminase}} \text{Inosine} \xrightarrow{\textit{Nucleoside phosphorylase}} \text{Hypoxanthine} \qquad (10)$$

$$O_2 + \text{Hypoxanthine} \xrightarrow{\textit{xanthine oxidase}} \text{uric acid} + H_2O_2 \qquad (11)$$

A wide variety of techniques can be used to achieve enzyme immobilization on a microelectrode while preserving enzymatic activity, rapid response time over a wide linear range, and providing stability for at least several hours of monitoring *in vivo*. These techniques can be divided into (1) cross-linking methods involving covalent bonds, and (2) entrapment into polymer matrices (Vasylieva and Marinesco, 2013).

16.5.1 Cross-linking

Cross-linking is a process by which macromolecules are covalently bound, providing stable long-term binding to the microelectrode. It is usually achieved by bifunctional agents that react with amine and carboxylic groups on the surface of proteins, creating a ramified network. Glutaraldehyde (pentane-1,5-dialdehyde) is the most widely used cross-linking agent in biosensor technology. Although it is a small, toxic molecule, it provides stable immobilization a reaction time of a few minutes (Burmeister and Gerhardt, 2001; Pernot *et al.*, 2008; Wassum *et al.*, 2008). Another bifunctional molecule, PEGDE, is also frequently used in microelectrode biosensors. It was first proposed by Heller *et al.* for use in implantable glucose biosensors for diabetic patients, but was later adapted for brain monitoring (Gregg and Heller, 1991; Kulagina *et al.*, 1999; Heller and Feldman, 2008; Vasylieva *et al.*, 2011; von Bornstadt *et al.*, 2015). A non-toxic cross-linker bearing two epoxy groups on each end of a poly(ethylene glycol) chain, PEGDE reacts preferentially with amine groups on proteins. The reaction process, however, is slower than that of glutaraldehyde, and requires several hours with moderate heating (Vasylieva *et al.*, 2011). Many redox hydrogels used in second generation biosensors incorporate PEGDE as a cross-linker in the electron-conducting matrix. One of the first successful polymers of this type employed a polymer backbone of polyvinyl pyridine and metal-organic complexes of $Os^{2+/3+}$ with substituted and unsubstituted pyridines cross-linked with PEGDE to achieve electron transfer between the electrode and the enzyme (Gregg and Heller, 1991; Ohara *et al.*, 1993; Heller, 2006).

16.5.2 Enzyme entrapment

In enzyme entrapment, enzyme molecules are incorporated in a polymer layer by mixing them with the monomer solution during polymerization. Enzymes are embedded in the polymer via electrostatic interactions or simply adsorbed through weak chemical bonds that preserve enzyme activity (Vasylieva and Marinesco, 2013; Chatard *et al.*, 2018). Sol-gel matrices are especially suited for providing a porous environment capable of entrapping enzymes at a microelectrode surface. This method involves the hydrolysis of alkoxysilanes into reactive products that can polymerize on the electrode surface. Such reactions are often facilitated at alkaline pH. Negatively charged electrodes can trigger the reduction of H^+ to H_2

or nitrate ions to nitrite, thereby locally increasing pH around the electrode and triggering sol-gel polymerization (Shacham *et al.*, 1999; Tian *et al.*, 2009). This is a useful property for locally addressing enzymes to specific microelectrodes on multi-channel devices. Sol-gel-derived, silica-based materials are well-suited for such enzyme entrapment (Rozhanchuk *et al.*, 2009; Tian *et al.*, 2009).

Enzymes can also be entrapped during electropolymerization of a matrix such as polypyrrole or poly-phenylenediamine. Pyrrole-derived molecules can readily polymerize at a potential allowing the generation of pyrrole-cation radicals that react with neighboring monomers (usually around –700 mV *versus* Ag/AgCl) (Foulds and Lowe, 1986; Cosnier, 1999; Llaudet *et al.*, 2003). Phenylenediamine, although highly carcinogenic, can also be readily polymerized into poly-phenylenediamine to incorporate enzymes such as glucose oxidase, glutamate oxidase, or D-amino acid oxidase (Lowry *et al.*, 1997; McMahon and O'Neill, 2005; Zain *et al.*, 2010).

16.5.3 Effect of enzyme immobilization on substrate specificity

Biosensor selectivity is dependent on the specificity of the enzyme immobilized on the microelectrode. Very often, the oxidases used in brain monitoring experiments are either very selective for one substrate, like glutamate oxidase, or can accommodate other molecules as secondary substrates that are not present *in vivo*. For example, glucose oxidase also displays low activity towards 2-deoxy glucose, mannose and lactose, but the concentrations of these secondary substrates *in vivo* are not sufficient to interfere significantly with glucose detection (Vasylieva and Marinesco, 2013). D-amino acid oxidase can degrade all neutral and basic amino acids, but only D-serine is present *in vivo* in detectable concentration (Pernot *et al.*, 2008). However, the immobilization procedure can significantly impact enzyme-substrate specificity. Glutaraldehyde, in particular, often results in decreased enzyme specificity. It can increase mannose and lactose detection by glucose oxidase, or make glutamate oxidase capable of oxidizing asparagine, aspartate, and glutamine. This can have profound effects on quantitative concentration estimates provided by microelectrode biosensors *in vivo* (Vasylieva and Marinesco, 2013). This effect probably results from the small size (less than 1 nm long) of the glutaraldehyde molecule, which could constrain enzyme movements and affect its three dimensional structure during cross-linking. In contrast the larger molecule PEGDE (over 4 nm) does not impact enzyme substrate specificity. Moreover, the effects of glutaraldehyde on enzyme substrate specificity are attenuated when it is incubated in a solution of BSA prior to enzyme immobilization (Burmeister *et al.*, 2013). It is possible that under these conditions, enzyme immobilization is achieved by glutaraldehyde molecules already bound to BSA, thereby providing a longer chain for cross-linking enzyme molecules. Entrapment techniques do not appear to modify enzyme specificity (Vasylieva and Marinesco, 2013).

16.6 Microelectrode biosensors for monitoring brain injury

Biosensors are promising tools for monitoring the brain after acute injury. They meet most of the ASSURED criteria proposed by the World Health Organization to select most the efficient diagnostic methods. According to these criteria, a diagnostic method should be **A**ffordable, **S**ensitive, **S**pecific, **U**ser-friendly, **R**apid and robust, **E**quipment-free and **D**eliverable to end-users (Peeling *et al.*, 2006). Microelectrode biosensors have the potential to fulfill these criteria, not only in pre-clinical research laboratories but, perhaps, in brain-injured patients in the future. In addition to their high sensitivity and specificity, they also offer low-cost and compact equipment compared, for example, to microdialysis which requires offline analysis with chromatography or mass spectrometry. They also provide real-time analysis, which can be instrumental for rapidly informing therapy in time-sensitive cases such as brain injury.

So far, microelectrode biosensors have only been implemented in animal models. Pathological situations that have been monitored with these devices include epilepsy, middle cerebral artery occlusion, traumatic brain injury, and cortical spreading depolarizations.

16.6.1 Epileptic seizures

Epilepsy is a chronic neurological disorder characterized by recurrent seizures during which a group of brain cells displays excessive hypersynchronized activity, resulting in involuntary movements ranging from isolated muscle jerks to full body convulsions and, sometimes, loss of consciousness (http://www.who.int/newsroom/fact-sheets/detail/epilepsy). A severe case of epilepsy is called *status epilepticus*, a situation of continuous intractable seizure "lasting more than 5 min or two or more seizures between which there is incomplete recovery of consciousness" (Lowenstein *et al.*, 1999). *Status epilepticus* induces neuronal injury and causes major neuronal loss particularly in the limbic system (Cavalheiro, 1995; Trinka *et al.*, 2012; Tyler *et al.*, 2012; Woeffler-Maucler *et al.*, 2014).

Microelectrode biosensors have helped understand the role of free radicals released during epileptic seizures in these neuronal injuries. Moon *et al.* focused on nitric oxide (NO) release in the rat cortex during pharmacologically-evoked seizures (Moon *et al.*, 2016). They designed an elegant dual-sensor microelectrode capable of measuring potassium and NO simultaneously. They detected a large NO release peaking around 0.2 μM. Because NO is a highly reactive nitrogen species, this concentration could contribute to neuronal injury induced by prolonged seizures (Moon *et al.*, 2016).

Neurochemical studies of shorter, pharmacologically-evoked seizures have also been performed using microelectrode biosensors. The high temporal resolution of these devices, on the order of a few seconds, is well suited for monitoring short epileptic seizures lasting only 30–60 s. Morales-Villagran *et al.* monitored glutamate extracellular concentration in the hippocampus of rats during epileptic seizures induced by the drug 4-aminopyridine (Morales-Villagran *et al.*, 2008). They detected an increase in glutamate level of about 5 μM within a few seconds of seizure onset, which supports a role for glutamate in seizure initiation. Evidence for increases in glutamate during epileptic seizures using microdialysis has been published by other teams in animals (Ueda and Tsuru, 1994; Kaura *et al.*, 1995; Khan *et al.*, 1999) and humans (During and Spencer, 1993). However, the temporal resolution of microdialysis is usually on the order of 2 to 5 min, which makes this technique ill-suited for analysis of neurotransmitter levels during seizures. Using capillary electrophoresis coupled to laser-induced fluorescence (CE-LIF), Parrot *et al.* achieved 20 s temporal resolution and detected a doubling of gamma-amino butyric acid (GABA) levels during seizures compared to inter-ictal periods (Parrot *et al.*, 2004). However, this method could not determine whether the increase occurred at the onset or the termination of the seizure. Therefore, the second-by-second temporal resolution achieved by microelectrode biosensors represents a major improvement in the field of epilepsy research.

This methodological improvement was particularly evident in the study of Van Gompel *et al.* (2014), in which adenosine extracellular concentration was monitored in pig and human cortex during epileptic seizures. Simultaneous ECoG and electrochemical recording revealed that adenosine levels rose during seizures to reach a peak concentration 260% of baseline (649 ± 513 nM), 2 to 8 s before seizure termination. This study confirmed that adenosine may play a prominent role in seizure termination and extended previous microdialysis studies showing an increase in adenosine levels within minutes of epileptic seizure. The high temporal resolution allowed by electrochemical techniques and microelectrode biosensors allowed researchers to tightly correlate increases in adenosine to the last seconds of epileptic seizure, which represented a major advance in our understanding of epilepsy (Van Gompel *et al.*, 2014).

Overall, microelectrode biosensors provide an important methodological breakthrough to investigate the neuronal mechanisms of epileptic seizures on a second-by-second basis.

16.6.2 Ischemia and stroke models

Brain ischemia results from a critical reduction in CBF that reduces the supply of oxygen and energy metabolites below the threshold necessary for cell survival. It is a recognized and common cause of irreversible brain damage (Raichle, 1983). Brain ischemia can be global, for example during cardiac arrest, or local, such as when a blood vessel is occluded in stroke patients. Ischemia can also be induced by SAH, when a major blood vessel bleeds, as in cases of ruptured aneurysm. Several animal models are available to study the mechanisms of ischemic brain injury *in vivo*. One of the most widely used models is occlusion of the middle cerebral artery (MCAO), a major blood vessel that provides blood supply to the whole anterior hemisphere. Five minutes of MCAO is sufficient to induce infarction and severe neurological deficits in rats but the precise mechanisms responsible for neuronal injury remain to be understood (Bederson *et al.*, 1986).

As in epilepsy research, the role of free radicals and reactive oxygen and nitrogen species in infarct formation is a subject of intense research. Malinski and colleagues succeeded in implanting an NO sensor (carbon fiber microelectrodes covered with a metallo-porphyrin and a screening layer of Nafion) in the brain of rats subjected to 2 h MCAO. They could monitor NO levels during the ischemic phase as well as during reperfusion and detected a large NO release peaking around 3 μM during ischemia, but not reperfusion (Malinski *et al.*, 1993). In a brain slice model, Ganesana monitored the formation of superoxide using biosensors based on gold wires covered with cytochrome *c*. When the slice was deprived of oxygen and glucose, superoxide production was doubled (Ganesana *et al.*, 2012). Overall, these studies indicate that free radicals are indeed released in response to ischemia and that they are likely to induce oxidative stress in neuronal cells and to contribute to infarct formation in the ischemic brain.

16.6.3 Traumatic brain injury (TBI)

TBI encompasses all alterations in brain function caused by an external force (Menon *et al.*, 2010). With an average incidence of about 262/100,000 inhabitants per year in Europe, TBI often affects young individuals and represents the primary cause of disability below age 45 (Tagliaferri *et al.*, 2006; Peeters *et al.*, 2015). The economic cost of TBI is immense, but remains difficult to estimate because it involves both direct costs of treatment, rehabilitation, and long-term assistance, but also the much higher indirect costs of productivity loss for patients, relatives, and society (Langlois *et al.*, 2006; Kayani *et al.*, 2009). In addition, there is often a long delay between the initial insult and the onset of neurological and cognitive deficits, even after mild or moderate TBI (Peeters *et al.*, 2015).

From a neurobiological point of view, TBI produces an initial insult via tissue compression or shearing. These primary lesions are later followed by a sequence of secondary complications that take place in the hours to days following the trauma. Such complications include marked hypoperfusion in surrounding areas of the brain, sometimes referred to as traumatic penumbra (von Oettingen *et al.*, 2002). In a recent study, Balança *et al.* induced moderate and severe TBI to rats using a

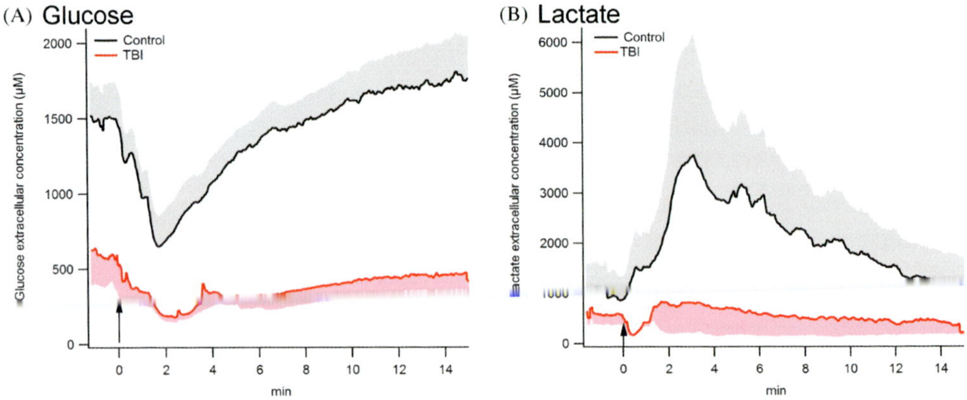

Figure 16.4 Changes in glucose (A) and lactate extracellular concentrations (B) in response to cortical spreading depolarizations evoked by KCl in control rats or spontaneously after severe traumatic brain injury (TBI). In TBI rats, brain metabolic disturbances are characterized by a decreased glucose basal concentration and a lack of lactate release evoked by spreading depolarization. Data are expressed as mean ± SEM (shading).

lateral fluid percussion model, in which a shockwave is applied to the dura through a small craniotomy. They observed a reproducible penumbra area encompassing the temporal cortex, area CA3 of the hippocampus, and the dorsal thalamus (Balança *et al.*, 2016). Using implanted enzyme microelectrode biosensors in penumbral region to monitor glucose and lactate extracellular concentration, as well as $PbtO_2$, they detected a persistently decreased glucose concentration that accompanied a drop in local brain perfusion (Figure 16.4A).

Glutamate microelectrode biosensors have also been implanted in the rat brain after mild or moderate TBI. Although glutamate basal levels were unchanged in frontal cortex, they were almost doubled in hippocampus and striatum two days after TBI (Hinzman *et al.*, 2010). Moreover, the increase in tonic glutamate in striatum depended on glial glutamate release as well as impaired glutamate reuptake (Hinzman *et al.*, 2012). These changes in glutamate neurotransmission after TBI could contribute to ongoing development of excitotoxic insult and neurological lesions over time.

16.6.4 Spreading depolarizations

Spreading depolarizations (SDs) represent an interesting feature of brain injury. They are waves of abrupt, sustained, near-complete breakdown of cellular transmembrane ion gradients that travel through the cortex at a speed of 2–4 mm/min (Somjen, 2001; Dreier, 2011; Lauritzen *et al.*, 2011; Balança *et al.*, 2017). SDs induce a massive, transient depolarization of brain cells and are observed in 56% of patients with severe TBI (Fabricius *et al.*, 2006; Hartings *et al.*, 2011b), 60–70%

of patients with intracerebral hemorrhage (Lauritzen *et al.*, 2011), 70–80% of patients with poor-grade aneurysmal SAH (Dreier *et al.*, 2006, 2012), and nearly all patients with malignant hemispheric stroke (Dohmen *et al.*, 2008; Woitzik *et al.*, 2013). In neurointensive care, SDs are considered to be markers of severe injury and poor prognosis (Dreier *et al.*, 2006; Hartings *et al.*, 2011a; Dreier and Reiffurth, 2015), correlated with cytotoxic oedema and new injuries (Dreier *et al.*, 2018).

The biochemical mechanisms responsible for SD induction in the injured brain are largely unknown. It is possible that SDs result from acute increases in neuronal excitability and/or a mismatch between high energy demand and low supply, such as in the peri-infarct zones in stroke models (Dreier *et al.*, 2002; von Bornstadt *et al.*, 2015). It is thought that SDs can also occur in healthy individuals, in which they are rather innocuous (e.g. during migraine aura), whereas they can aggravate neuronal injury in the already injured brain (Dreier and Reiffurth, 2015). For example, in healthy individuals, SDs induce a wave of hyperemia through neurovascular coupling mechanisms. It is hypothesized that in brain injured patients, neurovascular coupling mechanisms are impaired and thus SDs induce a spreading ischemia that can aggravate brain injury (Dreier, 2011).

SDs have been studied using microelectrode biosensors in both stroke and TBI models. Hinzman *et al.* implanted glutamate microelectrode biosensors in the ischemic core and penumbra in a rat MCAO model. They detected a transient increase in glutamate extracellular concentration for each spontaneous SD. Glutamate peaked around 6–12 μM and the increase lasted about 2–3 min, significantly longer than in healthy animals. Such prolonged elevation in glutamate extracellular level might underlie the mechanism for excitotoxic damage in the penumbral zone of stroke (Hinzman *et al.*, 2015).

Using a similar approach, Balança *et al.* (2017) studied the metabolic profile of spontaneous SDs observed a few hours after severe TBI in rats. They detected a hypermetabolic response to SDs in healthy rats, characterized by a drop in glucose extracellular concentration and a large increase in lactate levels. Interestingly, this hypermetabolic pattern was not observed in rats with severe TBI, in which lactate failed to increase in response to SD (Figure 16.4). The loss of this lactate response to SDs might reflect an early metabolic disturbance in brain injured rats that could be partially responsible for the delayed development of neurological lesions (Balança *et al.*, 2017).

The neurochemical monitoring of SDs has also provided an opportunity to compare the *in vivo* performance of microelectrode biosensors to that of microdialysis probes. Microdialysis probes are usually much larger than microelectrode biosensors. Their diameter ranges between 200 and 500 μm whereas microelectrode biosensors can be as small as 40–50 μm in diameter (Chatard *et al.*, 2018). Consequently, brain microdialysis probes are typically surrounded by compromised tissue, as evidenced by electron microscopic studies that observed neuronal cell death up to 400 μm away from the probe, and tissue disruption including swollen axons and degenerating cell bodies up to 1.2 mm (Clapp-Lilly *et al.*, 1999). To better understand the local effects of microdialysis probe implantation, Borland *et al.* detected dopamine release near a microdialysis probe with an external diameter of 220 μm with a carbon fiber microelectrode just 7 μm in diameter. Although the carbon fiber microelectrode detected a large dopamine release in response to neuronal stimulation in healthy tissue, dopamine release was no longer detected near the microdialysis probe (Borland *et al.*, 2005; Wintermark *et al.*, 2005). These data indicate that microdialysis probes can disrupt local neuronal transmission, probably in response to tissue lesions during implantation.

SD neurochemical monitoring confirms this notion. In healthy animals, microelectrode biosensors revealed a large hypermetabolic response to SDs during which glucose levels decreased by 70–80% and lactate increased by 60–100% (Balança *et al.*, 2017; Lourenco *et al.*, 2017). Microdialysis can also detect neurochemical changes in response to SDs, provided that it is performed at high temporal resolution (Rogers *et al.*, 2013; Varner *et al.*, 2017; Gowers *et al.*, 2018). Interestingly, the dynamic changes in glucose and lactate appear dampened when measured with this technique (Figure 16.5). The decrease in glucose levels was only 10–20% (Rogers *et al.*, 2013; Varner *et al.*, 2017), while the increase in lactate extracellular

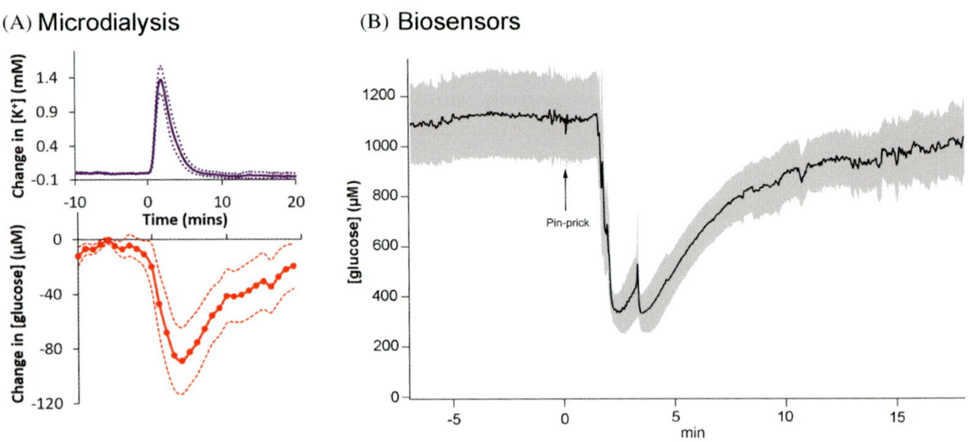

Figure 16.5 Changes in glucose extracellular concentration in response to a cortical spreading depolarization (SD). (A) Glucose changes monitored using continuous online microdialysis enhanced by local infusion of dexamethasone (reproduced with permission from Varner *et al.*, 2017). Top trace represents potassium concentration changes; bottom trace represents glucose. (B) Glucose changes monitored with microelectrode biosensors. Data are expressed as mean ± SEM. Microdialysis detected a change in glucose concentration of about −100 μM, which represents a 30–35% decrease from baseline, whereas with microelectrode biosensors, glucose dropped from 1.1 mM to 0.3 mM (−73 %). This comparison suggests that microelectrode biosensors are more sensitive to rapid changes in concentrations, probably due to the absence of a layer of compromised tissue acting as a low-pass filter.

concentration was 30% (Rogers *et al.*, 2013). It should be noted, however, that in brain injured patients the drop in glucose extracellular level evoked by SDs can be much larger (Feuerstein *et al.*, 2010; Hutchinson *et al.*, 2014; Rogers *et al.*, 2017). This comparison supports the view that microdialysis probes measure from a local microenvironment of compromised brain tissue that acts as a low-pass filter on dynamic concentration changes spontaneously occurring in the brain. Because of their smaller size, microelectrode biosensors have the potential to diminish the extent of brain tissue lesion around the site of probe implantation and, therefore, to gather more sensitive and more reliable data to understand brain neurochemistry.

16.7 Conclusion

Microelectrode biosensors represent a promising methodology for improved neurochemical monitoring of the injured brain. Their main advantages are (1) their excellent second-by-second time resolution, and (2) their potential for miniaturization to diameters of a few micrometers. These advantages could open new opportunities for recording neurochemical changes related to neurological lesion development after injury with improved sensitivity and temporal resolution. Although these devices are not yet approved for clinical use, it is likely that they will be in the future and become part of multimodal monitoring technologies in brain-injured patients.

Acknowledgments

CA is the recipient of a doctoral fellowship by Région Auvergne-Rhône-Alpes. This study was supported by Fondation Gueules Cassées "Sourire Quand Même" (grants FGC49-2016 and FGC34-2018), CNRS mission à l'interdisciplinarité, and Agence Nationale pour la Recherche (PALSE project ANR-11-IDEX-0007-02). BB is the recipient of a CCA Inserm-Bettencourt fellowship. CA, AM, and SM are supported by Inserm U1028, CNRS UMR5292, and University of Lyon.

References

Arima J, Tamura T, Kusakabe H, Ashiuchi M, Yagi T, Tanaka H, Inagaki K (2003) Recombinant expression, biochemical characterization and stabilization through proteolysis of an L-glutamate oxidase from Streptomyces sp. X-119-6. *J Biochem*, 134: 805–812.

Balança B, Bapteste L, Lieutaud T, Ressnikoff D, Guy R, Bezin L, Marinesco S (2016) Neuronal loss as evidenced by automated quantification of neuronal density following moderate and severe traumatic brain injury in rats. *J Neurosci Res*, 94: 39–49.

Balança B, Dailler F, Boulogne S, Ritzenthaler T, Gobert F, Rheims S, Andre-Obadia N (2018) Diagnostic accuracy of quantitative EEG to detect delayed cerebral ischemia after subarachnoid hemorrhage: A preliminary study. *Clin Neurophysiol*, 129: 1926–1936.

Balança B, Meiller A, Bezin L, Dreier JP, Marinesco S, Lieutaud T (2017) Altered hyper-metabolic response to cortical spreading depolarizations after traumatic brain injury in rats. Journal of cerebral blood flow and metabolism. *J Cereb Blood Flow Metab*, 37: 1670–1686.

Barcelos GK, Tholance Y, Grousson S, Renaud B, Perret-Liaudet A, Dailler F, Zimmer L (2013) Outcome of poor grade subarachnoid hemorrhage as determined by biomarkers of glucose cerebral metabolism. *Neurocrit Care*, 18: 234–244.

Bederson JB, Pitts LH, Tsuji M, Nishimura MC, Davis RL, Bartkowski H (1986) Rat middle cerebral artery occlusion: Evaluation of the model and development of a neurologic examination. *Stroke*, 17: 472–476.

Belay A, Collins A, Ruzgas T, Kissinger PT, Gorton L, Csoregi E (1999) Redox hydrogel based bienzyme electrode for L-glutamate monitoring. *J Pharm Biomed Anal*, 19: 93–105.

Bilgin S, Guclu-Gunduz A, Oruckaptan H, Kose N, Celik B (2012) Gait and Glasgow Coma Scale scores can predict functional recovery in patients with traumatic brain injury. *Neural Regen Res*, 7: 1978–1984.

Bindra DS, Zhang Y, Wilson GS, Sternberg R, Thévenot DR, Moatti D, Reach G (1991) Design and *in vitro* studies of a needle-type glucose sensor for subcutaneous monitoring. *Anal Chem*, 63: 1692–1696.

Bolger FB, Lowry JP (2005) Brain tissue oxygen: *In vivo* monitoring with carbon paste electrodes. *Sensors*, 5: 473–487.

Bollella P, Gorton L, Antiochia R (2018) Direct electron transfer of dehydrogenases for development of 3rd Generation biosensors and enzymatic fuel cells. *Sensors*, 18: Pii:E1316.

Booth MA, Gowers SAN, Leong CL, Rogers ML, Samper IC, Wickham AP, Boutelle MG (2018) Chemical monitoring in clinical settings: Recent developments toward real-time chemical monitoring of patients. *Anal Chem*, 90: 2–18.

Borgmann S, Schulte A, Neugebauer S, Schuhmann W (2012) Amperometric Biosensors, In *Advances in Electrochemical Science and Engineering*, Vol. 13 (Alkire RC, Kolb DM, Lipkowski J, eds.)

Borland LM, Shi G, Yang H, Michael AC (2005) Voltammetric study of extracellular dopamine near microdialysis probes acutely implanted in the striatum of the anesthetized rat. *J Neurosci Methods*, 146: 149–158.

Botteri M, Bandera E, Minelli C, Latronico N (2008) Cerebral blood flow thresholds for cerebral ischemia in traumatic brain injury. A systematic review. *Crit Care Med*, 36: 3089–3092.

Bouzat P, Marques-Vidal P, Zerlauth JB, Sala N, Suys T, Schoettker P, Bloch J, Daniel RT, Levivier M, Meuli R, Oddo M (2015) Accuracy of brain multimodal monitoring to detect cerebral hypoperfusion after traumatic brain injury. *Crit Care Med*, 43: 445–452.

Bratton SL, Chestnut RM, Ghajar J, McConnell Hammond FF, Harris OA, Hartl R, Manley GT, Nemecek A, Newell DW, Rosenthal G, Schouten J, Shutter L, Timmons SD, Ullman JS, Videtta W, Wilberger JE, Wright DW (2007) Guidelines for the management of severe traumatic brain injury. VIII. Intracranial pressure thresholds. *J Neurotrauma*, 24 Suppl 1: S55–58.

Bungay PM, Newton-Vinson P, Isele W, Garris PA, Justice JB (2003) Microdialysis of dopamine interpreted with quantitative model incorporating probe implantation trauma. *J Neurochem*, 86: 932–946.

Burmeister JJ, Davis VA, Quintero JE, Pomerleau F, Huettl P, Gerhardt GA (2013) Glutaraldehyde cross-linked glutamate oxidase coated microelectrode arrays: Selectivity and resting levels of glutamate in the CNS. *ACS Chem Neurosci*, 4: 721–728.

Burmeister JJ, Gerhardt GA (2001) Self-referencing ceramic-based multisite microelectrodes for the detection and elimination of interferences from the measurement of L-glutamate and other analytes. *Anal Chem*, 73: 1037–1042.

Burmeister JJ, Moxon K, Gerhardt GA (2000) Ceramic-based multisite microelectrodes for electrochemical recordings. *Anal Chem*, 72: 187–192.

Burmeister JJ, Pomerleau F, Huettl P, Gash CR, Werner CE, Bruno JP, Gerhardt GA (2008) Ceramic-based multisite microelectrode arrays for simultaneous measures of choline and acetylcholine in CNS. *Biosens Bioelectron*, 23: 1382–1389.

Carpenter KL, Young AM, Hutchinson PJ (2017) Advanced monitoring in traumatic brain injury: Microdialysis. *Curr Op Crit Care*, 23: 103–109.

Cavalheiro EA (1995) The pilocarpine model of epilepsy. *Ital J Neurol Sci*, 16: 33–37.

Chatard C, Meiller A, Marinesco S (2018) Microelectrode biosensors for in vivo analysis of brain interstitial fluid. *Electroanalysis*, 30: 977–998.

Chefer VI, Thompson AC, Zapata A, Shippenberg TS (2001) Overview of brain microdialysis. *Curr Protoc Neurosci*, 47: 7.1.1–7.1.28.

Chesnut RM, Temkin N, Carney N, Dikmen S, Rondina C, Videtta W, Petroni G, Lujan S, Pridgeon J, Barber J, Machamer J, Chaddock K, Celix JM, Cherner M, Hendrix T (2012) Global neurotrauma research. A trial of intracranial-pressure monitoring in traumatic brain injury. *New Engl J Med*, 367: 2471–2481.

Claassen J, Hirsch LJ, Frontera JA, Fernandez A, Schmidt M, Kapinos G, Wittman J, Connolly ES, Emerson RG, Mayer SA (2006) Prognostic significance of continuous EEG monitoring in patients with poor-grade subarachnoid hemorrhage. *Neurocrit Care*, 4: 103–112.

Clapp-Lilly KL, Roberts RC, Duffy LK, Irons KP, Hu Y, Drew KL (1999) An ultrastructural analysis of tissue surrounding a microdialysis probe. *J Neurosci Methods*, 90: 129–142.

Clark LC, Lyons C (1962) Electrode systems for continuous monitoring in cardiovascular surgery. *Annal New York Acad Sci*, 102: 29–45.

Cosnier S (1999) Biomolecule immobilization on electrode surfaces by entrapment or attachment to electrochemically polymerized films. A review. *Biosens Bioelectron*, 14: 443–456.

Cunningham AS, Salvador R, Coles JP, Chatfield DA, Bradley PG, Johnston AJ, Steiner LA, Fryer TD, Aigbirhio FI, Smielewski P, Williams GB, Carpenter TA, Gillard JH, Pickard JD, Menon DK (2005) Physiological thresholds for irreversible tissue damage in contusional regions following traumatic brain injury. *Brain*, 128: 1931–1942.

Davies DJ, Su Z, Clancy MT, Lucas SJ, Dehghani H, Logan A, Belli A (2015) Near-infrared spectroscopy in the monitoring of adult traumatic brain injury: A review. *J Neurotrauma*, 32: 933–941.

De Georgia MA (2015) Brain tissue oxygen monitoring in neurocritical care. *J Intensive Care Med*, 30: 473–483.

Dohmen C, Sakowitz OW, Fabricius M, Bosche B, Reithmeier T, Ernestus RI, Brinker G, Dreier JP, Woitzik J, Strong AJ, Graf R, Co-Operative Study of Brain Injury D (2008) Spreading depolarizations occur in human ischemic stroke with high incidence. *Ann Neurol*, 63: 720–728.

Dreier JP (2011) The role of spreading depression, spreading depolarization and spreading ischemia in neurological disease. *Nature Med*, 17: 439–447.

Dreier JP, Kleeberg J, Petzold G, Priller J, Windmuller O, Orzechowski HD, Lindauer U, Heinemann U, Einhaupl KM, Dirnagl U (2002) Endothelin-1 potently induces Leao's cortical spreading depression in vivo in the rat: A model for an endothelial trigger of migrainous aura? *Brain*, 125: 102–112.

Dreier JP, Lemale CL, Kola V, Friedman A, Schoknecht K (2018) Spreading depolarization is not an epiphenomenon but the principal mechanism of the cytotoxic edema in various gray matter structures of the brain during stroke. *Neuropharm*, 134: 189–207.

Dreier JP, Major S, Manning A, Woitzik J, Drenckhahn C, Steinbrink J, Tolias C, Oliveira-Ferreira AI, Fabricius M, Hartings JA, Vajkoczy P, Lauritzen M, Dirnagl U, Bohner G, Strong AJ (2009) Cortical spreading ischaemia is a novel process involved in ischaemic damage in patients with aneurysmal subarachnoid haemorrhage. *Brain*, 132: 1866–1881.

Dreier JP, Major S, Pannek HW, Woitzik J, Scheel M, Wiesenthal D, Martus P, Winkler MK, Hartings JA, Fabricius M, Speckmann EJ, Gorji A, group Cs (2012) Spreading convulsions, spreading depolarization and epileptogenesis in human cerebral cortex. *Brain*, 135: 259–275.

Dreier JP, Reiffurth C (2015) The stroke-migraine depolarization continuum. *Neuron*, 86: 902–922.

Dreier JP, Woitzik J, Fabricius M, Bhatia R, Major S, Drenckhahn C, Lehmann TN, Sarrafzadeh A, Willumsen L, Hartings JA, Sakowitz OW, Seemann JH, Thieme A, Lauritzen M, Strong AJ (2006) Delayed ischaemic neurological deficits after subarachnoid haemorrhage are associated with clusters of spreading depolarizations. *Brain*, 129: 3224–3237.

During MJ, Spencer DD (1993) Extracellular hippocampal glutamate and spontaneous seizure in the conscious human brain. *Lancet*, 341: 1607–1610.

Enroth C, Eger BT, Okamoto K, Nishino T, Nishino T, Pai EF (2000) Crystal structures of bovine milk xanthine dehydrogenase and xanthine oxidase: Structure-based mechanism of conversion. *Proc Natl Acad Sci USA*, 97: 10723–10728.

Euser SM, Hofman A, Westendorp RG, Breteler MM (2009) Serum uric acid and cognitive function and dementia. *Brain*, 132: 377–382.

Fabricius M, Fuhr S, Bhatia R, Boutelle M, Hashemi P, Strong AJ, Lauritzen M (2006) Cortical spreading depression and peri-infarct depolarization in acutely injured human cerebral cortex. *Brain*, 129: 778–790.

Feuerstein D, Manning A, Hashemi P, Bhatia R, Fabricius M, Tolias C, Pahl C, Ervine M, Strong AJ, Boutelle MG (2010) Dynamic metabolic response to multiple spreading depolarizations in patients with acute brain injury: An online microdialysis study. *J Cereb Blood Flow Metab*, 30: 1343–1355.

Foulds NC, Lowe CR (1986) Enzyme entrapment in electrically conducting polymers — immobilization of glucose-oxidase in polypyrrole and its application in amperometric glucose sensors. *J Chem Soc Farad*, T 82: 1259–1264.

Frew JE, Hill HA (1988) Direct and indirect electron transfer between electrodes and redox proteins. *Eur J Biochem*, 172: 261–269.

Frey O, van der Wal P, de Rooij N, Koudelka-Hep M (2007) Development and characterization of choline and L-glutamate biosensor integrated on silicon microprobes for *in-vivo* monitoring. *Conf Proc IEEE Eng Med Biol Soc*, 2007: 6040–6043.

Frey O, Holtzman T, McNamara RM, Theobald DE, van der Wal PD, de Rooij NF, Dalley JW, Koudelka-Hep M (2010) Enzyme-based choline and L-glutamate biosensor electrodes on silicon microprobe arrays. *Biosens Bioelectron*, 26: 477–484.

Ganesana M, Erlichman JS, Andreescu S (2012) Real-time monitoring of superoxide accumulation and antioxidant activity in a brain slice model using an electrochemical cytochrome *c* biosensor. *Free Radic Biol Med*, 53: 2240–2249.

Gerhardt GA, Oke AF, Nagy G, Moghaddam B, Adams RN (1984) Nafion-coated electrodes with high selectivity for CNS electrochemistry. *Brain Res*, 290: 390–395.

Gonon F, Buda M, Cespuglio R, Jouvet M, Pujol JF (1980) *In vivo* electrochemical detection of catechols in the neostriatum of anaesthetized rats: Dopamine or DOPAC? *Nature*, 286: 902–904.

Gowers SAN, Hamaoui K, Cunnea P, Anastasova S, Curto VF, Vadgama P, Yang GZ, Papalois V, Drakakis EM, Fotopoulou C, Weber SG, Boutelle MG (2018) High temporal resolution delayed analysis of clinical microdialysate streams. *Analyst*, 143: 715–724.

Grasso G, Alafaci C, Macdonald RL (2017) Management of aneurysmal subarachnoid hemorrhage: State of the art and future perspectives. *Surg Neurol Int*, 8: 11.

Gregg BA, Heller A (1991) Redox polymer films containing enzymes. 1. A redox-conducting epoxy cement: Synthesis, characterization, and electrocatalytic oxidation of hydroquinone. *J Phys Chem*, 95: 5970–5975.

Griffiths JR, Robinson SP (1999) The OxyLite: A fibre-optic oxygen sensor. *Br J Radiol*, 72: 627–630.

Guiza F, Depreitere B, Piper I, Citerio G, Chambers I, Jones PA, Lo TY, Enblad P, Nillson P, Feyen B, Jorens P, Maas A, Schuhmann MU, Donald R, Moss L, Van den Berghe G, Meyfroidt G (2015) Visualizing the pressure and time burden of intracranial hypertension in adult and paediatric traumatic brain injury. *Intensive Care Med*, 41: 1067–1076.

Hall SB, Khudaish EA, Hart AL (1999) Electrochemical oxidation of hydrogen peroxide at platinum electrodes. Part III: Effect of temperature. *Electrochim Acta*, 44: 2455–2462.

Hamdan SK, Mohd Zain A (2014) *In vivo* electrochemical biosensor for brain glutamate detection: A mini review. *Malays J Med Sci*, 21: 12–26.

Hartings JA, Bullock MR, Okonkwo DO, Murray LS, Murray GD, Fabricius M, Maas AI, Woitzik J, Sakowitz O, Mathern B, Roozenbeek B, Lingsma H, Dreier JP, Puccio AM, Shutter LA, Pahl C, Strong AJ (2011b) Spreading depolarisations and outcome after traumatic brain injury: A prospective observational study. *Lancet Neurol*, 10: 1058–1064.

Hartings JA, Watanabe T, Bullock MR, Okonkwo DO, Fabricius M, Woitzik J, Dreier JP, Puccio A, Shutter LA, Pahl C, Strong AJ (2011a) Spreading depolarizations have prolonged direct current shifts and are associated with poor outcome in brain trauma. *Brain*, 134: 1529–1540.

Hebb MO, McArthur DL, Alger J, Etchepare M, Glenn TC, Bergsneider M, Martin N, Vespa PM (2007) Impaired percent alpha variability on continuous electroencephalography is associated with thalamic injury and predicts poor long-term outcome after human traumatic brain injury. *J Neurotrauma*, 24: 579–590.

Heller A (2006) Electron-conducting redox hydrogels: Design, characteristics and synthesis. *Curr Opin Chem Biol*, 10: 664–672.

Heller A, Feldman B (2008) Electrochemical glucose sensors and their applications in diabetes management. *Chem Rev*, 108: 2482–2505.

Hinzman JM, DiNapoli VA, Mahoney EJ, Gerhardt GA, Hartings JA (2015) Spreading depolarizations mediate excitotoxicity in the development of acute cortical lesions. *Exp Neurol*, 267: 243–253.

Hinzman JM, Thomas TC, Burmeister JJ, Quintero JE, Huettl P, Pomerleau F, Gerhardt GA, Lifshitz J (2010) Diffuse brain injury elevates tonic glutamate levels and potassium-evoked glutamate release in discrete brain regions at two days post-injury: An enzyme-based microelectrode array study. *J Neurotrauma*, 27: 889–899.

Hinzman JM, Thomas TC, Quintero JE, Gerhardt GA, Lifshitz J (2012) Disruptions in the regulation of extracellular glutamate by neurons and glia in the rat striatum two days after diffuse brain injury. *J Neurotrauma*, 29: 1197–1208.

Hu Y, Wilson GS (1997) A temporary local energy pool coupled to neuronal activity: Fluctuations of extracellular lactate levels in rat brain monitored with rapid-response enzyme-based sensor. *J Neurochem*, 69: 1484–1490.

Hu Y, Wilson GS (1997) Rapid changes in local extracellular rat brain glucose observed with an *in vivo* glucose sensor. *J Neurochem*, 68: 1745–1752.

Hutchinson P, O'Phelan K *et al.* (2014) International multidisciplinary consensus conference on multimodality monitoring: Cerebral metabolism. *Neurocrit Care*, 21 Suppl 2: S148–158.

Jordan KG (2004) Emergency EEG and continuous EEG monitoring in acute ischemic stroke. *J Clin Neurophysiol*, 21: 341–352.

Karyakin AA, Gitelmacher OV, Karyakina EE (1995) Prussian blue based first-generation biosensor — a sensitive amperometric electrode for glucose. *Anal Chem*, 67: 2419–2423.

Karyakin AA, Karyakina EE, Gorton L (2000) Amperometric biosensor for glutamate using Prussian blue-based "artificial peroxidase" as a transducer for hydrogen peroxide. *Anal Chem*, 72: 1720–1723.

Katsounaros I, Schneider WB, Meier JC, Benedikt U, Biedermann PU, Auer AA, Mayrhofer KJ (2012) Hydrogen peroxide electrochemistry on platinum: Towards understanding the oxygen reduction reaction mechanism. *Phys Chem Chem Phys*, 14: 7384–7391.

Kaura S, Bradford HF, Young AMJ, Croucher MJ, Hughes PD (1995) Effect of amygdaloid kindling on the content and release of amino acids from the amygdaloid complex: *In vivo* and *in vitro* studies. *J Neurochem*, 65: 1240–1249.

Kawagoe KT, Jankowski JA, Wightman RM (1991) Etched carbon-fiber electrodes as amperometric detectors of catecholamine secretion from isolated biological cells. *Anal Chem*, 63: 1589–1594.

Kayani NA, Homan S, Yun S, Zhu BP (2009) Health and economic burden of traumatic brain injury: Missouri, 2001–2005. *Public Health Rep*, 124: 551–560.

Keilin D, Hartree EF (1952) Specificity of glucose oxidase (notatin). *Biochem J*, 50: 331–341.

Khan GM, Smolders I, Lindekens H, Manil J, Ebinger G, Michotte Y (1999) Effects of diazepam on extracellular brain neurotransmitters in pilocarpine-induced seizures in rats. *Eur J Pharmacol*, 373: 153–161.

Korbakis G, Vespa PM (2017) Multimodal neurologic monitoring. *HandbClinical Neurol*, 140: 91–105.

Kulagina NV, Shankar L, Michael AC (1999) Monitoring glutamate and ascorbate in the extracellular space of brain tissue with electrochemical microsensors. *Anal Chem*, 71: 5093–5100.

Kurtz P, Hanafy KA, Claassen J (2009) Continuous EEG monitoring: Is it ready for prime time? *Curr Op Crit Care*, 15: 99–109.

Langlois JA, Rutland-Brown W, Wald MM (2006) The epidemiology and impact of traumatic brain injury: A brief overview. *J Head Trauma Rehab*, 21: 375–378.

Lauritzen M, Dreier JP, Fabricius M, Hartings JA, Graf R, Strong AJ (2011) Clinical relevance of cortical spreading depression in neurological disorders: Migraine, malignant stroke, subarachnoid and intracranial hemorrhage, and traumatic brain injury. *J Cereb Blood Flow Metab*, 31: 17–35.

Le Roux P *et al.* (2014) The International Multidisciplinary Consensus Conference on Multimodality Monitoring in Neurocritical Care: Evidentiary tables: A statement for healthcare professionals from the Neurocritical Care Society and the European Society of Intensive Care Medicine. *Neurocrit Care*, 21 Suppl 2: S297–361.

Lee I, Loew N, Tsugawa W, Lin CE, Probst D, La Belle JT, Sode K (2018) The electrochemical behavior of a FAD dependent glucose dehydrogenase with direct electron transfer subunit by immobilization on self-assembled monolayers. *Bioelectrochem*, 121: 1–6.

Lietsche J, Gorka J, Hardt S, Karas M, Klein J (2015) Custom-made microdialysis probe design. *JoVE*: e53048.

Lindgren A, Ruzgas T, Gorton L, Csoregi E, Ardila GB, Sakharov IY, Gazaryan IG (2000) Biosensors based on novel peroxidases with improved properties in direct and mediated electron transfer. *Biosens Bioelectron*, 15: 491–497.

Llaudet E, Botting NP, Crayston JA, Dale N (2003) A three-enzyme microelectrode sensor for detecting purine release from central nervous system. *Biosens Bioelectron*, 18: 43–52.

Lourenco CF, Ledo A, Gerhardt GA, Laranjinha J, Barbosa RM (2017) Neurometabolic and electrophysiological changes during cortical spreading depolarization: Multimodal approach based on a lactate-glucose dual microbiosensor arrays. *Sci Rep*, 7: 6764.

Lowenstein DH, Bleck T, Macdonald RL (1999) It's time to revise the definition of status epilepticus. *Epilepsia*, 40: 120–122.

Lowry JP, Boutelle MG, Fillenz M (1997) Measurement of brain tissue oxygen at a carbon past electrode can serve as an index of increases in regional cerebral blood flow. *J Neurosci Methods*, 71: 177–182.

Lowry JP, Oneill RD (1994) Partial characterization in-vitro of glucose oxidase-modified poly(phenylenediamine)-coated electrodes for neurochemical analysis *in-vivo*. *Electroanalysis*, 6: 369–379.

Lugo-Morales LZ, Loziuk PL, Corder AK, Toups JV, Roberts JG, McCaffrey KA, Sombers LA (2013) Enzyme-modified carbon-fiber microelectrode for the quantification of dynamic fluctuations of nonelectroactive analytes using fast-scan cyclic voltammetry. *Anal Chem*, 85: 8780–8786.

Malinski T, Bailey F, Zhang ZG, Chopp M (1993) Nitric oxide measured by a porphyrinic microsensor in rat brain after transient middle cerebral artery occlusion. *J Cereb Blood Flow Metab*, 13: 355–358.

Malitesta C, Palmisano F, Torsi L, Zambonin PG (1990) Glucose fast-response amperometric sensor based on glucose oxidase immobilized in an electropolymerized poly (o-phenylenediamine) film. *Anal Chem*, 62: 2735–2740.

Maucler C, Pernot P, Vasylieva N, Pollegioni L, Marinesco S (2013) In vivo D-serine hetero-exchange through alanine-serine-cysteine (ASC) transporters detected by microelectrode biosensors. *ACS Chem Neurosci*, 4: 772–781.

McMahon CJ, Hopkins S, Vail A, King AT, Smith D, Illingworth KJ, Clark S, Rothwell NJ, Tyrrell PJ (2013) Inflammation as a predictor for delayed cerebral ischemia after aneurysmal subarachnoid haemorrhage. *J Neurointerv Surg*, 5: 512–517.

McMahon CP, O'Neill RD (2005) Polymer-enzyme composite biosensor with high glutamate sensitivity and low oxygen dependence. *Anal Chem*, 77: 1196–1199.

Menon DK, Schwab K, Wright DW, Maas AI (2010) Position statement: Definition of traumatic brain injury. *Arch Phys Med and Rehabil*, 91: 1637–1640.

Mergenthaler P, Lindauer U, Dienel GA, Meisel A (2013) Sugar for the brain: The role of glucose in physiological and pathological brain function. *Trends Neurosci*, 36: 587–597.

Meunier CJ, Mitchell EC, Roberts JG, Toups JV, McCarty GS, Sombers LA (2018) Electrochemical selectivity achieved using a double voltammetric waveform and partial least squares regression: Differentiating endogenous hydrogen peroxide fluctuations from shifts in pH. *Anal Chem*, 90: 1767–1776.

Miele M, Fillenz M (1996) *In vivo* determination of extracellular brain ascorbate. *J Neurosci Methods*, 70: 15–19.

Moon J, Ha Y, Kim M, Sim J, Lee Y, Suh M (2016) Dual electrochemical microsensor for real-time simultaneous monitoring of nitric oxide and potassium ion changes in a rat brain during spontaneous neocortical epileptic seizure. *Anal Chem*, 88: 8942–8948.

Morales-Villagran A, Medina-Ceja L, Lopez-Perez SJ (2008) Simultaneous glutamate and EEG activity measurements during seizures in rat hippocampal region with the use of an electrochemical biosensor. *J Neurosci Methods*, 168: 48–53.

Naylor E, Aillon DV, Barrett BS, Wilson GS, Johnson DA, Harmon HP, Gabbert S, Petillo PA (2012) Lactate as a biomarker for sleep. *Sleep*, 35: 1209–1222.

Netchiporouk LI, Shulga AA, Jaffrezicrenault N, Martelet C, Olier R, Cespuglio R (1995) Properties of carbon-fiber microelectrodes as a basis for enzyme biosensors. *Anal Chim Acta*, 303: 275–283.

O'Neill RD, Chang SC, Lowry JP, McNeil CJ (2004) Comparisons of platinum, gold, palladium and glassy carbon as electrode materials in the design of biosensors for glutamate. *Biosens Bioelectron*, 19: 1521–1528.

Ohara TJ, Rajagopalan R, Heller A (1993) Glucose electrodes based on cross-linked $[Os(bpy)_2Cl]^{+/2+}$ complexed poly(1-vinylimidazole) films. *Anal Chem*, 65: 3512–3517.

Ohara TJ, Rajagopalan R, Heller A (1994) "Wired" enzyme electrodes for amperometric determination of glucose or lactate in the presence of interfering substances. *Anal Chem*, 66: 2451–2457.

Okonkwo DO, Shutter LA, Moore C, Temkin NR, Puccio AM, Madden CJ, Andaluz N, Chesnut RM, Bullock MR, Grant GA, McGregor J, Weaver M, Jallo J, LeRoux PD, Moberg D, Barber J, Lazaridis C, Diaz-Arrastia RR (2017) Brain oxygen optimization in severe traumatic brain injury Phase-II: A Phase II randomized trial. *Crit Care Med*, 45: 1907–1914.

Oldenziel WH, Dijkstra G, Cremers TI, Westerink BH (2006a) Evaluation of hydrogel-coated glutamate microsensors. *Anal Chem*, 78: 3366–3378.

Oldenziel WH, Dijkstra G, Cremers TI, Westerink BH (2006b) *In vivo* monitoring of extracellular glutamate in the brain with a microsensor. *Brain Res*, 1118: 34–42.

Oldenziel WH, Westerink BH (2005) Improving glutamate microsensors by optimizing the composition of the redox hydrogel. *Anal Chem*, 77: 5520–5528.

Otite F, Mink S, Tan CO, Puri A, Zamani AA, Mehregan A, Chou S, Orzell S, Purkayastha S, Du R, Sorond FA (2014) Impaired cerebral autoregulation is associated with vasospasm and delayed cerebral ischemia in subarachnoid hemorrhage. *Stroke*, 45: 677–682.

Parrot S, Sauvinet V, Riban V, Depaulis A, Renaud B, Denoroy L (2004) High temporal resolution for in vivo monitoring of neurotransmitters in awake epileptic rats using brain microdialysis and capillary electrophoresis with laser-induced fluorescence detection. *J Neurosci Methods*, 140: 29–38.

Peeling RW, Holmes KK, Mabey D, Ronald A (2006) Rapid tests for sexually transmitted infections (STIs): The way forward. *Sex Transm Infect*, 82 Suppl 5: v1–6.

Peeters W, van den Brande R, Polinder S, Brazinova A, Steyerberg EW, Lingsma HF, Maas AI (2015) Epidemiology of traumatic brain injury in Europe. *Acta Neurochirurgica*, 157: 1683–1696.

Pernot P, Maucler C, Tholance Y, Vasylieva N, Debilly G, Pollegioni L, Cespuglio R, Marinesco S (2012) d-Serine diffusion through the blood-brain barrier: Effect on d-serine compartmentalization and storage. *Neurochem Int*, 60: 837–845.

Pernot P, Mothet JP, Schuvailo O, Soldatkin A, Pollegioni L, Pilone M, Adeline MT, Cespuglio R, Marinesco S (2008) Characterization of a yeast D-amino acid oxidase microbiosensor for D-serine detection in the central nervous system. *Anal Chem*, 80: 1589–1597.

Perumal V, Hashim U (2014) Advances in biosensors: Principle, architecture and applications. *J Appl Biomed*, 12: 1–15.

Pluta RM, Hansen-Schwartz J, Dreier J, Vajkoczy P, Macdonald RL, Nishizawa S, Kasuya H, Wellman G, Keller E, Zauner A, Dorsch N, Clark J, Ono S, Kiris T, Leroux P, Zhang JH (2009) Cerebral vasospasm following subarachnoid hemorrhage: Time for a new world of thought. *Neurological Research*, 31: 151–158.

Pochay P, Wise KD, Allard LF, Rutledge LT (1979) A multichannel depth probe fabricated using electron-beam lithography. IEEE *Trans Biomed Eng*, 26: 199–206.

Pollegioni L, Falbo A, Pilone MS (1992) Specificity and kinetics of Rhodotorula gracilis D-amino acid oxidase. *Biochim Biophys Acta*, 1120: 11–16.

Proust F, Callonec F, Clavier E, Lestrat JP, Hannequin D, Thiebot J, Freger P (1999) Usefulness of transcranial color-coded sonography in the diagnosis of cerebral vasospasm. *Stroke*, 30: 1091–1098.

Quinto M, Losito I, Palmisano F, Zambonin CG (2000) Needle-type glucose microbiosensor based on glucose oxidase immobilised in an overoxidised polypyrrole film (an in-vitro study). *Fres J Anal Chem*, 367: 692–696.

Ragland J, Lee K (2016) Critical care management and monitoring of intracranial pressure. *J Neurocrit Care*, 9: 105–112.

Raichle ME (1983) The pathophysiology of brain ischemia. *Ann Neurol*, 13: 2–10.

Reinstrup P, Stahl N, Mellergard P, Uski T, Ungerstedt U, Nordstrom CH (2000) Intracerebral microdialysis in clinical practice: Baseline values for chemical markers during wakefulness, anesthesia, and neurosurgery. *Neurosurgery*, 47: 701–709; discussion 709–710.

Rogers ML, Boutelle MG (2013) Real-time clinical monitoring of biomolecules. *Ann Rev Anal Chem*, 6: 427–453.

Rogers ML, Feuerstein D, Leong CL, Takagaki M, Niu X, Graf R, Boutelle MG (2013) Continuous online microdialysis using microfluidic sensors: Dynamic neurometabolic changes during spreading depolarization. *ACS Chem Neurosci*, 4: 799–807.

Rogers ML, Leong CL, Gowers SA, Samper IC, Jewell SL, Khan A, McCarthy L, Pahl C, Tolias CM, Walsh DC, Strong AJ, Boutelle MG (2017) Simultaneous monitoring of potassium, glucose and lactate during spreading depolarization in the injured human brain — proof of principle of a novel real-time neurochemical analysis system, continuous online microdialysis. *J Cereb Blood Flow Metab*, 37: 1883–1895.

Roh D, Park S (2016) Brain multimodality monitoring: Updated perspectives. *Curr Neurol Neurosci Rep*, 16: 56.

Roos YB, de Haan RJ, Beenen LF, Groen RJ, Albrecht KW, Vermeulen M (2000) Complications and outcome in patients with aneurysmal subarachnoid haemorrhage: A prospective hospital based cohort study in the Netherlands. *J Neurol Neurosur Ps*, 68: 337–341.

Rozhanchuk T, Tananaiko O, Mazurenko I, Etienne M, Walcarius A, Zaitsev V (2009) Electroanalytical properties of haemoglobin in silica-nanocomposite films electrogenerated on pyrolitic graphite electrode. *J Electroanal Chem*, 625: 33–39.

Ryan MR, Lowry JP, O'Neill RD (1997) Biosensor for neurotransmitter l-glutamic acid designed for efficient use of l-glutamate oxidase and effective rejection of interference. *Analyst*, 122: 1419–1424.

Sakadzic S, Roussakis E, Yaseen MA, Mandeville ET, Srinivasan VJ, Arai K, Ruvinskaya S, Devor A, Lo EH, Vinogradov SA, Boas DA (2010) Two-photon high-resolution measurement of partial pressure of oxygen in cerebral vasculature and tissue. *Nat Methods*, 7: 755–759.

Sanford AL, Morton SW, Whitehouse KL, Oara HM, Lugo-Morales LZ, Roberts JG, Sombers LA (2010) Voltammetric detection of hydrogen peroxide at carbon fiber microelectrodes. *Anal Chem*, 82: 5205–5210.

Sarrafzadeh AS, Smoll NR, Unterberg AW (2014) Lessons from the intracranial pressure-monitoring trial in patients with traumatic brain injury. *World Neurosurg*, 82: e393–395.

Sato M, Ohishi N, Yagi K (1979) Identification of a covalently bound flavoprotein in rat liver mitochondria with sarcosine dehydrogenase. *Biochem Biophys Res Comm*, 87: 706–711.

Schulz MK, Wang LP, Tange M, Bjerre P (2000) Cerebral microdialysis monitoring: Determination of normal and ischemic cerebral metabolisms in patients with aneurysmal subarachnoid hemorrhage. *J Neurosurg*, 93: 808–814.

Schuvailo OM, Soldatkin OO, Lefebvre A, Cespuglio R, Soldatkin AP (2006) Highly selective microbiosensors for in vivo measurement of glucose, lactate and glutamate. *Anal Chim Acta*, 573–574: 110–116.

Shacham R, Avnir D, Mandler D (1999) Electrodeposition of methylated sol-gel films on conducting surfaces. *Adv Mater*, 11: 384–388.

Smith SK, Lugo-Morales LZ, Tang C, Gosrani SP, Lee CA, Roberts JG, Morton SW, McCarty GS, Khan SA, Sombers LA (2018) Quantitative comparison of enzyme immobilization strategies for glucose biosensing in real-time using fast-scan cyclic voltammetry coupled with carbon-fiber microelectrodes. *Chem Phys Chem*, 19: 1197–1204.

Somjen GG (2001) Mechanisms of spreading depression and hypoxic spreading depression-like depolarization. *Physiol Rev*, 81: 1065–1096.

Sreenivas G, Ang SS, Fritsch I, Brown WD, Gerhardt GA, Woodward DJ (1996) Fabrication and characterization of sputtered-carbon microelectrode arrays. *Anal Chem*, 68: 1858–1864.

Stahl N, Mellergard P, Hallstrom A, Ungerstedt U, Nordstrom CH (2001) Intracerebral microdialysis and bedside biochemical analysis in patients with fatal traumatic brain lesions. *Acta Anaesthes Scand*, 45: 977–985.

Stein SC, Levine JM, Nagpal S, LeRoux PD (2006) Vasospasm as the sole cause of cerebral ischemia: How strong is the evidence? *Neurosurg Focus*, 21: E2.

Stiefel MF, Spiotta A, Gracias VH, Garuffe AM, Guillamondegui O, Maloney-Wilensky E, Bloom S, Grady MS, LeRoux PD (2005) Reduced mortality rate in patients with severe traumatic brain injury treated with brain tissue oxygen monitoring. *J Neurosurg*, 103: 805–811.

Tagliaferri F, Compagnone C, Korsic M, Servadei F, Kraus J (2006) A systematic review of brain injury epidemiology in Europe. *Acta Neurochirurgica*, 148: 255–268; *discussion* 268.

Thévenot DR, Toth K, Durst RA, Wilson GS (2001) Electrochemical biosensors: Recommended definitions and classification. *Biosens Bioelectron*, 16: 121–131.

Tholance Y (2014) Suivi du métabolisme énergétique cérébral chez les patients victimes d'hémorragies sous-arachnoïdiennes graves: Intérêt pour le pronostic individuel et le diagnostic des complications ischémiques. Neurosciences [q-bio.NC]. Doctoral Thesis, Université Claude Bernard, Lyon, France.

Tholance Y, Barcelos GK, Perret-Liaudet A, Omar E, Carrillon R, Grousson S, Lieutaud T, Dailler F, Marinesco S (2017) Placing intracerebral probes to optimise detection of delayed cerebral ischemia and allow for the prediction of patient outcome in aneurysmal subarachnoid haemorrhage. *J Cereb Blood Flow Metab*, 37: 2820–2832.

Tian F, Dale N (2013) Hexacyanoferrates as mediators for microelectrode biosensors. In: Microelectrode Biosensors (Marinesco S, Dale N, eds), pp. 69–93. Totowa, NJ: Humana Press.

Tian F, Gourine AV, Huckstepp RT, Dale N (2009) A microelectrode biosensor for real time monitoring of L-glutamate release. *Anal Chim Acta*, 645: 86–91.

Tian F, Llaudet E, Dale N (2007) Ruthenium purple-mediated microelectrode biosensors based on sol-gel film. *Anal Chem*, 79: 6760–6766.

Trinka E, Hofler J, Zerbs A (2012) Causes of status epilepticus. *Epilepsia*, 53 Suppl 4: 127–138.

Tyler AL, Mahoney JM, Richard GR, Holmes GL, Lenck-Santini PP, Scott RC (2012) Functional network changes in hippocampal CA1 after status epilepticus predict spatial memory deficits in rats. *J Neurosci*, 32: 11365–11376.

Ueda Y, Tsuru N (1994) Bilateral seizure-related changes of extracellular glutamate concentration in hippocampi during development of amygdaloid kindling. *Epilepsy Research*, 18: 85–88.

Vajkoczy P, Roth H, Horn P, Lucke T, Thome C, Hubner U, Martin GT, Zappletal C, Klar E, Schilling L, Schmiedek P (2000) Continuous monitoring of regional cerebral blood flow: Experimental and clinical validation of a novel thermal diffusion microprobe. *J Neurosurgery*, 93: 265–274.

Van Gompel JJ, Bower MR, Worrell GA, Stead M, Chang SY, Goerss SJ, Kim I, Bennet KE, Meyer FB, Marsh WR, Blaha CD, Lee KH (2014) Increased cortical extracellular adenosine correlates with seizure termination. *Epilepsia*, 55: 233–244.

Varner EL, Leong CL, Jaquins-Gerstl A, Nesbitt KM, Boutelle MG, Michael AC (2017) Enhancing continuous online microdialysis using dexamethasone: Measurement of dynamic neurometabolic changes during spreading depolarization. *ACS Chem Neurosci*, 8: 1779–1788.

Vasylieva N, Marinesco S (2013) Enzyme immobilization on microelectrode biosensors. In: Microelectrode Biosensors (Marinesco S, Dale N, eds), pp. 95–114. Totowa, NJ: Humana Press.

Vasylieva N, Marinesco S, Barbier D, Sabac A (2015) Silicon/SU8 multi-electrode microneedle for *in vivo* neurochemical monitoring. *Biosens Bioelectron*, 72: 148–155.

Vasylieva N, Maucler C, Meiller A, Viscogliosi H, Lieutaud T, Barbier D, Marinesco S (2013) Immobilization method to preserve enzyme specificity in biosensors: Consequences for brain glutamate detection. *Anal Chem*, 85: 2507–2515.

Vasylieva N, Barnych B, Meiller A, Maucler C, Pollegioni L, Lin JS, Barbier D, Marinesco S (2011) Covalent enzyme immobilization by poly(ethylene glycol) diglycidyl ether (PEGDE) for microelectrode biosensor preparation. *Biosens Bioelectron*, 26: 3993–4000.

Venton BJ, Robinson TE, Kennedy RT, Maren S (2006) Dynamic amino acid increases in the basolateral amygdala during acquisition and expression of conditioned fear. *Eur J Neurosci*, 23: 3391–3398.

Vergouwen MD, Vermeulen M, Coert BA, Stroes ES, Roos YB (2008) Microthrombosis after aneurysmal subarachnoid hemorrhage: An additional explanation for delayed cerebral ischemia. *J Cereb Blood Flow Metab*, 28: 1761–1770.

Vergouwen MD, Vermeulen M, van Gijn J, Rinkel GJ, Wijdicks EF, Muizelaar JP, Mendelow AD, Juvela S, Yonas H, Terbrugge KG, Macdonald RL, Diringer MN, Broderick JP, Dreier JP, Roos YB (2010) Definition of delayed cerebral ischemia after aneurysmal subarachnoid hemorrhage as an outcome event in clinical trials and observational studies: Proposal of a multidisciplinary research group. *Stroke*, 41: 2391–2395.

Vespa P, Bergsneider M, Hattori N, Wu HM, Huang SC, Martin NA, Glenn TC, McArthur DL, Hovda DA (2005) Metabolic crisis without brain ischemia is common after traumatic brain injury: A combined microdialysis and positron emission tomography study. *J Cereb Blood Flow Metab*, 25: 763–774.

von Bornstadt D, Houben T, Seidel JL, Zheng Y, Dilekoz E, Qin T, Sandow N, Kura S, Eikermann-Haerter K, Endres M, Boas DA, Moskowitz MA, Lo EH, Dreier JP, Woitzik J, Sakadzic S, Ayata C (2015) Supply demand mismatch transients in susceptible peri-infarct hot zones explain the origins of spreading injury depolarizations. *Neuron*, 85: 1117–1131.

von Oettingen G, Bergholt B, Gyldensted C, Astrup J (2002) Blood flow and ischemia within traumatic cerebral contusions. *Neurosurgery*, 50: 781–788; discussion 788–790.

Wang J (2008) Electrochemical glucose biosensors. *Chem Rev*, 108: 814–825.

Wassum KM, Tolosa VM, Wang J, Walker E, Monbouquette HG, Maidment NT (2008) Silicon wafer-based platinum microelectrode array biosensor for near real-time measurement of glutamate in vivo. *Sensors*, 8: 5023–5036.

Welch ME, Doublet T, Bernard C, Malliaras GG, Ober CK (2015) A glucose sensor via stable immobilization of the GOx enzyme on an organic transistor using a polymer brush. *J Polym Sci Pol Chem*, 53: 372–377.

Weltin A, Kieninger J, Enderle B, Gellner AK, Fritsch B, Urban GA (2014) Polymer-based, flexible glutamate and lactate microsensors for in vivo applications. *Biosens Bioelectron*, 61: 192–199.

Weltin A, Kieninger J, Urban GA (2016) Microfabricated, amperometric, enzyme-based biosensors for in vivo applications. *Anal Bioanal Chem*, 408: 4503–4521.

Willner I, Heleg-Shabtai V, Blonder R, Katz E, Tao GL, Buckmann AF, Heller A (1996) Electrical wiring of glucose oxidase by reconstitution of FAD-modified monolayers assembled onto Au-electrodes. *J Am Chem Soc*, 118: 10321–10322.

Wilson GS, Zhang Y (2009) Introduction to the Glucose Sensing Problem, in *In Vivo* Glucose Sensing (Cunningham DD, Stenken JA, eds). Vol. 174, pp. 1–27: Wiley, Hoboken, NJ.

Wintermark M, Sesay M, Barbier E, Borbely K, Dillon WP, Eastwood JD, Glenn TC, Grandin CB, Pedraza S, Soustiel JF, Nariai T, Zaharchuk G, Caille JM, Dousset V, Yonas H (2005) Comparative overview of brain perfusion imaging techniques. *Stroke*, 36: e83–99.

Wise KD, Angell JB, Starr A (1970) An integrated-circuit approach to extracellular microelectrodes. IEEE *Trans Biomed Eng*, 17: 238–247.

Woeffler-Maucler C, Beghin A, Ressnikoff D, Bezin L, Marinesco S (2014) Automated immunohistochemical method to quantify neuronal density in brain sections: Application to neuronal loss after status epilepticus. *J Neurosci Methods*, 225: 32–41.

Woitzik J, Hecht N, Pinczolits A, Sandow N, Major S, Winkler MK, Weber-Carstens S, Dohmen C, Graf R, Strong AJ, Dreier JP, Vajkoczy P, group Cs (2013) Propagation of cortical spreading depolarization in the human cortex after malignant stroke. *Neurology*, 80: 1095–1102.

Woodroofe MN, Sarna GS, Wadhwa M, Hayes GM, Loughlin AJ, Tinker A, Cuzner ML (1991) Detection of interleukin-1 and interleukin-6 in adult rat brain, following mechanical injury, by in vivo microdialysis: Evidence of a role for microglia in cytokine production. *J Neuroimmunol*, 33: 227–236.

Yorita K, Janko K, Aki K, Ghisla S, Palfey BA, Massey V (1997) On the reaction mechanism of L-lactate oxidase: Quantitative structure-activity analysis of the reaction with para-substituted L-mandelates. *Proc Natl Acad Sci USA*, 94: 9590–9595.

Yu B, Long N, Moussy Y, Moussy F (2006) A long-term flexible minimally-invasive implantable glucose biosensor based on an epoxy-enhanced polyurethane membrane. *Biosens Bioelectron*, 21: 2275–2282.

Zain ZM, O'Neill RD, Lowry JP, Pierce KW, Tricklebank M, Dewa A, Ab Ghani S (2010) Development of an implantable D-serine biosensor for in vivo monitoring using mammalian D-amino acid oxidase on a poly (o-phenylenediamine) and Nafion-modified platinum-iridium disk electrode. *Biosens Bioelectron*, 25: 1454–1459.

Zhu L, Yang R, Zhai J, Tian C (2007) Bienzymatic glucose biosensor based on co-immobilization of peroxidase and glucose oxidase on a carbon nanotubes electrode. *Biosens Bioelectron*, 23: 528–535.

Zoerle T, Carbonara M, Zanier ER, Ortolano F, Bertani G, Magnoni S, Stocchetti N (2017) Rethinking neuroprotection in severe traumatic brain injury: Toward bedside neuroprotection. *Front Neurol*, 8: 354.

INDEX